Forests in Time

Edited by **David R. Foster** and **John D. Aber**

Forests in Time

The Environmental Consequences of
1,000 Years of Change in New England

Yale University Press   New Haven & London

Frontispiece: Stone walls running through the woods of New England provide an enduring reminder of the great historical transformations in land use and land cover that have occurred in this landscape over the past few centuries. Photograph by D. R. Foster.

Copyright © 2004 by Yale University.

All rights reserved.

This book may not be reproduced, in whole or in part, including illustrations, in any form (beyond that copying permitted by Sections 107 and 108 of the U.S. Copyright Law and except by reviewers for the public press), without written permission from the publishers.

Designed by James J. Johnson and set in Melior and Syntax types by The Composing Room of Michigan, Inc. Printed in the United States of America.

Library of Congress Cataloging-in-Publication Data

Forests in time : the environmental consequences of 1,000 years of change in New England / edited by David R. Foster and John D. Aber.

     p. cm.

Includes bibliographical references.

  ISBN 0-300-09235-0 (hardcover : alk. paper)

  1. Forest ecology—New England—History. I. Foster, David R., 1954– II. Aber, John D.

  QH104.5.N4.F67 2004

  577.3'0974—dc21   2003012715

A catalogue record for this book is available from the British Library.

The paper in this book meets the guidelines for permanence and durability of the Committee on Production Guidelines for Book Longevity of the Council on Library Resources.

10 9 8 7 6 5 4 3 2 1

This book is dedicated to the vision and leadership of

John G. Torrey

and to the Harvard Forest legacies of

Richard T. Fisher

Albert C. Cline

Stephen H. Spurr

Hugh M. Raup

Ernest M. Gould

Walter H. Lyford

and

Martin H. Zimmermann

When I walk . . . I forget that this which is now Concord was once Muske-taquid, and that the *American race* has had its destiny also. Everywhere . . . the earth is strewn with the relics of a race which has vanished as completely as if trodden in with the earth. I find it good to remember the eternity behind me as well as the eternity before. Wherever I go, I tread in the tracks of the Indian.
    —H. D. Thoreau, 1842

[New England is] cloathed with infinite thick Woods.
    —John Josselyn, 1672

The forests are not only cut down, but there appears little reason to hope that they will ever grow again.
    —Timothy Dwight, 1804

What is a New England landscape this sunny August day? A weather-painted house and barn, with an orchard by its side, in midst of a sandy field sur-rounded by green woods, with a small blue lake on one side. A sympathy be-tween the color of the weather-painted house and that of the lake and sky.
    —H. D. Thoreau, August 26, 1856

Except in the towns there was evidence of a population which had once lived here and farmed and had its being and had been driven out. The forests were marching back, and where farm wagons once had been only the big logging trucks rumbled along. And the game had come back, too; deer strayed on the roads and there were marks of bear.
    —John Steinbeck, 1962

This unintentional and mostly unnoticed renewal of the rural and mountain-ous east—not the spotted owl, not the salvation of Alaska's pristine ranges—represents the great environmental story of the United States, and in some ways of the whole world.
    —Bill McKibben, 1995

Although writing at very different times in the historical trajectory of the New England landscape, each of these authors share a common trait: in their own style and context each recognized the enormity of the past and ongoing changes that shaped the land and its forests. Each clearly underscored the role of human activity in these changes and recognized the clear linkages between the history of the land and its people and its modern character. For each author an appreciation for the past helped to clarify the significance and uniqueness of current conditions.

In like fashion, since 1907, when Harvard University established a center for forest research and education in central Massachusetts, ecologists at the Harvard Forest have based their studies on one central notion: that understanding landscape history provides essential insights into modern ecosystem structure, composition, and function. Research at the Forest has thus always combined a historical perspective with field studies in order to interpret current conditions and to manage forest ecosystems more effectively for the future. This Harvard Forest approach to forest ecology is based, in large part, on a series of very practical considerations. Trees are long-lived organisms, and in their very age span they impart a certain inertia, legacy, or historical memory to a forest. The maples, oaks, and hemlocks that constitute our woods establish, grow, and are shaped over time by a wide range of conditions; the effects of past windstorms, climates, or human effects become embodied in the shape and structure of the trees and appearance of the forest, which are then carried forward through time as the forest grows. In similar fashion, many important ecosystem processes such as soil development occur very slowly. Consequently, historical effects or ecological changes leave imprints on the structure and function of ecosystems that persist for decades or centuries. Changes in soil nutrients induced by Indian burning or colonial farming practices, physical impacts like the uprooting of trees or the plowing of a field, or changes in plant and animal composition induced by climate change, disease, or hunting practices all set in motion profound ecological changes that continue to ripple through ecosystems over time. A massive hemlock falling from a windstorm in an old-growth forest may lie on the ground for a century as it slowly molders into the soil. The challenge for the ecologist is to integrate biology and environmental science with an understanding of the complexities of landscape history in order to forge meaningful interpretations of the present.

To understand processes that have shaped current plant and animal assemblages through time and to document the slow progress of important ecological processes, we need to extend the time frame for our studies and our thinking by conducting truly long-term research and by using a creative array of historical approaches. Whether we are interested in the abundance of large mammals, the distribution of rare plants, the fate of carbon storage in our forests, or the response of all ecosystems to the nitrogen and other pollutants released by fossil fuel combustion, we cannot make sound interpretations of the modern landscape solely by examining current conditions. We must always invoke a detailed knowledge of the history of the land and its people. Similarly, our attempts to project or guide future changes in these ecosystems are always improved by considering the ecological processes that have operated in the past.

In this volume we convey our integrated understanding of forest pattern and process in New England couched within a historical perspective of landscape changes. At the heart of the research is a series of intensive long-term observations and measurements of modern forest structure and function. These measurements are paralleled by a series of experiments in which we simulate important stresses and disturbances that have been or are projected to become important in shaping the forest ecosystems of New England. We then interpret these results in the context of the landscape history of natural and human-induced changes that are captured by the quotations of Josselyn, Dwight, Thoreau, Steinbeck, and McKibben. This history, in turn, is investigated in great detail for the ecological lessons that it conveys, for the guidance that it may offer for conservation and land policy, and for the simple enjoyment that it brings to each of us as it reveals delightful details of our cultural and natural heritage. The result is a millennial perspective into the forest ecosystems and landscape of New England.

Acknowledgments

Research at the Harvard Forest is a collaborative enterprise that pushes far beyond the bounds of the Long Term Ecological Research Program to draw on the talents of many colleagues and the support of diverse institutions and agencies. As indicated in our dedication, the research that we present here builds on the superb legacy of studies by the students, faculty, and staff who preceded us and who made the Harvard Forest a unique scientific resource.

Special thanks go to the following people who assisted our efforts in many ways and made our studies possible and productive: P. Bakwin, S. Barry, A. Bazzaz, J. Bellemare, A. Bright, N. Brokaw, D. Bryant, J. Budney, J. Burk, K. Chamberlin, E. Chilton, E. Cline, B. Coleman, A. Colman, S. Cromley, B. Daube, B. Donahue, E. Doughty, M. Downs, N. Drake, A. Dunn, P. Dunwiddie, R. Eberhardt, E. Ellin, A. Elseroad, E. Faison, S.-M. Fan, K. Fetherston, M. Fisher, M. Fluet, B. Flye, J. Fontenault, D. Francis, R. Garrish, F. Gerhardt, F. Gimelfarb, A. Goldstein, D. Goodwin, E. Gottlieb, M. Goulden, J. Hadley, B. Hall, E. Hammond-Pyle, B. Hansen, J. Harrod, A. Hirsch, C. Horii, L. Hutyra, S. Jackson, J. Jenkins, C. Jones, J. Jones, S. Katt, D. Kittredge, M. Kizlinski, C. Le Cat, M. Lefer, R. Lent, A. Lezberg, C. Mabry, D. MacDonald, K. May, T. McLellan, J. Merriam, K. Metzler, M. Miliefsky, J. Muckenhoupt, M. Mulholland, S. Newman, J. O'Keefe, F. Paillet, C. Parker, T. Parshall, W. Patterson, D. Pelletier, W. Peterjohn, G. Peterken, J. Peterson, C. Pinney, L. Plourde, G. Quigley, G. Rapalee, T. Rawinski, M. Riddell, K. Rolih, T. Rudnicky, E. Russell, J. Sanderman, L. Schaider, L. Schmitt, P. Schoonmaker, M. Serrano, A. Simoneau, B. Slater, D. Sperduto, M. Sperduto, C. Spooner, J. Stone, S. Stratton, D. Sutton, F. Swanson, P. B. Tomlinson, K. Touvinen, S. Urbanski, L. Waller, W. Warren, P. Weatherbee, A. Wilson, C. Wilson, J. Wisnewski, and N. Wurzburger.

Support for the research described in this book came in many forms from diverse agencies and organizations, including the National Science Foundation programs in Long Term Ecological Research, Research Experience for Undergraduates, Ecology, Ecosystems, and Atmospheric Science and the International Program; the A. W. Mellon Foundation; the U.S. Department of Energy Biological and Environmental Research Program and the Northeast Regional Center of the National Institute for Global Environmental Change (NIGEC); the U.S. Environmental Protection Agency's Global Change Program and Athens Environmental Research Laboratory; NASA—Terrestrial Ecosystem Program; The Nature Conservancy; Exxon Mobil Corporation; Conservation Research Foundation; Sweetwater Trust; Mount Grace Land Conservation Trust; Electric Power Research Institute; Connecticut Department of Environmental Protection; in Massachusetts, the Natural Heritage and Endangered Species Program, Department of Environmental Management, Department of Fisheries and Wildlife, Executive Office of Environmental Affairs, Massachusetts District Commission, and the University of Massachusetts Herbaria; and the USDA Competitive Grants program.

Finally, we would like to make special mention of the extremely talented group of individuals who helped in the development and production of this book, including Dottie Recos Smith, Julie Pallant, Jean E. Thomson Black, Mary Pasti, Audrey Barker Plotkin, and Jessie Dolch. In many ways they made this effort an enjoyable and exciting enterprise.

I

*A Research Program on
Forest Ecology and Change*

CHAPTER 1

Background and Framework for Long-Term Ecological Research

D. FOSTER and **J. ABER**

In 1907 Harvard University acquired nearly 3,000 acres of land in the central Massachusetts town of Petersham to establish the Harvard Forest as a center for research and education in forest ecology, conservation, and management. In the ensuing century of investigations, students, faculty, and visiting researchers came to rely heavily on accumulated and continuing historical studies as a complement to intensive field and laboratory work and as a source of insight into important processes that have shaped the land, its people, and its biota. By developing long-term studies of the past and present, we can uncover events and processes that are infrequent in occurrence, we can examine physical and biological processes that unfold over long periods of time, and we can sift through the many changes and factors that have operated in the landscape over time in order to identify those that are critical for interpreting modern conditions and dynamics (Figure 1.1).

This long-term approach to ecological research was a central driver in the selection of research directions when we teamed together with colleagues from several Harvard departments, the University of New Hampshire, the Ecosystem Center at the Marine Biological Laboratory, and the University of Massachusetts in 1988 to form the Harvard Forest Long Term Ecological Research (LTER) program. In particular, we applied our understanding of the history of the land, modern forest dynamics, and projections for future changes in the regional and global environment to select a suite of important disturbances, stresses, and forest ecosystem processes to investigate in detail. The broad objective of these studies was to develop information and approaches that will answer fundamental ecological questions and to generate data and perspectives that have broad application to major environmental and conservation issues.

A sketch of the history of New England's land and people highlights the major changes that shape the present landscape and the key objectives of our investigations.

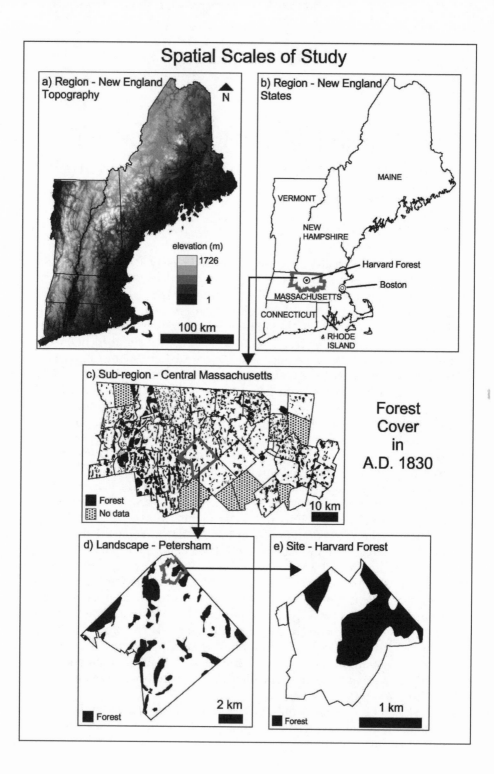

Spatial Scales of Study

a) Region - New England Topography

elevation (m)
1726

1

100 km

b) Region - New England States

MAINE

VERMONT

NEW HAMPSHIRE

Harvard Forest

Boston

MASSACHUSETTS

CONNECTICUT

RHODE ISLAND

c) Sub-region - Central Massachusetts

■ Forest
▨ No data

10 km

Forest
Cover
in
A.D. 1830

d) Landscape - Petersham

■ Forest

2 km

e) Site - Harvard Forest

■ Forest

1 km

Overview of Environmental and Forest Dynamics in Central New England

Over the past few thousand years, New England's landscape has been remarkably dynamic in response to varied changes in environmental, cultural, and biological factors that control the structure, composition, and function of forest ecosystems (Figure 1.2). The broadscale geological and physiographic structure of the region has been relatively unaltered since the last wave of glaciation overran the gentle hill and valley topography and blanketed it with shallow deposits of silt, sand, clay, and rock some 14,000 years ago. In contrast, the region's climate has evolved continually on a time frame of decades to millennia. Relatively warm conditions in the mid-Holocene period approximately 5,000 years ago gave way to cooler conditions about 2,000 years ago that persisted until the recent past. Subtle changes in the composition of New England forests throughout this time and increasingly in the few hundred years before European settlement suggest that significant shifts occurred in the amount and seasonal distribution of precipitation and temperature. As a consequence, the abundance of important trees, including beech, hemlock, spruce, and chestnut, has shifted in pronounced though poorly understood ways.

When we examine any forest's history more closely, it is apparent that natural disturbances ranging from frequent, small events to infrequent, broadscale and intense impacts are important in structuring the New England landscape. Nearly 5,000 years ago, an insect pest caused hemlock to decline abruptly throughout its range in eastern North America. Over the ensuing thousand-year period, forest and lake ecosystems changed as this long-lived and shade-tolerant tree species declined and then gradually rebounded, albeit with considerable geographic variation in the rate and extent of recovery. The details of this episode continue to emerge through new studies and are clearly relevant to modern questions concerning the importance of individual species to ecosystem function and the potential impacts of introduced organisms, such as the hemlock woolly adelgid, on our forests.

In contrast to the solitary occurrence of a natural pest, the historical record of powerful hurricanes leads us to suspect that tropical storms episodically disrupt and shape New England forests (Figures 1.3 and

Figure 1.1. Primary spatial scales of investigation in the Harvard Forest Long Term Ecological Research program. The New England–wide maps depict topography and state boundaries as well as the location of the Harvard Forest. Other maps show the distribution of forest area (black) remaining in the mid-nineteenth century at the height of agricultural activity and deforestation. Areas in white were predominantly open pastureland, other agricultural fields, and village areas. Compiled and modified from U.S. Geological Survey (unpublished), Foster, Motzkin, and Slater 1998, Foster 1992, and Hall et al. 2002.

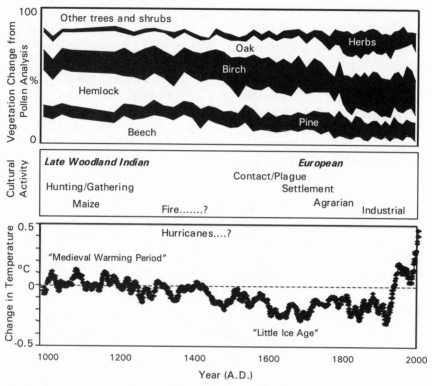

Historical Setting for Research

Figure 1.2. The temporal setting for long-term studies in New England in relation to important biotic, cultural, and environmental changes. Biotic (vegetation) change (top) is illustrated by the varying percentages of pollen of major plant taxa from Aino Pond in north central Massachusetts (Fuller et al. 1998). Major cultural changes (middle) highlight the shift from native hunting, gathering, and horticulture to European agriculture and industry. Climate dynamics (bottom) are depicted by the long-term change in Northern Hemisphere temperature over the past 1,000 years, as reconstructed from tree-ring records and other proxies (adapted from Crowley 2000, data archived at the World Data Center for Paleoclimatology, Boulder, Colorado, USA).

1.4). Severe storms tend to follow similar paths across New England, and therefore it is possible that distinct regional gradients and repeatable landscape patterns of windthrow and damage may exist. Long-term records and historical studies also suggest that less intense disturbance resulting from northeasterlies, downbursts, ice storms, and late-season snowstorms are important drivers of forest gap dynamics, processes that may diversify forest landscape patterns over time. Understanding the relative role of these different types and scales of physical disturbance and their distribution across New England has remained an elusive goal of ecologists.

Figure 1.3. The 1938 hurricane damaged more than 70 percent of the volume of timber on the Harvard Forest and exerted a profound effect on ecological thought, forest planning, and interpretations of New England's environmental history. Photograph of Albert Cline, Harvard Forest director who was faced with coping with this unanticipated event, from the Harvard Forest Archives.

Fire is less frequent in the temperate forests of New England than in the boreal region, midwestern grasslands, or western woodlands. However, fire may have a long-lasting effect on forest composition and function, so interpreting its frequency, intensity, type, and variability through time remains a great challenge. This subject is especially relevant to modern conservation and land management as fire is increasingly viewed as a critical tool for maintaining important plant or animal assemblages. Discussion of fire, which is often placed in the category of "natural" disturbances, invariably introduces the issue of cultural influences on our landscape as most New England fires are ignited by people. Indeed, fire

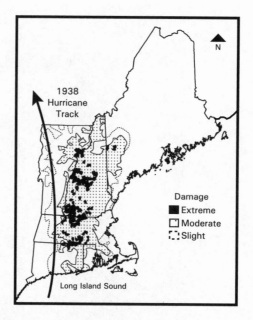

Figure 1.4. The 1938 hurricane damaged forests in a 100-kilometer-wide swath that extended from Long Island Sound to northern Vermont. Damage was concentrated to the east of the storm track due to the counterclockwise rotation and forward velocity of the storm. The track was reconstructed from meteorological records, whereas damage estimates are based on town-wide assessments of salvaged timber by the New England Timber Salvage Administration (unpublished data). The low damage values in southern New England are due to the relatively small amount of merchantable timber-size trees in that region in 1938. Modified from Foster 1988a, published by Blackwell Scientific Publications, by permission of the British Ecological Society.

represents the major means by which a relatively small population of Native Americans might have exerted a widespread effect on natural vegetation (Figure 1.5).

For New England, the general patterns of Indian activity are broadly understood: concentrated coastal and river valley populations and much lower densities in upland areas; cultural patterns, seasonal activities, and densities that varied historically with the availability of resources; the late arrival of maize (corn) horticulture circa A.D. 1000; and a transformation of subsistence patterns and culture with the arrival of European people. However, many major questions remain concerning the type and spatial extent of human effects on natural ecosystem patterns, and these motivate our ongoing studies.

Over the nearly 400 years since European settlement, the rate of ecosystem change has accelerated. Across New England, extensive de-

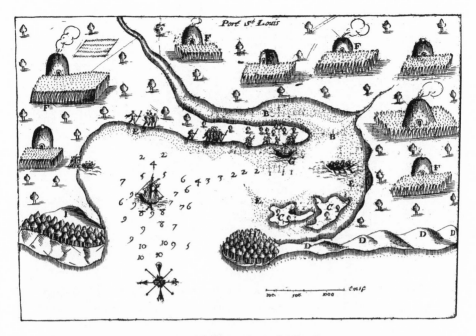

Les chifres montrent les braſſes d'eau.

A Monſtre le lieu'ou poſent les vaiſſeaux.
B L'achenal.
C Deux iſles.
D Dunes de ſable.
E Baſſes.

F Cabannes où les ſauuages labourent la terre.
G Le lieu où nous fuſmes eſchouer noſtre barque.
H vne maniere d'iſle tem-

plie de bois tenant aux du-nes de ſable.
I Promontoire aſſez haut qui paroiſt de 4. a 5. lieux à la mer.

Figure 1.5. Champlain's map of Indian lodges, cornfields, and the landscape around Plymouth Bay, Massachusetts, in 1605. Although evidence for Indian encampments and local effects on the landscape is especially abundant near the coast and in major river valleys, the details of this activity and the extent of Indian influence on the interior countryside remain a matter of debate. In particular, archaeological evidence for large agricultural fields, extensive corn consumption, fortified settlements, and sizable villages in the region is lacking. Map originally printed in "The Voyages of Samuel de Champlain" (S. de Champlain, 1613).

forestation for agriculture peaked after 1850 and was followed by broad-scale abandonment of farming and reforestation through natural succession (Figures 1.6 and 1.7). Vast areas that supported scattered woodlots in a panorama of fields and pastures only 150 years ago are now covered with a mosaic of maturing second-growth forests that support natural ecosystem processes and native wildlife populations (Figure 1.8). Consequently, a central question looms over all environmental studies in the region: what are the enduring consequences of land-use history?

During the past century the forests and environment of New England

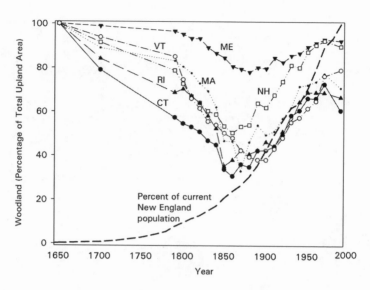

Figure 1.6. Historical changes in the New England population (heavy dashed line) and forest cover for each of the individual states. Excluding Maine, where the northern forest region was never settled and remained forested, the states share a common history of deforestation for agriculture followed by farm abandonment and natural reforestation. The extent to which these secondary forests differ in structure and function from permanently wooded areas or the forests of the presettlement period forms a major research emphasis of Harvard Forest studies. Population data compiled from U.S. censuses; forest cover data from Baldwin 1942 and U.S. Forest Service 1958, 1965, 1973, and 1990.

have also been exposed to new stresses induced through human activity. A series of introduced insects and diseases—chestnut blight, Dutch elm disease, gypsy moth, beech bark disease, and hemlock woolly adelgid— has selectively weakened, defoliated, or decimated major tree species across the region. Industrialization has changed the earth's atmosphere, which among other things has increased the deposition of nitrogen, a major limiting nutrient in most terrestrial ecosystems, as well as sulfur, in forms that acidify the rain and the region's ecosystems. Photochemical processes in the earth's upper atmosphere are depleting the tropospheric ozone (O_3) layer that shields nature and humans from ultraviolet radiation. Meanwhile, during the summer growing season, stagnant air circulation along the industrialized eastern seaboard brings damaging ozone and other pollutants to New England forests.

Increases in major greenhouse gases—including carbon dioxide (CO_2), methane (CH_4), and nitrous oxide (N_2O)—may be initiating a global increase of temperature that could reach 3° to 4°C in the north-

Figure 1.7. A stone wall built in the eighteenth century to separate a tilled field from an adjacent pasture winds through a second-growth hardwood stand in central Massachusetts. Despite the natural appearance of the vast forest area in New England, most of the region bears subtle clues to its cultural past. Photograph by D. R. Foster.

eastern United States by the end of the century. Increases in CO_2, nitrogen, and ozone in the atmosphere may also have subtle, though important, consequences for plant growth and competition, plant-animal interactions, and ecosystem processes. Evaluating the comparative effects of these novel stresses and the potential interactions that they may have with historically important disturbances is a challenge for ecologists and an important focus for conservationists and natural resource managers.

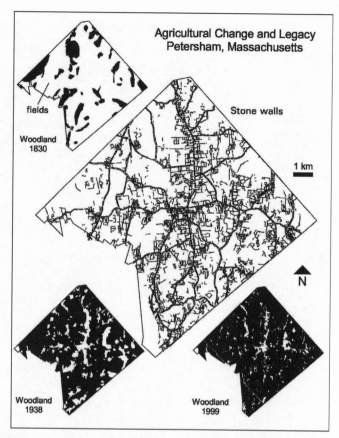

Figure 1.8. The pattern of stone walls bordering colonial fields and roads in the central Massachusetts town of Petersham (center) documents the extent and intensity of agriculture in New England's history. The approximately 380 kilometers of stone walls run through a landscape that is now nearly completely forested (bottom right) as a result of forest succession on abandoned farmland (top and bottom left). Maps compiled from the Worcester County White Pine Blister Rust project (1938–39; Harvard Forest Archives), 1830 land-use maps (Massachusetts State Archives, Harvard Forest Archives, and Hall et al. 2002), 1938 aerial photographs, and MassGIS 2002.

Broad Ecological Questions Concerning New England Forests

Our brief historical overview highlights many changes in the physical, biotic, and human environment of New England that have initiated a range of forest dynamics and have set the vegetation and landscape on a long-term trajectory. This overview also raises broad questions that drive our research as we seek to understand, conserve, and manage the modern landscape and anticipate future changes. Not surprisingly, these questions address fundamental issues relevant to many natural ecosystems worldwide.

1. *How does the array of environmental factors and disturbance processes interact to shape forest ecosystems over time?* The preceding section identified major uncertainties concerning the ways in which climate change, natural disturbance, and human activities have operated at local to regional scales through time. Of specific interest are details of the natural disturbance regimes, especially variations in wind, pathogens, and fire and the way in which these have interacted with prehistoric and historical human activity and environmental change.

2. *What are the contrasting effects of natural, physical disturbance versus novel, anthropogenic stress on the function of forest ecosystems?* Increasingly, forest ecosystems are exposed to chemical and climatic stresses that are qualitatively different from the types of impacts that forests have experienced for millennia. Recognizing that forest species evolved within a context of natural disturbance and environmental change, it is important to assess whether forests retain the same degree of control over ecosystem processes (for example, nutrient cycling, hydrology, and forest growth) under these novel conditions as they do under historically important stresses. Specifically, we are interested in contrasting the relative effects of physical disturbance, such as hurricanes, with important new stresses such as nitrogen additions and rapid climate change.

3. *What changes in forest patterns and processes were generated by the history of intensive land use since European settlement, and how persistent are the physical and biological legacies of this historical disturbance in New England's reforested landscape?* Large areas of northwestern Europe, Latin America, and eastern North America have experienced or are currently undergoing landscape transformations analogous to the forest–deforestation–reforestation history of New England, and thus lessons from our region should have general relevance. Major questions remain concerning the initial effects of colonial land-use activity, the ability of forest ecosystems to return to predisturbance conditions, and the legacy of historical changes on modern forest characteristics. For example, is the history of sites that were in agriculture 150 years ago reflected in the modern forest composition, soils, and fertility, or in the way that the forest will respond to the next hurricane or to acid rain?

4. *What application do answers to these questions have for environmental policy issues such as (a) designing effective local and regional conservation strategies, (b) anticipating forest ecosystem response to modern stresses and disturbances, and (c) developing global strategies to mitigate future climate change?* As we develop an improved understanding of modern forest ecosystems and their history of change, we can bring this information to bear on fundamental ecological questions concerning the patterns and processes of natural ecosystem organization and dynamics. We can also assist in the application of this information to education and the management of our natural environment and resources.

Design of the Harvard Forest Research Program

In order to address the broad ecological questions raised above, our research effort has been organized to integrate studies across disciplines, scientific approaches, and a range of spatial and temporal scales. By augmenting the lengthy record of ecosystem change that had developed over nearly a century of study at the Harvard Forest, we have selected historically important and relevant processes for investigation. We have also expanded existing programs in order to make education and public outreach a major goal of our program (Table 1.1).

We use a suite of scientific approaches in order to identify important ecological processes, to create a long-term series of measurements, and to assess ecosystem response and dynamics.

Ecological history provides information on the range of environmental conditions and types of natural and human disturbance processes that have been historically important in a landscape (Figure 1.9). This information identifies processes that are critical to study in order to understand ecosystem structure and function, and it contributes to our understanding of the relative role of historical factors versus environmental factors in controlling modern conditions. In addition, many key

Table 1.1. Design of the Harvard Forest Long Term Ecological Research Program

Research Approaches

1. Reconstruction of ecosystem dynamics using paleoecology, historical ecology, and modeling to evaluate long-term trends, to study infrequent processes, and to interpret the development of modern conditions
2. Measurement of modern ecosystem structure, composition, processes, and dynamics on permanent plots, through remote sensing and through eddy flux measurements of atmosphere-biosphere exchanges to define current conditions and rates
3. Experimental manipulations of ecosystems to evaluate and compare patterns of response and to collect integrated measurements on multiple processes
4. Controlled environment studies of plant and population responses to specific environmental changes
5. Integration through modeling, comparative studies, monthly meetings, annual symposia, and synthetic products, including coordination with other research programs
6. Application to ecological theory, conservation biology, environmental policy, and forest management

Spatial Scales of Investigation

1. Plot: 0.1 km
2. Site: 1 km—Harvard Forest
3. Landscape: 10 km—Petersham, Massachusetts
4. Subregion: 100 km—e.g., central Massachusetts, Cape Cod and islands, White Mountains
5. Region: 1,000 km—New England and adjacent New York

Disturbances, Stresses, and Environmental Processes Investigated

1. Climate change
2. Windstorms and other environmental extremes
3. Fire
4. Native and introduced pathogens and pests
5. Land use: aboriginal, historical, and current
6. Changes in atmospheric chemistry

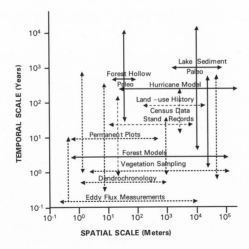

Figure 1.9. The relationship between the temporal and spatial scales of many of the ecological methods used in the Harvard Forest research program. Approaches include field studies (dotted lines), historical research (dashed lines), paleoecological and other retrospective studies (solid lines), and modeling (also solid). For each approach, vertical and horizontal lines approximate the temporal and spatial coverage achieved. Based on Foster et al. 1990.

ecological processes, such as succession, ecosystem development, large disturbances, species invasion, and ecosystem response to environmental change, operate on decadal to millennial timescales or with such great temporal variability that they are difficult or impossible to measure through conventional studies. Historical techniques enable the evaluation of such processes and allow these observations to be placed within the context of long-term trends and cycles of environmental change.

Experimental field manipulations allow us to evaluate infrequent though historically important processes as well as to anticipate future ecosystem response to predicted changes in climate or chemical stresses. At the Harvard Forest, we have focused large experiments on a subset of important though contrasting disturbances and stresses, including simulation of windthrow from an intense hurricane, timber harvesting, chronic nitrogen amendments to simulate enhanced deposition of nitrogen, soil warming as a component of climate change, and alteration of organic matter inputs to soils to examine basic soil processes linked to carbon and nitrogen dynamics. In the case of historically important processes such as hurricanes and forest logging, results from the experimental manipulations are compared directly with long-term studies of "natural experiments," such as the 1938 hurricane or historical logging that occurred throughout central New England. Many of the experiments are compared with parallel studies in different ecosystems. For example, the nitrogen saturation experiment has counterparts at the

Bear Brooks watershed in Maine maintained by the U.S. Forest Service and in the enhanced deposition of nitrogen that is occurring at high elevations in New England and in central Europe; the soil warming experiment has been replicated in the subarctic region at the Abisko Research Station in northern Sweden; organic matter manipulation experiments have been undertaken at the University of Wisconsin and other sites within the U.S. LTER network; and results from the experimental hurricane have been compared with those from natural events in many temperate and tropical forests. In all cases, the integrated measurements of ecosystem structure and pattern enable comparison among these important manipulations.

Long-term measurements of ecosystem patterns and processes in a range of natural forests carry forward observations of current conditions and results from reconstructive studies. In particular, measurements continued over decades assess seasonal and interannual variation, long-term trends and trajectories, and ecosystem function under highly varied sets of conditions. Permanent plots and repeated sampling of forest stands at five- to ten-year intervals continue long-term experiments and observations that were initiated in the early 1900s by the first faculty and students at the Harvard Forest. Remote sensing, using modern and historical aerial photographs or satellite images, increases the coverage of many measurements across two or more of our spatial scales of observation (for example, from plots in a forest to the landscape or region) and over many decades. Control areas, which are the undisturbed but monitored parts of our experimental manipulations, provide baseline measurements that may be linked with other data sets, such as the studies of atmosphere-biosphere exchange at our Environmental Measurement Station. The coupling of retrospective historical studies and long-term measurements of intact and experimentally manipulated ecosystems provides an integrated assessment of ecosystem dynamics and function under a range of historical, modern, and simulated conditions.

Harvard Forest research operates at four primary spatial scales of investigation: site (approximately 1 kilometer), landscape (approximately 10 kilometers), subregion (approximately 100 kilometers), and region (approximately 1,000 kilometers) (Figure 1.1). Intensive site studies at the scale of individual organisms, forests, or sample plots represent the heart of our long-term effort. Most of our studies at this scale occur on Harvard Forest land (approximately 1,200 hectares) in central Massachusetts, where varied vegetation, site conditions, and history, along with nearly 100 years of study, provide an ideal setting for long-term measurements and experiments. Infrastructure improvements, such as access to electrical and telecommunications cables at major experiments, erection of a series of canopy access and environmental measurement towers, the use of mobile lifts to reach into the crowns of mature trees, and surveyed grid points, enable diverse studies. Geographic information system–based data-management systems allow field sampling to be integrated with other sources of information such as low-elevation aerial photography, satellite imagery, radiotelemetry, historical surveys, and vegetation maps.

Many important processes, including natural and human disturbance, wildlife movement, and hydrologic flows, occur at a *landscape* scale, where physiography, slope position, vegetation structure, and soil variation interact to form complex patterns. In central New England, the area of an individual town (often approximately 10 by 10 kilometers) captures substantial landscape variation of the characteristic hill and valley topography. Consequently, the town of Petersham, Massachusetts, serves as one focus for many landscape studies because it includes the main tracts of the Harvard Forest and represents a typical rural village in the New England uplands. Given the politically independent structure of New England town governments, much of the geographical, social, and environmental data relevant to ecological studies are collected or aggregated by public agencies at a town level, making this political unit a particularly convenient scale of study.

To place site- and landscape-level studies in a broader context and to examine variation in environmental, social, and biotic processes, we conduct a considerable amount of research at the *subregional* scale (for example, central Massachusetts, Cape Cod and the Islands, the White Mountains of New Hampshire, the Connecticut Valley), and the *regional* scale of New England, oftentimes including adjacent New York. Selection of these areas is based on ecological, cultural, and pragmatic considerations. For example, the central Massachusetts subregion (approximately 5,000 square kilometers) extends 100 kilometers east from the Connecticut Valley Lowland through the Central Uplands physiographic region to the Eastern Lowlands west of Boston, and 50 kilometers south from the New Hampshire border approximately halfway to the Connecticut border. Petersham and the Harvard Forest lie directly in the center of this diverse subregion, which encompasses a wide range of the physical and biological variation in central New England, as well as a substantial amount of the cultural variation that has occurred from precolonial to modern times. The ability to place intensive studies within the context of these major cultural and environmental gradients is extremely useful for interpreting the generality of results and for understanding the broadscale controls over major ecological processes. On the practical side, this subregion consists of fifty townships in four counties of one state, which presents a manageable, though considerable, challenge for the collection and archiving of archaeological, historical, environmental, and biological data. Information for this region comes in three primary forms—continuous spatial coverage (for example, elevation, land-cover maps, and remote sensing imagery), township-level data (for example, population, agricultural, and forestry statistics), and networks of site-specific data (for example, sample plots, archaeological sites, and intensive measurement locations).

Considerably greater variation in environmental and cultural conditions occurs across the New England region, and the dynamics and effects of many of the broadscale disturbances and anthropogenic stresses can be understood only at this scale. In order to evaluate processes that are relevant at the regional scale, we conducted select studies utilizing diverse historical, modern, and modeling approaches. These studies yield data that may be continuous, aggregated at the county scale, or site specific. Importantly, these studies also enable us to see how well our approaches and results translate to other areas.

Thus, the research approach followed by the Harvard Forest LTER program is a continuation of the long-standing approach to understanding the New England landscape that Harvard Forest researchers have used for nearly a century. We use historical studies to understand the development of modern forests and to study infrequent and variable events and slow processes; we integrate our understanding of modern measurements and experiments with results from retrospective studies; we emphasize long-term experiments with an informative and secure data management structure; and we attempt to synthesize the results of all of these studies such that they address fundamental ecological questions and provide insights into societally relevant management issues.

CHAPTER 2

The Physical and Biological Setting for Ecological Studies

D. FOSTER

In the late 1890s, Nathaniel Shaler, dean of the Lawrence School of Science at Harvard University, and Richard Fisher, professor of forestry, agreed on Harvard's need to establish a research forest in the New England countryside to serve as a laboratory for investigations in forest biology and management and as an outdoor classroom where students could learn the emerging discipline of forest science. After considerable travel and with encouragement and support from local alumni, Fisher selected a series of tracts in the town of Petersham in north-central Massachusetts as the center for the new Harvard Forest. He discovered in this rolling and picturesque upland landscape a large range of the biological and environmental conditions that were characteristic of New England and that fit his criteria for a field laboratory and classroom. The land captured much of New England's considerable variation in soils and vegetation as it stretched from lowlands such as the Tom Swamp valley—where acid spruce and larch bogs, red maple swales, and sedge and shrub marshes prevailed—to mesic forests of hemlock and hardwoods on gentle slopes, and up to dry and rocky oak knolls (Figure 2.1). Equally important for studies of land management, the landscape that Fisher surveyed embraced a wide variety of land-use history, from intensely cultivated agricultural sites on the broad upland ridges, to pasture and woodlots on the rocky slopes, to small pockets of older growth forest on moist, rocky, and less-accessible sites. Thus, with the establishment of the Harvard Forest in central Massachusetts and with subsequent acquisitions like the old-growth Pisgah Forest in southern New Hampshire, the base was created for the intensive and long-term studies that form most of our research.

However, all researchers at the Forest—including Fisher, ecologists and foresters such as Hugh Raup, Steve Spurr, Ernie Gould, and Walter Lyford, and their modern successors—have recognized the need to expand on this modest center of intensive studies and to place the results from Petersham in a broader geographical context. Thus, through time, the research laboratory and classroom grew in size beyond Harvard

Figure 2.1. Diorama from the Fisher Museum of the old-growth hemlock forest adjoining Harvard Pond in Petersham, Massachusetts. Depicted in the foreground are Nathaniel Shaler, Harvard dean, geology professor, and first director of the U.S. Geological Survey, and Richard Fisher, first director of the Harvard Forest. Shaler was instrumental in promoting the concept of a field laboratory and classroom for forest studies at Harvard, and Fisher located the property and then initiated the ecological and historical studies that remain the hallmark of Harvard Forest research. The old forest was subsequently damaged by the 1938 hurricane. Photograph by J. Green, from Foster and O'Keefe 2000.

lands to embrace a range of sites throughout Petersham and north-central Massachusetts, across much of the New England region, and progressively to many distant places around the globe. In the integrated studies that we present in this volume, the hierarchy of geographical scales from site to landscape to subregion to region and ultimately to the globe is embraced as an explicit framework for research. To facilitate discussions in subsequent chapters, much of which crosses scales rather fluidly, we provide a brief overview of the physical environment and biological characteristics of these settings for our studies.

New England Region

Our regional studies focus on most of New England (Maine, New Hampshire, Vermont, Massachusetts, Connecticut, and Rhode Island) because of a broad similarity across this six-state area in vegetation and flora, natural disturbance regimes, and cultural history. Overall, the physiographic divisions, the broad patterns of topography, and regional variation in soil characteristics in New England are closely tied to geology. Local relief has been strongly modified by some twenty episodes of glaciation over the past 2 million years, and the landscape patterns of soils, stream drainage, and topography are largely a consequence of interactions between the bedrock geology and the erosional and depositional history of the most recent glacial period. New England is a predominantly hilly region of broad highlands ranging from 200 to 500 meters above sea level (a.s.l.), with narrow valleys and a few broad lowlands that extend below an elevation of 200 meters. Six major physiographic areas include the Champlain Valley, Green Mountain Uplands, Connecticut River Valley, Central or Inland Uplands, and the Coastal Lowlands, which grade into a narrow Coastal Plain (Figure 2.2). Additional physiographic variation is found in the Taconic Highlands and Vermont Valley in the west and the New Brunswick Highlands and St. John Lowlands in Maine. At both the broad scale and across local landscapes, much of this variation occurs through alternating valleys and uplands that trend north to south because of the structure of the underlying bedrock.

Physiographic Divisions

The *Champlain Valley* of Vermont is part of the St. Lawrence Lowland that ranges from 15 to 130 kilometers in width and extends 700 kilometers from Newburgh, New York, to Quebec City. Composed of sedimentary rocks (shales, dolomites, and limestones) overlain by sedi-

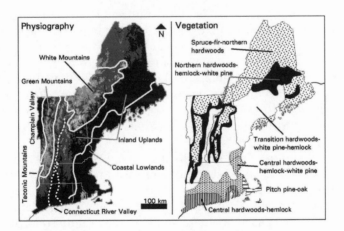

Figure 2.2. Elevation, physiography, and vegetation of New England. The broad lowland and gentle relief of the southern and eastern coasts gives way to more rugged upland terrain to the north and west. The Connecticut River Valley to the south and Champlain Valley to the northwest form prominent lowlands that interrupt this general pattern. Modified from Wright 1933 and Jorgensen 1977. Major vegetation zones are adapted from Westveld 1956.

ments from the Champlain Sea, the lowland has relatively rich soils, gentle relief, and elevations ranging from sea level to 100 meters. To the east, the *Green Mountain Uplands* emerge abruptly from the valley to reach elevations 400 to 1,200 meters a.s.l. and extend the length of New England. Metasedimentary and metavolcanic rocks comprise the bulk of these uplands, which include such familiar landmarks as the Hudson Highlands of New York, the Berkshire Hills of Massachusetts, and the Green Mountains of Vermont. Discontinuous areas of calcareous and alkaline rocks provide locally rich soils, high floristic diversity, and unusual plant assemblages. In general, this region has more productive soils and therefore a more diverse flora than the Central Uplands and the White Mountains.

The *Connecticut River Valley* separates the two large upland regions in New England and ranges from 2 to 35 kilometers in width in its 250-kilometer extent from Long Island Sound on the Connecticut coast to its headwaters beyond northern Vermont and New Hampshire. Underlain primarily by sandstone and shale to the south and metasedimentary and metavolcanic rocks to the north, the gentle relief and low elevation (largely less than 100 meters a.s.l.) is broadly controlled by glacial lake sediments and alluvial deposits interrupted by occasional bedrock ridges and domes. To the east and north, the *Central Uplands* comprise the largest physiographic region and include the White Mountains, with the tallest peaks in New England (Mount Washington at 1,886 meters a.s.l. and other peaks in the Presidential Range), and the intensive study areas of the Harvard Forest LTER program (Figure 2.3). The rocks of this region are variable but tend to be of metasedimentary and metavolcanic origin and produce acidic soils of low nutrient status. Elevations range from 200 meters in the south to 500 meters and higher in the north.

The *Coastal Lowlands* form a 60- to 100-kilometer-wide belt that extends from coastal New Jersey to central Maine. Relief is generally low, the bedrock is highly variable, and the contact with the adjoining highland areas to the north and west is typically abrupt. The extensive Coastal Plain of New England is largely submerged off the Atlantic Coast, where it forms the Continental Shelf that includes Georges Bank and the Gulf of Maine.

Morainal and outwash deposits that have been modified by coastal processes since the last glaciation comprise the areas from Cape Cod southward through the islands of Nantucket and Martha's Vineyard in Massachusetts, Block Island in Rhode Island, and Long Island in New York state. These coastal areas are dominated by sandy soils, low elevation, and varied relief.

Across New England, strong gradients in precipitation, temperature, and length of growing season are driven largely by elevation and latitude and to a lesser extent by a moderating coastal influence that extends inland 5 to 20 kilometers. Mean annual temperature ranges from

Figure 2.3. View looking east across the margin of the Connecticut River Valley toward the Central Uplands of northern Massachusetts where the Harvard Forest is located. Photograph by D. R. Foster.

11°C in southern Connecticut to 4°C in the northern highlands of Vermont, New Hampshire, and Maine, whereas precipitation ranges from 88 to 125 centimeters and is distributed fairly evenly through the year. Gradual environmental gradients are restricted to areas of gentle relief such as the sloping uplands from southern Connecticut through the Harvard Forest and northward into the southern White Mountains. Elsewhere, relatively sharp transitions occur from lowland areas such as the Champlain and Connecticut valleys to the adjoining hills. These abrupt physiographic changes produce a tight juxtaposition of climate zones, geology, vegetation, and land-use history.

The forests of this environmentally complex region are described fairly well by a classification and map developed in 1956 by a group headed by Marinius Westveld for the Society of American Foresters. Vegetation was described in terms of its natural "undisturbed" composition, as well as the important successional taxa in each region that had increased in abundance because of land-use activities or other disturbances. A simplified version of this Westveld classification includes five forest vegetation zones: Spruce–Fir–Northern Hardwoods, Northern Hardwoods–Hemlock–White Pine, Transition Hardwoods, Central Hardwoods, and Pitch Pine–Oak (Figure 2.2). Interestingly, Westveld's map agrees quite well with recent reconstructions based on witness tree data and led by ecologist Charlie Cogbill as a collaborative effort of the Hubbard Brook and Harvard Forest research groups (Figure 2.4).

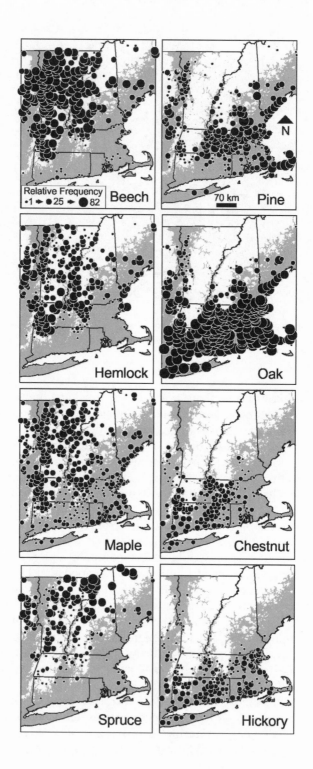

The *Spruce-Fir* and *Northern Hardwoods* zones cover northern and eastern Maine, extending south and west through the higher elevations of the Green and White Mountains of Vermont and New Hampshire and the Berkshires of Massachusetts. Characteristic hardwood species are beech, yellow birch, and sugar maple, with paper birch, white ash, red maple, and aspen on cut-over, wet, or high-elevation sites. Red and white spruce and balsam fir predominate at higher elevations, in old fields, and along the Maine coast, whereas black spruce occupies poorly drained sites. In southern and lower areas, hemlock and white pine are more common than spruce and fir.

Transition Hardwood forest covers low elevations in New Hampshire and Vermont, southern Maine, central Massachusetts, and northwestern Connecticut where Northern Hardwood species overlap with the oaks and hickories of the Central Hardwood region. White pine and hemlock are common conifers, and indeed in earlier classifications this area was called the "white pine region" because of the extensive white pine on old abandoned farm fields. However, when this first generation of pines is cut or otherwise removed, it is generally succeeded by a mixture of hardwood species. This zone includes Petersham, which helps to explain the relatively diverse vegetation across the Harvard Forest. Transition Hardwoods grade into the *Central Hardwood forest* of Connecticut, Rhode Island, eastern Massachusetts, and the Connecticut River Valley of southern New England. Typical tree species include black, red, and white oaks; red maple; black birch; and pignut and shagbark hickory.

On the sandy soils of Cape Cod and the Islands, along with scattered sand plains elsewhere in southern New England, *Pitch Pine–Oak forests* occur. Dry site conditions, long and intensive human land use, and relatively frequent fires control the composition and growth of these forests. In addition to pitch pine, typical trees include white oak, scarlet oak, black oak, and sassafras, with scrub oak and heath species in the understory.

Central Massachusetts Subregion

The central Massachusetts area constitutes portions of four counties in the north-central part of the state and encompasses physical, biological, and cultural gradients that vary across three major physio-

Figure 2.4. Distribution and abundance of major tree taxa across New England at the time of European settlement. The witness tree data that were used to compile these maps were derived from property surveys conducted for individual towns at the time of their settlement. Massachusetts was the location of abundant white pine and a relatively sharp shift from southern taxa (hickory, chestnut, oaks) to northern taxa (beech, maple, hemlock, and spruce). White areas are uplands higher than 225 meters in elevation. Based on data from Cogbill et al. 2002.

Figure 2.5. Nearly unbroken forest fills the view of the Central Uplands region, including the northern part of the Harvard Forest and town of Petersham on the distant ridge. Photograph by D. R. Foster.

graphic regions that differ in relief, geology, soils, land-use history, and climate. The broadest physiographic area is the Central Uplands (Figure 2.5), which in north-central Massachusetts is characterized by north-south trending hills and narrow valleys from 150 to 430 meters a.s.l. The bedrock of gneisses, schists, and granites is overlain by thin glacial till on the uplands and deeper and more level outwash, alluvial deposits, and peats in the narrow valleys. Soils are acid sandy loams of low nutrient status.

With the exception of developed areas, lakes, and marshes, the Central Uplands are predominantly forested, with few remaining farms. Upland villages with low population density are scattered across the forested hills, whereas larger industrial towns and small cities border some of the major streams. The Quabbin Reservoir, which was created in the 1930s to provide drinking water for more than one-quarter of the Massachusetts population living in the Boston metropolitan area, forms a 10,000-hectare water body surrounded by approximately 23,000 hectares of land owned and managed by the Metropolitan District Commission, a state agency. The reservoir occupies the historic Swift River

Valley and portions of the former towns of Dana, Prescott, Greenwich, and Enfield. The tops of bedrock hills that emerged through the old valley floor now persist as parallel north-south islands in the reservoir. The Quabbin Reservation forms the largest piece of an extensive, though loosely affiliated, conservation partnership, the North Quabbin Regional Land Partnership (NQRLP). Lands in the North Quabbin Region are protected from commercial and residential development and are managed by federal and state agencies, nonprofit conservation organizations, and educational institutions including the Harvard Forest. Most of the land in the area and, indeed, in New England is in small private ownership. Much of the forested land is actively managed for wood products and other values.

To the west, the Massachusetts portion of the Connecticut River Valley is easily distinguished by topography, vegetation, and land use. The level to rolling plains at 30 to 75 meters a.s.l. are underlain by sedimentary sandstone and shale and support level deposits of outwash, alluvium, and glacial lake sediments. Soils range from excessively well-drained, sandy outwash to poorly drained, silty, flood-plain sediments. A series of bedrock ridges composed mainly of volcanic basalt ("traprock") emerge through the valley bottom and reach a maximum height of 400 meters a.s.l. in the Pocumtuck Range. The rich and fertile soils, level terrain, ease of river navigation, and long settlement history by Native Americans and Europeans have led to a modern cover of extensive farmland; concentrated urban, industrial, and residential areas; and discontinuous forests. In contrast, the traprock ridges remain wooded. The diverse environments offered by the flood plain, rich alluvial soils, sandy outwash, and basaltic and sedimentary outcrops support a remarkable array of unusual plant assemblages. These threatened communities and the great cultural heritage make the Connecticut River Valley a priority area for state and national conservation.

To the east, the Central Uplands grade gradually into the Eastern Lowland, which is part of the extensive Coastal Lowland. This area of hills, gentle relief, and meandering rivers set in broad valleys ranges from 40 to 200 meters a.s.l. Acid bedrock is overlain by till, broad glacial-lake sediments, alluvium, and marine deposits. The region grades eastward from rural, agricultural, and forested areas into the densely populated suburban and high-technology region adjoining Boston.

The physiographic variation across the central Massachusetts subregion yields subtle gradients in environment, history, and vegetation. The Connecticut River Valley and Eastern Lowland have low elevations, gentle relief, and mild climates, with summer and winter temperatures that average 2° to 3°C warmer than the intervening Central Uplands. In response to this climatic variation, southern plant species decline on the Uplands, and the Northern Hardwoods-Hemlock forest extends southward from New Hampshire onto this area of higher elevation. Broadly, the Connecticut River Valley is lower, has more nutrient-rich soils, and is more agricultural than the Eastern Lowlands.

Petersham and the Harvard Forest

The township of Petersham, Massachusetts, contains most of the Harvard Forest and is typical of many rural hill towns in the Central Uplands (Figure 2.6). Embracing an original area of approximately 10,000 hectares, Petersham was nearly doubled in extent in 1938 upon dissolution of four towns in the Swift River Valley to form the Quabbin Reservoir and its forest reservation. The town lies in an undulating part of northern Worcester County where broad north-south trending hills of schists and gabbros overlain by a thin mantle of glacial till are separated by narrow valleys with rocky slopes and small streams, wetlands, and mill ponds. Elevations range from 150 to 300 meters a.s.l, but even this minor variation can lead to substantial differences in local environment (Figure 2.7).

Like much of central New England, Petersham supports extensive

Figure 2.6. The town common (village green) and Unitarian Church in the center of Petersham, Massachusetts. Photograph by D. R. Foster.

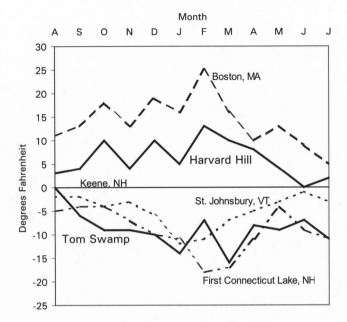

Figure 2.7. Variation in topographic position and vegetation in New England can lead to extreme differences in temperature between sites less than 5 kilometers apart. The figure shows deviations in monthly minimum temperature from that at Keene, New Hampshire (zero line), for two sites at the Harvard Forest (Harvard Hill—an upland site above Tom Swamp; Tom Swamp—a valley bottom site; see Figure 2.8), and areas in southern (Boston), northern (St. Johnsbury, VT), and extreme northern (First Connecticut Lake, NH) New England. Because of cold air drainage and radiative cooling caused by a thin canopy, the wetland site is markedly cooler than the adjoining upland forest. The differences between the local sites at the Harvard Forest are of the same magnitude as regional differences among sites separated latitudinally by more than 200 kilometers. Modified from Rasche 1958.

forests that are 50 to 100 years in age and that arose following succession on old abandoned agricultural fields in the late nineteenth or early twentieth century, logging, or blowdown and salvage logging due to the 1938 hurricane (see Figures 1.3, 1.4). Forest composition varies with history and site conditions with mixed hardwood stands of red oak, black oak, white oak, red maple, white ash, black birch, and paper birch on sites that were cut or blown down; white pine on old-field sites that were not cut or blown down; red maple on poorly drained lowland swales; and hemlock, often with white pine and hardwoods, in older forests that were cut historically as woodlots but never cleared for agriculture. Shrub-dominated and forested wetlands occupy muck and peat soils. A limited amount of agricultural land remains on the broad ridges in the central and eastern part of town that support the most productive soils in this landscape.

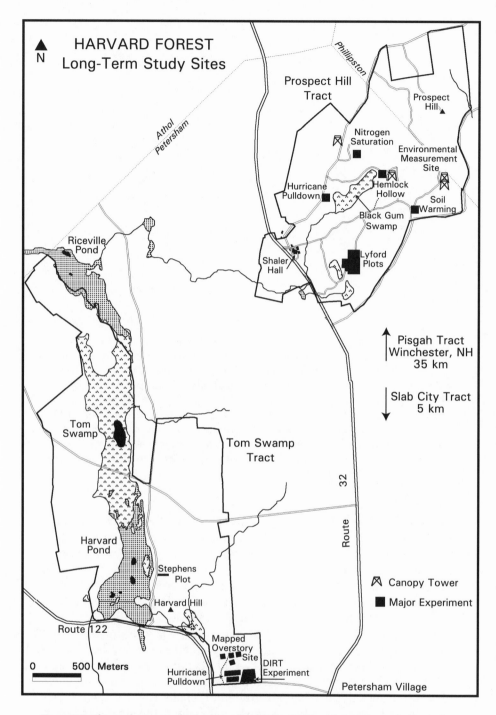

Figure 2.8. The northern part of Petersham showing the location of major research sites and experiments on the Prospect Hill and Tom Swamp tracts of the Harvard Forest.

The Harvard Forest has maintained a research and educational program focused on forest ecology, biology, and management since its establishment in 1907. The continuity of mission and data collection provides a unique long-term record on the more than 1,200 hectares of land that compose the three main parcels of land. These areas—the Prospect Hill, Tom Swamp, and Slab City tracts—encompass the broad range of site conditions, vegetation, and land-use history that characterize the Central Uplands (Figure 2.8). The Prospect Hill tract occupies a predominantly upland location in the northern part of Petersham and the adjacent town of Phillipston and includes the highest elevation in the area (Prospect Hill, 417 meters a.s.l.) as well as the headquarters and main research facilities at the Forest. As described so well in Hugh Raup's "The View from John Sanderson's Farm," the gentle, upland topography was extensively cleared and farmed through the early 1900s except for the large Black Gum Swamp, an adjoining hemlock woodlot, and smaller wetlands in the center of the tract.

The Tom Swamp tract extends downslope and westward from Prospect Hill and the center of Petersham to include Harvard Pond and large areas of sedge, shrub, and black spruce swamp in the bottom of the valley. The wetland areas and steep and rocky hill slopes were never cleared for agriculture, which was largely confined to the gentler terrain toward the hill crests on either side of the valley. The Slab City tract extends in a narrow series of properties south from the center of Petersham and down a steep and narrow tributary to the East Branch of the Swift River, which is the largest stream feeding the Quabbin Reservoir. Old-field forests on the gentle hills give way to large second-growth hardwood forests on the valley slopes and extensive areas of older-growth hemlock, white pine, and hardwoods on the protected slopes and valley bottoms adjacent to the river. This large, older-growth area, along with an old mixed-hardwood and hemlock forest on the slopes above Tom Swamp, make up the major research natural areas in the Forest.

CHAPTER 3

Biogeochemistry

The Physiology of Ecosystems

J. ABER

Topography, species composition, and land use are characteristics that can be directly viewed. Patterns of change across the regional landscape are accessible to an interested observer. In contrast, important processes that determine ecosystem function, and the role of ecosystems in regional environmental quality, are invisible. Leaves exchange water for carbon through small pores, but both water vapor and CO_2 are invisible gases. Pollutants such as ozone can alter the rates of exchange, but both the pollutant and the process are inaccessible to our vision and can be "seen" only with various instruments. Because of this, important ecosystem processes are less intuitive than structural elements and may require some introduction.

Biogeochemistry is one term that has come to be associated with the measurement of those functions of ecosystems that we study at the Harvard Forest. As the name implies, it is an interdisciplinary endeavor, combining elements of biology, geology, and chemistry. Several other disciplines could be added as well (hydrology, microbiology, and physics), making the name completely intractable! Biogeochemistry can be thought of as the metabolism or physiology of the landscape. It is the study of the movements of energy and materials through ecosystems, which are the units of study within the landscape.

Many of the important ecosystem services through which environmental quality is maintained rely upon exchanges between ecosystems and the atmosphere, or ecosystems and streams or groundwater. Forest stands at the Harvard Forest now take up more CO_2 through photosynthesis than they give off through respiration, thereby reducing atmospheric concentrations of this greenhouse gas. Excess nitrogen in rainfall derived from air pollution is strongly retained within these same forests, providing protection against acidification of soils and streams. Because of the historical resurgence of forests over the past century, the physiology of the New England landscape is determined largely by the biogeochemistry of ecosystems like those we study at the Harvard Forest. Will those functions change over time? One part of predicting our regional environmental future requires that we understand ecosystem function (or biogeochemistry) and its dynamics over time.

Energy, Carbon, Nitrogen, and Water: Interactions among Resources

Plants require energy (sunlight), carbon (CO_2), nitrogen (mostly in simple inorganic forms), and water as well as a number of other elements and compounds to produce new organic matter (also called *biomass*). A central concept in biogeochemistry is that we should not think of these resources individually, but should focus on interactions among them. As an example, we can present the interacting cycles of water, carbon, and nitrogen that form the basis of most of the biogeochemical research at the Harvard Forest.

The energy for all processes occurring in the forest comes ultimately from solar radiation. Two key processes driven by sunlight are *photosynthesis* and *transpiration* (Figure 3.1). Photosynthesis converts CO_2 in the atmosphere to simple sugars, which are then used in the plant for both energy and as the first building blocks in producing plant tissues. Photosynthesis requires that the leaves be open to the atmosphere so that CO_2 can diffuse into the intercellular spaces in the leaves where the actual fixation of carbon occurs. Carbon dioxide enters the leaf through pores called *stomates,* and the rate of this exchange is a function of how many of these pores are present and how open they are (summarized in a term called *conductance,* or the opposite of resistance to gas exchange between the atmosphere and internal leaf air spaces).

Cell surfaces within the leaf are always moist, so water evaporates into the air spaces within the leaf. This water vapor will diffuse out through the stomates at a rate related to both the conductance of the leaf and the relative humidity of the atmosphere. This process of evaporation from within leaves is called transpiration. Water lost in this way must be replaced by uptake from the soil if the plant is to avoid wilting. This creates a large demand for water, which is ultimately provided by precipitation. If this demand is not met, and water stress develops within the plants, then the stomates will close to reduce water loss. This also reduces the exchange of CO_2 with the atmosphere, shutting off photosynthesis.

The reciprocal exchange of CO_2 and water in Figure 3.1 is summarized in the term *water-use efficiency* (the amount of carbon fixed per unit of water transpired) and is one example of a term that captures the interaction between the cycles of two resources, in this case water and carbon. Because of the low concentration of CO_2 in the atmosphere and the high humidity inside leaves, forests can lose much more water than they gain carbon. Depending on the humidity of the atmosphere, water loss can range from 20 to 500 grams per gram of carbon fixed (water-use

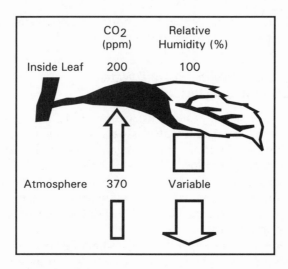

	CO₂ (ppm)	Relative Humidity (%)

Inside Leaf 200   100

Atmosphere 370   Variable

Figure 3.1. Leaves gain carbon dioxide (CO_2) from the atmosphere at the expense of water lost through transpiration. The concentration of CO_2 in the atmosphere is relatively constant day to day (although is it rising rapidly year to year), as is CO_2 inside the leaf. The lower concentrations in the leaf derive from fixation of CO_2 by photosynthesis and drive the diffusion of CO_2 into the leaf through stomatal pores. Similarly, plant cell surfaces are saturated with water, which evaporates into the internal leaf spaces and then out through the leaf pores. The rate of diffusion through the stomates into the atmosphere (transpiration) is controlled by the degree to which the stomates are open and the relative humidity of the atmosphere outside the leaf.

efficiency values of 0.002 to 0.05). This value will be lower still if it is calculated using the amount of biomass produced (photosynthesis minus plant respiration) rather than carbon fixed in photosynthesis.

Dark green leaves tend to fix carbon through photosynthesis more rapidly than lighter leaves. Darker leaves have more chlorophyll and also tend to have more of the proteins and enzymes required to carry out photosynthesis. As nitrogen is a major component of chlorophyll and proteins, the rate at which both photosynthesis and transpiration occur in leaves is partially a function of their nitrogen content, resulting in a strong three-way linkage among water, carbon, and nitrogen in ecosystems (Figure 3.2).

Very little of the nitrogen in this year's leaves came from the atmosphere this year. Although 78 percent of the atmosphere is nitrogen gas, the symbiotic relationship between plants and soil microbes that results in the direct fixation of nitrogen gas from the atmosphere into organic materials is relatively unimportant in the forests of New England. Atmospheric pollution may have increased the nitrogen content of precipitation by a factor of four or more. Even so, nitrogen deposition in rain

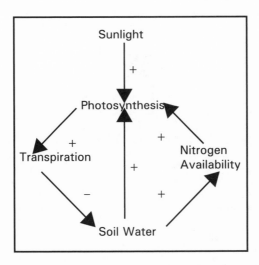

Figure 3.2. Schematic diagram of the interactions among energy, water, carbon, and nitrogen cycles (+ indicates a positive effect on the rate of a process, − a negative effect). Sunlight drives photosynthesis and transpiration. Transpiration reduces soil water content. Since high soil water content increases photosynthesis (and the lack of soil water decreases photosynthesis), continued photosynthesis in the absence of precipitation will eventually reduce both photosynthesis and transpiration. Reduced soil water content can also reduce nitrogen mineralization, and hence nitrogen availability, reducing nitrogen content in foliage the next year and the maximum rate at which photosynthesis can occur.

and snow is still only 5 to 15 percent of the annual requirement of nitrogen for forest growth.

Most of the nitrogen plants use comes from the large reserves of nitrogen in soil organic matter, which is released very slowly through microbial decay. Therefore, the rates at which carbon is fixed and water is evaporated from forests are both determined in part by the rate at which nitrogen is mineralized (converted from organic to inorganic forms available for plant uptake) in soils. In turn, the relative demand for nitrogen in the production of biomass (the nitrogen concentration in plant tissues) also varies between systems and can be expressed as a ratio of carbon to nitrogen (C:N ratio) in those materials, one measure of nitrogen-use efficiency, which again captures the interaction between two resources.

We can complete the complex set of interactions among these three resources by noting that nitrogen mineralization rates in soils are controlled by the relative content of carbon and nitrogen in soil organic matter (derived from the C:N ratio of plant materials), as well as by the fraction of soil pore space occupied by water.

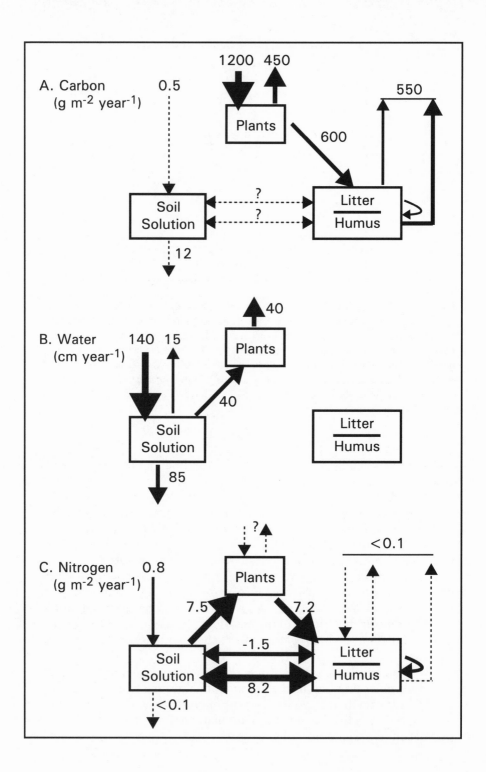

A. Carbon
(g m⁻² year⁻¹)

B. Water
(cm year⁻¹)

C. Nitrogen
(g m⁻² year⁻¹)

A General Comparison of Budgets

Keeping these interactions in mind, we can discuss the internal processes by which water, carbon, and nitrogen move through ecosystems. The three boxes in Figure 3.3 represent storage of elements in plants and in the soil, with two separate soil compartments shown. The soil solution contains both soil water and all the elements and chemicals, including organics, that are dissolved in that water at any one time. The Litter/Humus box contains elements stored in solid form, which, for carbon and nitrogen, is essentially soil organic matter. Microbial populations are included in this box. Arrows pointing up and down at the top of the diagrams indicate losses to and gains from the atmosphere. The arrow pointing down from the soil solution box indicates leaching of water below the rooting zone and into groundwater or streams. The arrows pointing from the soil solution to plants and from plants to litter capture plant uptake and shedding of senescent leaves and twigs, and the death of whole trees, respectively. The arrow from litter to humus is the production of long-term stable organics from freshly fallen materials. Nutrients are exchanged with the soil solution during this process of litter decay and are ultimately released by the decomposition of humus. In each case, the presence and width of the arrows suggest the relative importance of each process. Typical rates of annual fluxes are also shown in the figure.

The Carbon Cycle

Carbon inputs to forest ecosystems are dominated by a single process, *gross photosynthesis,* which is also called gross primary production (GPP) or gross carbon exchange (GCE). Carbon fixed by photosynthesis is partitioned between the release of carbon back to the atmosphere through plant respiration and the production of new plant biomass, which is called *net primary production* (NPP). After plant materials are shed as litter, microbial decay provides the second major pathway for return of carbon to the atmosphere. Decomposition, however, is generally incomplete and leads to the production of humus, an amorphous, complex organic material that is highly resistant to microbial action and decays only slowly. The total net carbon balance of the system, the difference between gross photosynthesis and the total respi-

Figure 3.3. A more complete diagram of the movement of carbon (A), water (B), and nitrogen (C) through forest ecosystems. See text for a description of pathways. Numbers represent annual fluxes for each pathway for a typical deciduous hardwood forest at the Harvard Forest in the units specified. Note that both carbon and nitrogen contents of these aggrading forests are increasing, and that nitrogen loss below the rooting zone is essentially zero.

ration by plants and soil organisms, is called *net ecosystem production* (NEP) or *net ecosystem exchange* (NEE).

The dashed lines into and out of the soil solution box in Figure 3.3 represent the transfers of carbon as dissolved organics in precipitation, leaching, and exchange with soil organic matter. While these transfers are relatively small in comparison with the processes described above, they are receiving increasing attention for the role they may play in the chemistry of both soils and waters. The composition of these compounds is still poorly known but in soils seems to be similar to that of humus. Rates of production, decomposition, and chemical exchange with soil solids are also poorly know.

Swamped by CO_2 exchanges and so lost in Figure 3.3A are exchanges of other carbon-based trace gases. Among the most important of these are methane (CH_4), carbon monoxide (CO), and a host of reduced carbon compounds produced by plants (for example, hydrocarbons like isoprene). Total fluxes of these compounds are small and rarely affect the total carbon balance of upland forests. However, either by acting as greenhouse gases or by altering critical reaction pathways in the atmosphere, these compounds may play significant roles in atmospheric chemistry and global change.

The Water Cycle

The water cycle (Figure 3.3B) is also dominated by a single input from the atmosphere, precipitation, which is partitioned into three major outputs: transpiration (from leaves), evaporation (from soil or plant surfaces), and leaching or runoff to streams and groundwater. Internal storage in the soil solution is small relative to the inputs and losses in most forest ecosystems. It is this relatively small storage of water combined with the large and continuing demands for transpiration that can result in intermittent droughts even in areas, such as the Harvard Forest, where annual precipitation is high. Droughts of even two to three weeks in duration can affect soil water content, reducing the water available for transpiration and resulting in stomatal closure, reduced photosynthesis, and eventually reduced NPP. Dry soils also lead to a reduction in microbial activity in soils, reducing CO_2 evolution from soils, so the net effect of drought on NEE can be either positive or negative.

The Nitrogen Cycle

In contrast to water and carbon, inputs to the nitrogen cycle from the atmosphere (Figure 3.3C) come in many forms and are relatively small in comparison with internal cycling rates, at least for stands at the Harvard Forest. Inputs can be as gases, as particles, or in dissolved form or through fixation of nitrogen from the atmosphere. At the Harvard Forest, most nitrogen input is in inorganic form as nitrate (NO_3^-) and ammonium (NH_4^+). Since total nitrogen inputs are much lower than those for carbon or water, small fluxes in novel forms generated by human activity can be an important part of the total nitrogen balance and have received increasing attention. Outputs can also occur as gaseous or dissolved forms, and again the relative importance is a function of site conditions, disturbance, and hydrologic regime, among other factors. All of this means that net input/output balances of nitrogen in forests are difficult to determine accurately. However, these fluxes are small relative to internal recycling. Internal nitrogen pools and rates of cycling tend to change slowly in the absence of disturbance. As we will see in later chapters, the ability of human disturbance to alter nitrogen pools and cycling rates drastically, and the slow nature of recovery from such disturbances, is a major factor affecting current forest ecosystem function at the Harvard Forest.

Note the large arrows in Figure 3.3C pointing in both directions between the soil solution and litter and humus boxes. In general, the flow of nitrogen is into litter and from humus, relative to the plant-available forms in the soil solution, but the two-way arrows capture the processes of gross mineralization and immobilization that make up the net fluxes. Mineralization is the release of nitrogen in inorganic forms (first ammonium, and then possibly nitrate) from organic matter. These are the residual products of microbial activity under conditions of carbon (or energy) limitations. Litter, which is relatively rich in fresh plant products such as sugars and starches and low in nitrogen, is a rich source of energy for microbes. As the microbes grow on these substrates, they require more nitrogen than is present in the decaying material and actually compete effectively against plants for available ammonium and nitrate, taking it out of the soil solution (a process called *immobilization*). Thus, while individual microbes grow and die, taking up and releasing inorganic nitrogen, the net flux in energy-rich litter tends to be toward net immobilization, or removal of inorganic nitrogen from the soil solution (resulting in the negative number in Figure 3.3C). In contrast, humus is a nitrogen-rich material composed of very complex compounds that are difficult to attack and do not yield much of a net energy gain to microbes. Decomposition of humus generates little microbial nitrogen demand and releases nitrogen contained in humus to the soil solution as ammonium (resulting in the positive number in Figure 3.3C).

As with carbon, the study of nitrogen in dissolved organic forms has increased markedly. In relatively unpolluted areas, inputs and losses of nitrate and ammonium are fairly low, and both inputs and losses of dissolved forms can affect the overall balance significantly. In addition, evidence is increasing that plants can take up dissolved organic forms of nitrogen directly from soils, bypassing the need for complete mineralization by microbes. The importance of this form of nitrogen cycling, and the role that the intriguing root/fungal symbiosis known as mycorrhizae plays in this and other forms of nitrogen cycling, is an active area of research and will be discussed again in a later chapter.

Regional History and Landscape Dynamics

CHAPTER 4

The Environmental and Human History of New England

D. FOSTER, G. MOTZKIN, J. O'KEEFE, E. BOOSE,
D. ORWIG, J. FULLER, and B. HALL

Although sketched in broad outline above, the history of environmental dynamics, natural disturbance, and cultural changes provides major ecological insights and is a subject of intensive study for Harvard Forest researchers. In particular, an overview based largely on results from these studies underscores the historical interactions of climate change, human activity, and natural disturbance that continue to control modern ecological patterns and processes.

Long-Term Climate and Vegetation Change

Interpretations based primarily on records of pollen and other fossils preserved in the sediments of lakes and wetlands confirm that the environment and vegetation of New England have changed continually through time. Although the rate and extent of change have varied since the last Ice Age, precipitation, temperature, storminess, and growing season length have all been dynamic as a consequence of long-term changes in solar, orbital, global, and atmospheric processes (Figure 4.1). The changing environment has initiated shifts, some subtle and others quite substantial, in the range and relative abundance of plant and animal species and in the composition, structure, and function of forest ecosystems. The magnitude of these changes over time and the different and unusual combinations of organisms that have grown together in the past support the notion that forest ecosystems are highly variable in composition and resilient to many natural perturbations.

This long-term perspective underscores the complexity of environmental change and ecosystem response and the challenges that we face as we seek to interpret past scenarios and to anticipate the future. Early environmental reconstructions were largely based on the notion that the climate changes in a relatively simple fashion—for example, from cool and wet conditions to drier and warmer ones. These interpretations often assumed that analogs to past conditions could generally be matched by modern environments elsewhere on Earth. However, we now recog-

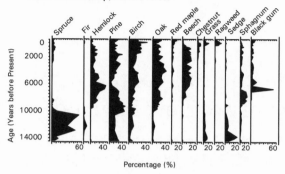

Black Gum Swamp, Harvard Forest

Figure 4.1. Pollen diagram from the Black Gum Swamp on the Prospect Hill tract of the Harvard Forest depicting the major changes in the vegetation over the past 14,000 years (see Figure 2.8 for location). Tundra vegetation was replaced by boreal spruce forest and then by temperate species, including pine, oak, hemlock, and beech approximately 9,000 years ago under a climate regime that was slightly warmer than today. A slight cooling over the past 2,000 years is apparent in the recent increase in spruce. Although most of the changes are driven by climate and species' migrations, the decline in hemlock about 5,000 years ago is the consequence of an insect pest. Pollen diagram, by T. Zebryk, adapted from Foster and Zebryk 1993, with permission from the Ecological Society of America.

nize that environmental changes are multifaceted and complex and often lack any modern equivalents. Not only the amount but also the seasonal and daily distribution of rainfall or temperature change through time, and these may vary as factors, such as storminess, atmospheric CO_2 concentrations, or animal and human populations, also change. As a consequence, the environmental conditions that have occurred in the past, and will develop in the future, may have no close parallels in any current landscape. In fact, the continual development of novel environmental settings through time is a major reason that fossil and other historical records attest to a long and changing sequence of unique and "nonanalog" plant and animal assemblages and ecosystem dynamics.

Changes in vegetation and the environment are most obvious over the relatively long time since the last glacial period. In the near-glacial environment that accompanied the wasting of the Laurentide ice sheet covering northern North America, the New England landscape stood in stark contrast to today. A major southward shift of the jet stream and dominance of the regional climate by arctic air masses favored a forest-tundra landscape in which herb, grass, and shrub communities dominated the uplands, and boreal forests of spruce, birch, and then pine were interspersed in sheltered sites. Large mammals, including mastodons, mammoths, giant beavers, and sloths, roamed the region, and extensive glacial lakes filled the valleys of the Connecticut, Champlain, Merrimack, and Nashua drainages. Broad coastal areas were exposed because of the drop in sea level, connecting Martha's Vineyard, Nantucket, and Block Island directly to the mainland. The resulting coastline configuration and detailed physiography were quite distinct from the present. Relatively rapid and large changes in precipitation, temperature, and wind conditions occurred about 10,000 years ago in response to regional and global shrinking of ice sheets and the northward shift of the warm Gulf Stream toward northern Canada, Greenland, and northwestern Europe.

By 8,000 years ago, temperate climatic conditions broadly similar to those of the present were established, and the major forest zones with which we are familiar today were in place—conifer forests in the mountains and across northern New England; mixed forest of broadleaf and conifer species in central New England; and oak-hardwood forest in southern New England (see Figure 2.2). Since that time, however, climate has continued to fluctuate, and important changes in forest composition and tree-species distributions have occurred. From 8,000 to approximately 5,000 years ago, conditions apparently 1° to 2°C warmer than today resulted in an expanded northern range and elevational extent of hemlock and white pine and a decrease in the abundance of boreal species, including spruce, across the region. Under warmer conditions many of our common tree species (for example, red maple, beech, and hickory) migrated into New England from southern ranges that they occupied through the glacial period.

Over the past 1,500 to 2,000 years, climate cooling across the Northeast has initiated significant changes in vegetation. A reduction in the latitudinal and elevational range of some trees was accompanied by a regional increase in spruce, presumably resulting from the expansion of populations that had persisted in local sites like wetlands (Figure 4.1). In one of the greater anomalies of New England vegetation history, at approximately the same time that spruce, a northern species, was increasing, chestnut, which has a southern and Appalachian distribution, made a delayed appearance and increased across Connecticut and Massachusetts. Subsequently, within the past 500 to 750 years, there has been a general decline in hemlock and beech, two trees that were abundant across central and northern New England and are locally important in many forests today.

This history of plant and animal migrations over thousands of kilometers in response to global climate change is one of the great biological stories of our landscape; it offers unusual insights into the selective pressures that have operated on species as well as their remarkable capacity for coping with major changes in the environment. It also underscores the humbling recognition that the conditions and ecosystems that we study today are only a minor subset of the range of possible or even typical conditions. Indeed, the species that live together in New England forests today occupied distinctly different distributions and environments from each other during the glacial period. As the climate warmed, these species, which differ in important life history traits such as longevity, seed production, and dispersal, migrated northward quite independently, along different routes and at different rates. The ability of tree, shrub, and herb species to move relatively rapidly (for example, at sustained rates of 400 to 1,000 meters per year) over a continental scale; the differential responses that they exhibit to climate change; and the fact that they may have undergone such migrations some twenty times over the past 2 million years present compelling evidence that biotic systems are composed of individualistic species highly adapted to environmental change.

Within the historical period since European settlement, the climate has continued to fluctuate, although the extent of alteration in vegetation and landscape by land-use activities makes it difficult to ascertain the role of climate in forest dynamics. Anecdotal and historical accounts of snow depth, ice cover, and temperature indicate that the New England climate was colder and more variable in the seventeenth through the nineteenth centuries than it is today. Ingenious approaches to climate reconstruction based on rigorous interpretations of the daily journals of farmers and rural residents confirm that this period witnessed much greater variation in the length of the growing season. In particular, late spring and early fall frosts were more frequent than today. These unpredictable conditions caused fluctuating crop yields and jeopardized human enterprises such as early settlements in coastal Virginia and New England, but their influence on native vegetation and environmental processes is largely conjectural. Similarly, we have limited information on rain or snowfall patterns before the first meteorological records of the mid-nineteenth century. One of the longest New England records comes from Amherst, Massachusetts, 45 kilometers to the south of Petersham. Here we see little overall change in precipitation since 1835 but notable droughts, especially in the 1960s. For temperature, there is a lengthy trend of an increase of approximately 1.5°C in mean annual temperature to the present (Figure 4.2). Interestingly, a comparison of the earliest forty-year period with an equivalent period a century later indicates a consistent increase in temperature for all months of the year. This long-term record roughly parallels the global average, which depicts rapid rises in the late 1800s, 1910–40, and since 1980 as well as many short-term fluctuations.

Thus, at a resolution of decades to thousands of years and on a landscape to regional scale, the record of vegetation and environment in New England is one of change. These changes have not involved a simple or progressive trend. Rather, they include complex alterations in interrelated environmental factors that trigger independent responses of individual plant and animal species. The relatively continuous and

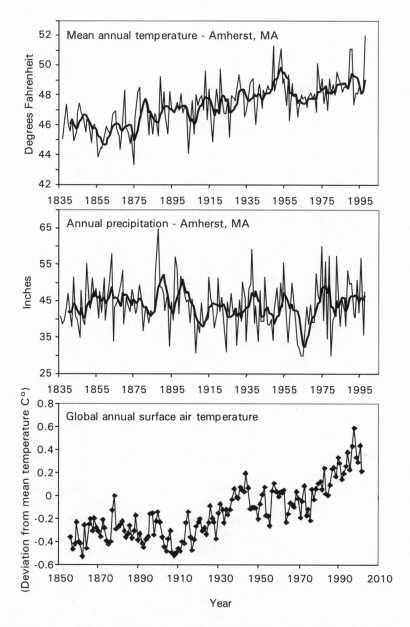

Figure 4.2. Long-term fluctuations in annual temperature (top) and precipitation (middle) as recorded at Amherst College, 40 kilometers southwest of the Harvard Forest, and global annual surface air temperature (bottom). The latter is depicted as the anomalies from the period 1961–90. Top panels, data from Bradley et al. 1987 and unpublished (used with permission of the author; Amherst data); bottom panel, data from the University of East Anglia Climate Research Unit.

complex nature of these changes is notable, as is the ability of organisms to shift in abundance and location in order to acclimate to them. A question that we will address in detail is, within this broad setting of progressively changing climate, what additional factors controlled the patterns of vegetation across the landscape at a finer scale?

Natural Disturbance Processes

Although the evidence remains geographically and historically incomplete, disturbances such as wind, pathogens, fire, and Native American activity were ecologically important factors in the pre-European landscape. The relative importance of each disturbance varied regionally and with changes in the climate and vegetation. For example, fire was apparently frequent and widespread during the period of boreal forest dominance approximately 10,000 years ago, for this was a time when flammable vegetation of spruce and pine and relatively dry weather created ideal conditions for large and intense crown fires. Fires were ignited by lightning and people and presumably created mosaics of vegetation and wildlife habitat, much as occur today in boreal wilderness areas from Alaska to Labrador. As temperatures warmed and hardwood species increased, the incidence of fire declined. High moisture levels in forests of broad-leaved trees keep flammability low until the fall or following spring when dry leaf litter provides fuel and open forest conditions increase air movement and decrease humidity. Consequently, dormant-season (early spring, late fall) surface fires dominate the broad-leaved forests of New England. Importantly, however, the potential for Indian activity to influence this regional fire regime will require us to reconsider fire in detail as we turn to prehistoric human activity later in this chapter.

It seems reasonable that meteorologically driven disturbances such as hurricanes, downbursts, thunderstorms, and ice and snow damage also varied through the postglacial period as the vegetation and climate changed. For example, research on tropical storms shows that the frequency and intensity of hurricanes along the Eastern seaboard are sensitive to such broadscale climatic factors as synoptic weather patterns in sub-Saharan Africa, surface temperatures of the Atlantic Ocean, and the position of the Bermuda high-pressure system. As these and other climatic parameters changed in the past, there should have been corresponding variation in the hurricane regime and its impact on our forest ecosystems. However, data to address this notion are scanty, and, in general, information regarding prehistoric disturbance regimes is poor. Consequently, as we discuss our findings on natural disturbance regimes here and in subsequent chapters, we try to highlight the questions that remain and that help to shape our ongoing studies.

Wind Damage

Wind damage was clearly important in the precolonial land-scape. Soil evidence for the uprooting of forest trees dates back nearly 1,000 years in sites across the Northeast, and the uneven mound and pit topography that characterizes old-growth and primary (that is, perma-nently forested) forests confirms the ubiquity of windthrows (Figure 4.3). However, the ecological role of wind varies greatly with the type, intensity, and frequency of the meteorological event; it is therefore criti-cal that we interpret the actual details of wind occurrence across New England as thoroughly as possible. For example, tornadoes create a nar-row track of intense damage tens to hundreds of meters wide that may skip irregularly over the ground to create a roughly linear pattern with little relationship to topography or vegetation structure. Downbursts, intense unidirectional winds often associated with frontal storms, may cover tens to hundreds of square kilometers and interact with vegetation and topography to form complex patterns. Meanwhile, tropical storms that reach hurricane intensity (greater than 74 miles per hour or 64 knots) may blow down extensive forest areas along tracks that are 50 to 100 kilometers in breadth and extend the length of New England. These extreme events contrast with our more typical winds that are generated by thunderstorms, northwesterlies, and frontal systems and break branches or blow down small groups of trees and confined patches of forest.

To assess the ecological importance of intense windstorms, two in-novative approaches were developed over time at the Harvard Forest: the historical reconstructive technique and meteorological modeling and reconstruction. Both have advanced our understanding of distur-bance and forest dynamics and both are suited for use in many different areas of the globe. Since intense wind events occur infrequently, these techniques seek to develop records that are centuries long.

THE HISTORICAL RECONSTRUCTIVE TECHNIQUE
The first approach that we use regularly in many studies is site based and provides a local record of disturbance and forest change. It was de-veloped in the 1940s by Earl Stephens, a doctoral student working with Hugh Raup. This technique goes beyond a simple determination of past wind events as it involves the assembly and interpretation of essentially all of the biological and physical clues to forest history contained in the vegetation, soils, and site itself. In many ways it formalizes the long tra-dition of historical observation and ecological detective work practiced by natural historians such as Henry Thoreau, John Muir, Frederick Clements, and George Nichols. Stephens termed it the *historical recon-structive technique,* and over more than five years he applied it in

Figure 4.3. Professor Hugh Raup next to a large black birch growing atop a mound created through the uprooting of an old-growth hemlock tree by a strong storm. As the soil mound erodes through time, the birch must continually adjust to changes in the substrate, producing large stiltlike roots. By determining the age of such trees, Earl Stephens, a doctoral student working with Raup, was able to ascertain a minimum date for storm events and assess the rate at which soils recovered after such major disturbances. The photo, from the Harvard Forest Archives, was taken adjacent to Stephens's intensive study site (see Figure 2.8).

painstaking detail to a 20-by-60-meter area in an old hemlock forest that occupies the wooded slope east of Harvard Pond and Tom Swamp.

From historical documents, including deeds and old maps, Stephens concluded that the site supported what we call *primary forest;* that is, it had been continuously wooded through the settlement period and had never been cleared for agriculture. Stephens's interpretation was corroborated by field reconnaissance, which showed that the rocky slope bore none of the evidence of past clearing such as stone walls, open-grown trees, or a soil surface smoothed by plowing or lengthy pasturing. After mapping all of the trees, he cut them down and sectioned them at 4-foot intervals. By counting the growth rings at each section, Stephens could reconstruct the tree's history of vertical and diameter growth and then assemble these individual records into maps and diagrams of forest stand development. Subsequently, he mapped, aged, and removed all fallen trees and woody debris, raked off the leaf litter, and surveyed the site topography down to a 6-inch contour. Having mapped all uproot mounds and pits appearing on the soil surface, he determined a minimum age for each feature from the age of trees growing on it and then dissected the pit and mound complex with deep soil trenches. Through careful study of the soil profiles, Stephens confirmed that the uprooting of trees mixes the soil substantially and that new soil horizons develop gradually over 500 years or more. A comprehensive record of forest dynamics in response to cutting as well as wind disturbance emerged from this compilation of historical, soil, and biotic records.

Stephens emerged from his classic study with a new technique and a history of the stand and its dynamics that was linked to hurricanes in the 1400s, the 1600s, 1815, and 1938 as well as a series of logging activities. Subsequently, David Henry, a graduate student working with Mark Swan in the 1960s, used Stephens's approach at the old-growth Pisgah Forest, a stand of 300-year-old white pine and hemlock that was windthrown in the 1938 hurricane (Figure 4.4). The outcome was a detailed chronology of fire, wind events, and forest dynamics for this virgin stand, set within a landscape perspective that had emerged from earlier studies by Richard Fisher, Al Cline, Steve Spurr, and their students. The studies by Stephens and Henry are landmarks in our understanding of northeastern forest process and history. For example, Henry and Swan documented that the old-growth forest that was windthrown in 1938 had itself developed after a major disturbance in the early 1600s, most likely a windstorm followed by fire. Both studies profoundly influenced ecological thinking concerning forest ecosystems by underscoring the importance of natural disturbance and forest dynamics.

METEOROLOGICAL MODELING AND RECONSTRUCTION
Although we use the historical reconstructive technique regularly in all of our field studies, we are also keenly interested in understanding

Figure 4.4. Standing snag of an old-growth white pine tree killed by the 1938 hurricane at the Harvard Pisgah tract in southwestern New Hampshire. Photograph by D. R. Foster.

broader patterns of wind disturbance. Recognizing the futility of pursuing detailed site reconstructions across the New England landscape, we turned to the more than 300-year-old regional historical and meteorological record to develop an approach for quantifying regional gradients and landscape variation in hurricane impacts. This approach is based on the fact that although tropical storm behavior is difficult to predict in advance, hurricane meteorology can be reconstructed and explained fairly simply after the event. Using this knowledge, Emery Boose developed an approach to reconstruct the hurricane disturbance regime for New England through modeling coupled with historical research. The ultimate objective was to compile a complete chronology of all historical hurricanes that caused damage to the forests. We used the technique to analyze sixty-one damaging hurricanes in great detail: thirty-seven storms during the modern period (1871–1997), when meteorological records are fairly complete, and twenty-four storms during the earlier period extending back to European settlement (1620–1870; Figure 4.5).

The historical analysis that initiated the reconstruction was based on contemporary accounts from sources across the region, primarily newspapers since the early 1700s and diaries for earlier storms. In particular, extensive research by David Ludlum was invaluable for locating sources before 1871. For each storm, reports of wind damage to trees, buildings, and other property were collected and indexed by town. Each report was assigned a damage rating on the Fujita or F-scale, a scheme proposed by Theodore Fujita at the University of Chicago for assessing tornado and hurricane damage. A regional map of observed wind damage for each hurricane was created using the maximum damage class for each town in the region. Of course, limitations are inherent in this method—for example, bias in the large numbers of observations from populated areas, fewer reports for earlier storms, variations in building construction practices from urban to rural areas and through time, difficulties separating wind from storm-surge damage in coastal towns or flooding elsewhere, and possible inaccuracies. Nonetheless, the resulting damage maps were geographically consistent with meteorological expectations, showing the greatest impacts to the right of the storm track (where the forward and rotational motions of the wind coincide) and a

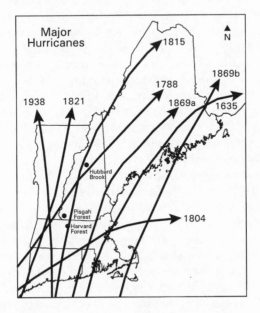

Figure 4.5. Tracks and dates of the major damaging hurricanes in New England since 1620. Modified from Boose et al. 2001, with permission from the Ecological Society of America.

lessening of damage across the region inland and to the north as hurricanes weakened over land (compare Figure 1.4). As expected, the completeness of the resulting maps was greater for severe and/or recent hurricanes.

To reconstruct the meteorology of each hurricane, we developed a simple model of hurricane surface winds (HURRECON) that utilized information on storm track, size, and intensity to predict the wind speed, direction, and damage. The model also takes into account whether the storm is over the land or water, as storms weaken and change in meteorology because of surface friction with the land and a loss of warm and moist sea air. The model generates estimates of wind characteristics for particular sites as well as regional maps of the characteristics and effects. The meteorological data that our reconstructions are based on come from the National Oceanic and Atmospheric Administration for the modern period (1871–1997) or from contemporary accounts of peak wind direction, storm surge, and wind damage for earlier storms (1620–1870). The reconstruction of each storm was checked against independent meteorological data and interpretations as well as the maps of actual wind damage. Results from individual storms were subsequently compiled into regional maps of hurricane frequency and intensity and time lines of events at particular locations.

Across New England there are strong gradients from the southeast to northwest in maximum storm intensity and the frequency of storms of a given intensity (Figure 4.6). The highest values of both occur along the shore of eastern Connecticut, Rhode Island, and southeastern Massachusetts, whereas the lowest values are in northern New England near the Canadian border. These gradients result from the rather consistent direction of the storm tracks (hurricanes approach New England from the south), the shape of the coastline (southern New England juts out into the path of these storm tracks), and the tendency for hurricanes to weaken rapidly over land or over the cold ocean water north of the Gulf Stream. Average return intervals across New England for F0 damage (defoliation, branch break, occasional blowdowns) range from 5 to 110 years; for F1 damage (isolated blowdowns), from about 10 years to none in 110 years; and for F2 damage (extensive blowdowns), from about 100 years to none in 375 years. Undoubtedly, the actual effects of each of the historical storms were influenced to some extent by differences in regional forest types, but there are few historical data to verify this.

On a landscape scale, local topography may modify wind flow considerably and exert a strong influence on damage patterns. In particular, hills or ridges may protect leeward areas and create sharp discontinuities in forest damage. In New England the most damaging hurricane winds normally come from the southeast because of the direction of the storm tracks (south to north), the rapid forward motion of the storms (which shifts the highest winds to the right of the storm track), and the

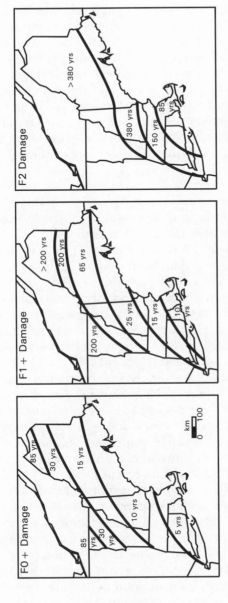

Figure 4.6. Gradients in hurricane frequency and intensity across New England since 1620. The Fujita scale provides an estimated range of damage from wind: F0—branches broken, trees damaged; F1—trees blown down; F2—extensive blowdowns; F3 (not shown)—most trees blown down. Reprinted from Boose et al. 2001, with permission from the Ecological Society of America.

inward, counterclockwise spiraling of winds at the land surface. To investigate these landscape patterns, Emery developed a simple model of exposure to wind (EXPOS) that utilizes digital elevation maps to predict which sites on a landscape are exposed to or protected from a given wind direction. The usefulness of this approach was demonstrated in two ways: by comparing predicted patterns of wind exposure with observed patterns of forest damage in two different settings—the Great 1938 Hurricane in New England and Hurricane Hugo in Puerto Rico— and by examining the distribution of white pine, a species that is highly susceptible to wind damage, in the town of Petersham before and after the 1938 hurricane. Results from the Petersham study showed that tall stands of mature white pine that survived the hurricane, and the large individual white pines that remain in the landscape today, are located almost exclusively in small areas that are topographically protected from southeast winds.

Forest response to winds of a given speed varies considerably as a function of the size, arrangement, and types of trees. Our early studies, based on data collected by graduate student Willett Rowlands in the windthrown landscape after the 1938 hurricane, showed that damage increased with forest height but that conifers (mostly white pine) sustained much greater damage than hardwoods of comparable height and exposure (Figure 4.7). Thus site factors (for example, geographic location, topographic position, soils) and prior disturbance history (such as wind, fire, disease, and land use, all of which influence forest age, height, and type) play a critical role in determining forest response to wind. In central New England, for example, the two most powerful hurricanes since European settlement (1815 and 1938) had strikingly different effects because of significant differences in land-use history and forest conditions. In 1815, most of New England was open agricultural land. The forests, which covered less than 30 percent of the area, were relatively young and short, cut-over stands with low susceptibility to wind damage. In contrast, in 1938, more than 60 percent of the landscape was forested, and an unusual abundance of highly susceptible white pine was established in abandoned pastures and fields. Although the two storms were of comparable strength (see Figure 4.8), the hurricane of 1938 thus caused much greater regional damage to forests.

The historical record shows considerable temporal variation in hurricane frequency (Figure 4.8). It is not unusual for New England to be struck by two or even three hurricanes in the same year, and at other times no hurricanes may occur for one or more decades. Longer-term trends on a scale of centuries are difficult to identify because there are fewer reports over a more restricted area in the earlier period. The total number of hurricanes reported does tend to increase somewhat over the entire 375-year period. This trend may simply reflect the bias of available historical data, but it may also indicate (at least in part) a real in-

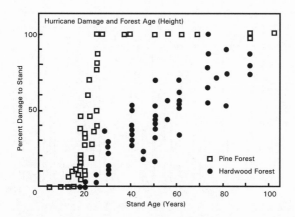

Figure 4.7. Relationship between damage from the 1938 hurricane and forest age (which is closely related to height) for white pine forests and oak–hickory–red maple forests at the Harvard Forest. White pine is among the most wind-prone tree species in the New England landscape. The intensity of the storm and the abundance of old-field white pine forests in 1938 due to succession on old farmland led to widespread forest damage. Modified from Foster 1988a, published by Blackwell Scientific Publications, by permission of the British Ecological Society.

crease in storm frequency caused by factors such as climate change, including warming associated with the end of the Little Ice Age in the nineteenth century. The most severe hurricanes are fairly evenly distributed in time, with the greatest number occurring in the 1800s; such major storms are unlikely to have escaped notice in the historical record.

At the local scale, hurricane damage is quite heterogeneous. Even the most severe storms, which produce broadscale blowdowns (F2 damage), leave extensive areas with only scattered tree falls, branch break, and defoliation. In fact, large hurricanes tend to produce a preponderance of small patches of varying damage intensity (Figure 4.9). Consequently, it appears that the role of hurricanes in producing frequent local and low-intensity damage has been largely overlooked by ecologists who have focused on the intensively damaged areas generated by occasional catastrophic storms. However, our results suggest that relatively frequent crown damage and small gap dynamics are also important attributes of the New England hurricane regime. The long-term impacts of forest damage and windthrow may include changes in structure, composition, coarse woody debris, soil topography, and susceptibility to fire (Figure 4.10). The effects of repeated minor damage (loss of leaves and small branches) are not well understood but may be significant when combined with other stresses such as drought or disease.

There is little doubt that the hurricane regime in New England has varied in the past and will vary in the future with changes in Earth's climate. If sea surface temperatures increase with global warming, for ex-

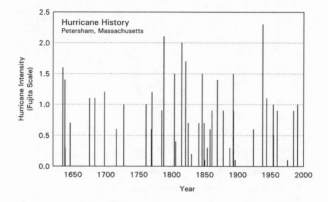

Figure 4.8. Reconstructed time line of hurricane occurrence and intensity at the Harvard Forest since 1620. Although hurricanes have had a frequent impact on the area, the 1938 storm stands out as the most intense in recorded history. See text and Figure 4.6 for a description of the Fujita scale of wind damage. Reprinted from Boose et al. 2001, with permission from the Ecological Society of America.

ample, then the theoretical upper limit on hurricane intensity will increase, but the effects on average hurricane intensity, frequency, and size remain unclear. Similarly, there is little evidence on which to predict whether the regions affected by hurricanes will expand or contract. In fact, some modeling results actually show a counterintuitive result of a decrease in predicted hurricane frequency under a doubled CO_2 (and therefore warmer climatic) regime. An alternative and empirical ap-

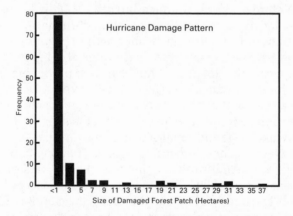

Figure 4.9. Size distribution of contiguous areas of similar forest damage resulting from the 1938 hurricane at the Harvard Forest. Although hurricanes are generally characterized as creating large windthrow areas, the fine-scale patchiness of the damage pattern is an important characteristic of wind disturbance. Redrawn from Foster and Boose 1992, published by Blackwell Scientific Publications, by permission of the British Ecological Society.

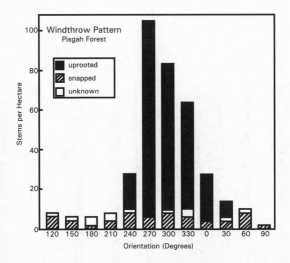

Figure 4.10. Orientation and damage type (uprooted, snapped, unknown origin of fall) of downed trees in the old-growth Pisgah Forest resulting from the 1938 hurricane. Whereas trees directly blown down by the storm are oriented to the west-northwest, trees that died standing and subsequently snapped and fell exhibit random orientations. Because of the prevalence of strong winds from the east and south in 1938, there is a distinct northwesterly orientation of old downed trees across much of New England. Modified from Foster 1988b, published by Blackwell Scientific Publications, by permission of the British Ecological Society.

proach to the problem is to expand our understanding of the hurricane and climate regimes of the past, thus lending new significance to studies of historical and prehistoric hurricanes.

Pests and Pathogens in Prehistory

The great number of insects and diseases that have been introduced in the Northeast in recent history raises many questions concerning the effect of natural and exotic pests and pathogens in controlling forest structure and function. To date, evidence exists for only a single major infestation by a forest pest in North America before European settlement: an insect outbreak on hemlock trees nearly 5,000 years ago (Figure 4.11). This event provides fascinating insights into ecosystem dynamics and landscape history as the insect inflicted broadscale and long-lasting impacts on forest composition and initiated many changes in forest and aquatic ecosystems. This mid-Holocene hemlock decline also yields a sobering backdrop for our efforts to assess the effects of the introduced hemlock woolly adelgid on northeastern forests.

Beginning approximately 4,800 years ago, a major and apparently synchronous decrease in eastern hemlock occurred across the tree's entire range in eastern North America. Although initially interpreted as a

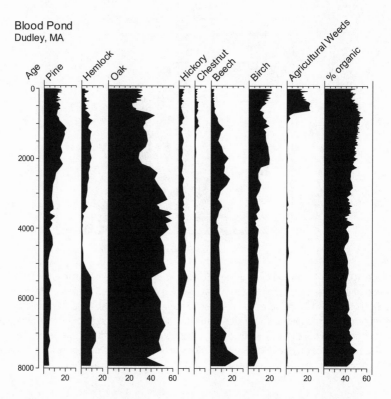

Figure 4.11. Pollen diagram from Massachusetts depicting the abrupt decline in hemlock and subsequent response of other species as a consequence of an insect outbreak approximately 5,000 years ago. Over the range of hemlock, this decline persisted for nearly 1,000 years before the species recovered. Unpublished diagram by B. Hansen (Harvard Forest Archives).

response to mid-Holocene climate warming, the rapidity and range-wide nature of the decline in a single species eventually led Margaret Davis to propose that the cause was a species-specific pest or pathogen. In fact, she eventually singled out the eastern hemlock looper, a native insect that occasionally irrupts in outbreaks that defoliate localized areas of hemlock today, as a likely culprit. Corroboration of Davis's ideas came in the 1990s when large numbers of looper fossils were identified by Najat Bhiry and Louise Filion in sediments that date to the time of the hemlock decline and contain chewed hemlock needles.

Following its abrupt decline, hemlock persisted at low though variable levels across its range for nearly 1,000 years before recovering. During this period, other taxa increased, notably sugar maple and beech in northern New England and oak to the south. What finally enabled hemlock to recover? Although we may never know for sure, it is possible that the species evolved a partial resistance to the looper through natural selection. Alternatively, the particular environmental conditions that enabled looper populations to thrive may have changed, allowing hemlock populations to rebound.

At the Harvard Forest, a paleoecological study that analyzed pollen records from paired sites, one a swamp that records vegetation dynamics from a broad area and the other a small vernal pool whose sedimentary record tracks local changes in the hemlock woodlot (see Figures 4.1, 6.2, and 6.6), enables us to evaluate many details of the hemlock decline. The swamp record indicates that across the region hemlock was replaced by oak, white pine, hickory, and sugar maple. Interestingly, the persistent though low levels of hemlock pollen through the decline suggests that a population of hemlock trees survived, perhaps in the forested swamp where the core was obtained. However, hemlock recovers only moderately in the ensuing millennia and never returns to its former abundance in the landscape. In contrast, in the hemlock woodlot, oak and pine increased as hemlock decreased, but subsequently hemlock recovered its prior abundance. Approximately 1,500 years after the initial decline, hemlock was again the dominant tree in this stand. In this record the decline in hemlock pollen also coincides with a discrete layer of charcoal in the sediments. This provides one unusual example of a phenomenon that is often discussed in the ecological literature: the linkage between fire and other disturbances, such as windstorm or pathogens. Evidently tree mortality and the accumulated leaf litter provided an abundance of fuel for an intense surface fire.

This unique and ancient history of forest response to an insect pest offers some interesting insights into the dynamics and history of forest ecosystems. Notably, (1) the response of forest communities to the hemlock decline differed across New England and the range of hemlock as various tree species benefited from hemlock's demise in different areas; (2) changes in forest composition continued for more than 500 years after hemlock began to decline, indicating a long-term adjustment of forest ecosystems to the loss of a dominant species; (3) in most areas hemlock populations recovered, but the species never resumed its former abundance, possibly because of slight environmental changes that occurred during the ensuing period; and (4) ecosystem-level responses in soil and stream chemistry and lake productivity were apparently minor, despite the loss of a major forest species. Finally, although hemlock populations did recover rather remarkably, this story offers little optimism for our modern forests that are under the onslaught of new pests and pathogens. The sobering reality is that the recovery process for hemlock required more than 1,000 years to play out.

The hemlock decline provides an important contrast to some natural disturbance processes, such as hurricanes and fire, in terms of the extended duration and geographical scope of its effect. It also underscores the importance of studying the consequences of modern pathogens and pests, such as the white pine blister rust, Dutch elm disease, chestnut blight, gypsy moth, and beech bark disease, in order to understand the changes that even unintentional human activity may bring to our forest ecosystems. The recent spread of the hemlock woolly adelgid, which is decimating hemlock stands across southern New England and the Appalachians, provides an unfortunate opportunity to evaluate insect spread to and impact on biotic systems (see below, Figure 4.27).

Native American Impacts and the Importance of Fire

One central but enigmatic issue in our quest to understand factors shaping the pre-European landscape is the effect of Native Americans on fire regimes, wildlife, and vegetation patterns. Since Indians did not have domesticated grazing animals and only used horticulture on a supplemental and localized basis, fire was the major mechanism through which they could affect forests on a broad scale. Understanding the prehistoric distribution and effect of Indians is directly pertinent to our interpretation of the "primeval" New England landscape and thereby figures into current discussions of conservation goals and management approaches. In fact, prescribed fire is increasingly used by organizations such as The Nature Conservancy, National Park Service, and state Natural Heritage and Endangered Species programs to manage lands for uncommon plants and animals, based largely on the notion that Indian fire was an important environmental factor structuring pre-European landscapes.

Recent major revisions in estimates of pre-Columbian Indian populations, coupled with new archaeological and paleoecological interpretations, have led to a reassessment of Indian impacts on American landscapes (Figure 4.12). Estimates for the population of the Americas in the fifteenth century range between 40 and 65 million inhabitants, nearly equal to that of the Old World at the time. These estimates exceed previous values, in large part because they account for the extraordinary effects that introduced disease exerted on Indian populations before they were chronicled by early explorers. However, this population was unevenly distributed across two massive continents, and its heterogeneity precludes generalization. Indeed, there were strong latitudinal gradients in human density and land-use impact, with approximately 20 million people spread across Mexico and Central America, 25 million in the Andes and lowland South America, 3 million in the Caribbean, and only 3.8 million (or fewer) in the entire North American continent. In North America the heaviest concentrations of people occurred in the South, Midwest, and Mississippi drainages. Consequently, whereas Aztec Mexico and the Mayan Lowlands supported cities such as Tenochtitlan, which in A.D. 200 exceeded the size of London a millennium later, northeastern Indians were scattered in small, seasonal encampments.

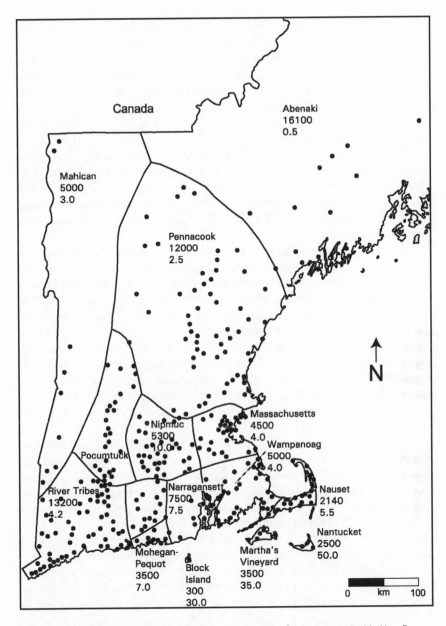

Figure 4.12. Distribution of major Indian groups at the time of European contact in New England. General locations of known sites are indicated along with an estimate of the group's population size and density (people per square mile). With the exception of some of the coastal islands, the population is estimated to be low (approximately 75,000 total) and dispersed in small groups. Compiled from Braun and Braun 1994, Cook 1976, and Whitney 1994.

Similar regional differences existed in the timing, development, and extent of important cultural innovations, including agriculture.

New England occupation extends over 10,000 years to the time when archaic people occupied the boreal landscape after the melting of the massive ice sheets. These and subsequent inhabitants were seasonally mobile, gathering diverse plant foods and hunting and fishing for a wide range of animals. Substantial shifts in regional populations and distribution occurred through time, presumably because of the changing availability of plants, such as nut-bearing trees, and wildlife and variation in climate. Streams, ponds, and coastal areas were important sources of fish and shellfish; freshwater and saltwater wetlands, which developed extensively 3,000 to 5,000 years ago, provided diverse waterfowl, small animals, and plants; and upland forests harbored nuts, plants, and game.

Important tools and foods that we often associate with New England Indians arrived or developed surprisingly recently and were accompanied by significant changes in culture and perhaps in ecological impact. In fact, the past 1,000 years embraced an especially dynamic period. In particular, so many fundamental changes are broadly coincident with European contact in the sixteenth century that a major challenge arises in reconciling archaeological and ethnographic sources of information for this complex period. Before 1,000 years ago, the spear and atlatl (spear thrower) were major hunting tools. Introduction of the bow and arrow since then improved hunting efficiency greatly. However, the most important cultural transition accompanied the adoption of corn, or maize, a Central American plant.

Archaeological sites before A.D. 1000 indicate limited horticulture based on the use of plants such as beans, squash, marsh elder, sunflowers, and knotwood. Evidence for maize appears sporadically across southern New England only about five centuries before extensive European contact. New England Indians evidently adopted corn agriculture slowly and unevenly, and people embracing agriculture may have become more sedentary. However, there is little evidence for large, semipermanent villages and large population centers even in broad valleys like the Connecticut or along the coast where marine, freshwater, and terrestrial resources were abundant. Meanwhile, across the interior uplands, populations remained small, and hunting and gathering appear to have predominated.

By the mid-1500s, a distinctive pattern of distribution and activity had developed, with recognizable subregional differences among coastal, riverine, and upland groups (Figure 4.12). Approximately 70,000 Indians occupied New England, with more than three-quarters concentrated in Connecticut, Massachusetts, Rhode Island, and southern New Hampshire. The densest populations occurred along the coast and major river valleys, but trail networks connected these regions with inland areas and seasonal encampments. The period from A.D. 1100 to 1500 may have represented the greatest extent of pre-European inhabitation of New England; however, by the time explorers and colonists provided detailed descriptions of Indian activities, the native population had been decimated by disease and was changing rapidly.

Forensic archaeology suggests that pre-European-contact Indians had a life expectancy of approximately thirty-seven years, comparable to that of contemporary Europeans; older Indians actually lived longer than their European counterparts. North America was largely free of the contagious diseases characteristic of the dense population centers of the Old World, and thus Indian history may not have included massive episodes of mortality such as associated with the Black Death of medieval Europe. That picture changed abruptly when Europeans arrived with, as Alfred Crosby has put it, "all of the microscopic parasites of humans, which had been collected from all parts of the known world." Infectious diseases transported from the Old to the New World include bubonic and pneumonic plagues, chicken pox, cholera, diphtheria, dysentery, influenza, measles, scarlet fever, tuberculosis, typhus, typhoid, whooping cough, and, among the most deadly, smallpox.

Diseases were transmitted so effectively across the ocean and landscape that some of the great epidemics in North America preceded the arrival in 1620 of the *Mayflower*. Indeed, the tribes that greeted the waves of colonists were only a shadow of the original populations and social organizations. Of note is the fact that the customs and demographic characteristics documented by such early colonists as William Bradford, John Josselyn, and William Wood were consequently affected by European contact and disrupted by widespread mortality. The most famous example of this phenomenon is the story of Squanto, who is known by most Americans as the Indian who taught the colonists the native agricultural practices they needed to survive the harsh New England conditions. In fact, the Squanto who readily greeted the colonists in English had already made a three-year voyage to England and back via Canada. By the time of his return, his entire village had died from disease, and the land was in ruins. Consequently, as we read of the "Indian" practices that he taught the colonists, we must wonder how much of his knowledge was indigenous and if any of it was influenced by his encounters and experience in the Old World.

Early epidemics presumably arose from the initial contact with explorers and fishermen. John Cabot's voyages to North America began in 1497, and French explorations by Giovanni da Verrazano in 1524 produced the earliest coastal maps, which were subsequently used by Jacques Cartier (1535–41) and Samuel de Champlain (1603 onward) and others. Meanwhile, the Basques established outposts in Labrador and the Maritime Provinces of Canada in 1536. The Indian propensity for using trade as a means of cultural contact and reduction of conflict and the European desire for furs provided ample opportunity for re-peated exposure and the spread of infection.

Epidemics occurred in 1535 in the St. Lawrence Valley and 1564–70 and 1586 in New England. During 1616–19 massive disease outbreaks reduced populations from Rhode Island to southern Maine by as much as 90 percent. The effects were clear to early explorers: as smallpox raged up the Connecticut River Valley to Vermont in the 1630s, it killed "thousands" of Indians but only two Europeans. From then on, history is replete with Indian diseases: influenza in 1647, diphtheria in 1659, smallpox again in 1662–63, a "strange disease" in 1675–76 during King Philip's War, smallpox reemerging in Vermont in 1684 and 1690, and so on. On the basis of the staggering loss of Native Americans during this time, some historians have come to call sixteenth- and seventeenth-cen-tury America the "widowed" rather than the "virgin" land.

This view is poignantly supported by Bradford and Winslow's de-scription of a 1620s scene in southeastern Massachusetts: "Thousands of men have lived there, which dyed in a great plague not long since; and pity it was and is to see, so many goodly fieldes, and so well seated, without men to dresse and manure the same."

In point of fact, however, the empty fields provided what the colo-nists viewed as a godsend, and Captain Thomas Dermer described (in 1626) as "ancient plantations, not long since populous, now utterly void." By reducing the number and strength of Indians and disrupting their social structure, disease aided the transformation of the landscape to European dominance. As elegantly documented by Alfred Crosby, disease assisted colonial expansion worldwide.

Perhaps in large part because of the impacts from European contact including disease, early descriptions of Indian settlements are difficult to resolve with archaeological results. Entering northern Vermont in 1604 along the lake now bearing his name, Champlain chronicled that "there is a great deal of land cleared up and planted with Indian corn" and that the open shores support "fertile fields of maize." In 1524 Ver-razano described New England as a populated agricultural landscape and noted fields in Rhode Island, eastern Massachusetts, Block Island, Nantucket, and Martha's Vineyard. In 1614, Captain John Smith counted forty Indian villages from Cape Cod to Penobscot Bay in southern Maine and wrote that "the sea coast as you pass shewes you all along large Corne fields." Captain Martin Pring spent six weeks in Plymouth, Mas-sachusetts, where he ate "Pease and Beanes" with Indians and noted fields that exceeded an acre in size, filled with vegetables and tobacco. When the Indians on Block Island were finally massacred in 1662 in re-taliation for the death of Captain John Oldman, the colonists proudly document that they destroyed more than 200 acres of Indian corn fields and sixty wigwams in the oak-sprout-covered forest.

Clearly, one possible Indian impact on the environment was the local clearing of forests for settlement and planting, a difficult task for people equipped with only fire and tools of wood and stone. Schematic drawings and chronicles by the early explorers suggest a pattern of short-fallow cultivation for maize and a landscape mosaic of gardens, forests picked of firewood, and brushy areas of abandoned old fields. In Boston Bay, Champlain documented that "there were also several fields entirely uncultivated, the land being allowed to remain fallow. When they wish to plant it, they set fire to the weeds, and then work it over with their wooden spades," suggesting that there was some alternation among cultivated and abandoned sites, presumably to allow the restoration of fertility.

The initial clearing of mature forests evidently involved the girdling of trees to kill them, burning and cutting to remove them, planting among the stumps, and eventual removal of stumps. In addition to the maintenance of villages and fields, Indians must have collected large quantities of firewood. Early historical records, by colonists arriving from a nearly deforested European landscape where firewood was scarce, include many references to the Indians' profligate waste of wood. Such observations suggest widespread wood gathering and clearing of forests around villages. However, the extent of Indian activity must have varied tremendously, and the intensive effects were undoubtedly focused on high-density areas in the major river valleys and coastal sites.

The Role of Fire

It has long been recognized that fire represents the mechanism by which Indians may have had the most pervasive effect on the landscape. Worldwide, native people have used fire to improve wildlife habitat, enhance agriculture, drive game, and open village sites. Nonetheless, the extent of fire-management practices in northeastern landscapes and their broadscale importance in controlling vegetation patterns continue to be widely debated. Conflicting interpretations emerge because of the scanty number of ethnographic sources, most of which are confined to the coastal region, and the potential for intentional or inadvertent bias. A few sources yield an impression of widespread or frequent burning that affected local forest conditions. For example, in Rhode Island Roger Williams indicated that Indians "burnt up all the underwoods in the Countrey, once or twice a yeare." Captain Martin Pring on Cape Cod, collecting sassafras, noted an Indian ignition "which we did behold to burne for a mile space," and in the early days of Plymouth the colonists viewed one "place where the savages had burnt the space of five miles in length."

The most often-cited references to the Indian practice of landscape management with fire come from Thomas Morton and William Wood, early inhabitants of Massachusetts Bay near present-day Boston. Although extremely limited in geographical scope, these observations provide specific detail on Indian motivations as well as the seasonal timing, behavior, and effect of fires on forests. Intriguingly, Wood explicitly links the precipitous decline of the Indian population from disease with decreased fire and corresponding increase in forest undergrowth.

> The Savages are accustomed, to set fire of the Country in all places where they come; and to burne it, twize a year, vixe at the Springe, and the fall of the leafe. The reason that mooves them to doe so, is because it would other wise be so overgrowne with underweedes, that it would be all a copice wood, and the people would not be able in any wise to passe through the Country out of a beaten path . . . for this custome hath bin continued from the beginninge. . . . For when the fire is once kindled, it dilates and spreads it selfe as well against, as with the winde; burning continually night and day, untill a shower of raine falls to quench it. And this custome of firing the Country is the meanes to make it passable, and by the meanes the trees growe here, and there as in our parks. (Thomas Morton, 1632)

> And whereas it is generally conceived, that the woods grow so thicke, that there is no more cleare ground than is hewed out by labour of man; it is nothing so; in many places, divers Acres being cleare, so that one may ride a hunting in most places of the land, if he will venture himselfe for being lost: there is no underwood saving in swamps and low grounds that are wet . . . for it being the custom of the Indians to burne the wood in November, when the grasse is withered, and leaves dryed, it consumes all the underwood, and rubbish, which otherwise would over grow the Country, making it unpassable, and spoil their much affected hunting; so that by this means in those places where the Indians inhabit, there is scarce a bush or bramble, or any cumbersome underwood to bee seene in the more champion ground. . . . In some places where the Indians dyed of the Plague some foureteene yeares agoe, is much underwood as in the mid way betwixt Wessaguscus and Plimouth, because it hath not been burned. (William Wood, 1634)

On the basis of such accounts it appears certain that Indians used fire knowledgeably to manage at least some parts of the forest landscape. The fires were probably low- to moderate-intensity surface fires, burning through the leaf litter and understory. In fact, the descriptions of Wood and others fit well with the current understanding of fuel loadings and fire dynamics that comes from the studies of Bill Patterson and others across the Northeast. Because of the high moisture content of green foliage and the humidity levels and low wind speeds in summer beneath a deciduous canopy, fires spread more easily when the plants are leafless. Consequently, woodland fires were most important during droughts in early spring and fall.

Broadleaf forests in New England will generally support only surface fires. However, with heavy loadings of fine or highly flammable fuels such as grasses, huckleberry, or scrub oak, intense fires and flame heights of 5 to 10 meters or more can occur in open oak forests. Dense pitch pine stands, which are abundant on Cape Cod, coastal islands like Martha's Vineyard, and inland outwash plains across the New England–New York region, can also support intense crown fires. Surface fires would clear the forest undergrowth and would selectively damage and kill smaller and thin-barked trees as well as less-fire-resistant species. The result would be an open understory and an increase in fire-tolerant and sprouting species such as oaks, hickories, birches, and pines over fire-sensitive hemlocks, maples, and beeches. Nonetheless, studies of prescribed burning on Cape Cod indicate that repeated fires rapidly consume the available fuel and reduce the frequency that fires can burn in any one-forested location to a few times per decade. The annual or twice-annual burning described by Wood and Morton is physically impossible.

Likewise, it is highly doubtful that fire was frequent or widespread across most of New England, especially in the upland areas where Indian activity was limited. Evidence obtained from charcoal fragments incorporated in wetlands, lake sediments, and small vernal pools indicates a regional gradient of fire with charcoal abundance and inferred fire frequency decreasing from southern coastal sites inland toward northern New England and higher elevations (Figure 4.13). This pattern parallels the broad climate and vegetation gradients, with more charcoal apparent in regions with warmer and longer growing seasons and on sandy, drought-prone soils, especially in coastal areas dominated by flammable vegetation. The fact that high charcoal abundance also coincides with regions with greater Indian populations makes disentangling environmental and human factors a great challenge.

Even in fire-prone areas, historical data suggest that fire frequency may have been only once every 10 to 100 years, whereas in the moist, rolling uplands, fire-free periods of more than 1,000 years clearly occurred. We note, however, that fire need not be frequent to exert important and long-lasting effects on forest structure, composition, and ecosystem function (Figure 4.14). At century-long intervals fire can still promote fire-resistant species and their associated effects on soils, wildlife, and productivity.

Our studies also show considerable landscape-level variation driven by local differences within regions. For example, the coastal area of Cape Cod and the Islands are not homogenous but are physiographically differentiated into morainal and outwash areas that are strikingly different in soil texture, topography, and vegetation. In general, the hilly moraines with finer textured soils and more mesic forests supported lower fire occurrence than the flatter, sandier, and more droughty outwash plains, with their flammable vegetation of oak, scrub oak, and huckleberry. An extreme example of this pattern is seen on the island of Martha's Vineyard. Harlock Pond, which lies in the moraine on the island's western flank, was surrounded by relatively mesic forests of

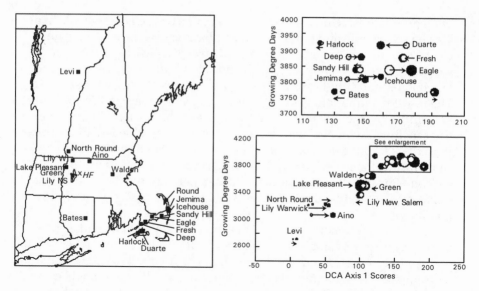

Figure 4.13. Historical importance of fire across the New England landscape as indicated by the occurrence of charcoal in lake sediments. Left: The distribution of lakes sampled. Right: The abundance of charcoal in the sediments of each lake during the pre-European (open circles) and European (black circles) periods. Arrows indicate position shift through time. Sites are displayed on axes representing climate (growing degree days, with low values indicating northern areas with a short growing season) and the first axis of a multivariate analysis (detrended correspondence analysis; DCA). The importance of fire clearly increases with longer growing season and warmer climate. Highest pre-European charcoal levels occur on Cape Cod and the southeastern island of Martha's Vineyard. For sites away from the coast, charcoal abundance and the importance of fire increase greatly after European settlement. Modified from Parshall and Foster 2002, with permission from Blackwell Science Ltd.

Figure 4.14. The long-term trajectory of vegetation change in the vicinity of Hemlock Hollow on the Prospect Hill tract (see Figures 2.8, 6.2, 6.8). Top panel shows vegetation changes over the past 11,000 years. Each dot represents one sediment sample, and its relative position indicates the vegetation composition. Through time the original forest of spruce changed to pine (approximately 9,350 years B.P.), then hemlock and hemlock–northern hardwood (approximately 8,350 years B.P.), and finally to hemlock–hardwood–spruce and chestnut (approximately 1,750 years B.P.), which persisted for the 1,500 years before European settlement. After colonization the forests were cut and burned, and the vegetation changed dramatically.

The four bottom panels provide details of the past 8,000 years from the top diagram and illustrate that within the general pattern of forest change driven by climate, there were periods of sudden compositional change resulting from fire (7,650; 6,650; 6,150; 4,700; 1,900 years B.P.), the hemlock decline (approximately 4,700 years B.P.), and European land-use activity. Modified from Foster and Zebryk 1993, with permission from the Ecological Society of America.

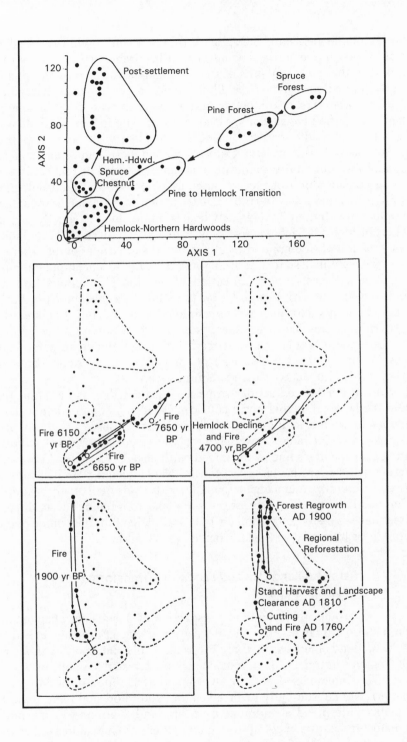

beech, maple, oak, and black gum. With the ocean upwind and vegetation of low flammability, the site had a fire frequency as low as that found in the mesic uplands of northern Massachusetts. In sharp contrast, Duarte Pond on the Great Plain, an oak- and scrub oak–dominated outwash plain just 8 kilometers to the east in the center of the island, had among the highest charcoal values of any site in New England and may have burned every few decades. Quite clearly, site and vegetation factors, as well as long-term feedbacks among these, can generate major spatial variation in fire regime on both landscape and regional scales.

In consideration of these results, as well as recent archaeological studies that downplay the role of maize agriculture and larger sedentary populations, Indian impacts may be best understood as locally important but highly variable across New England. These impacts included the creation of fine-grained local patchworks of vegetation of different age, successional status, and composition around encampments and seasonal villages, and modification of wildlife populations through hunting. Indian activities did alter the landscape that the early European explorers and colonists encountered and clearly influenced the pattern and practice of European settlement. However, no solid evidence remains that Native Americans influenced the broad patterns of vegetation or created sizable areas of open or early successional habitat. This scale of impact would await European arrival.

During the first century of European settlement, the colonists from the Old World were moving into a landscape shaped by depopulation. This land represented many things to the different observers and was described in a wide range of ways. It was, however, clearly diverse, striking in its beauty and resources, and already influenced by humans. Verrazano wrote in 1524, on the east coast toward New England, "We could see a stretch of country much higher than the sandy shore, with many beautiful fields and plains full of great forests, some sparse and some dense; and the trees have so many colors, and are so beautiful that they defy description. And these trees emit a sweet fragrance over a large area."

Historical Land Use and Landscape Transformations in New England

Through the long-term perspective of postglacial change and prehistoric human activity, the few centuries since European arrival to New England emerge as distinct. In a remarkably short period, much of the region was transformed from extensively forested to open and agrarian, with interspersed and cut-over woodlands. Equally rapidly, this pattern was substantially reversed as the land reforested naturally following widespread abandonment of farmland beginning in the mid-nineteenth century. The rapid rate and extent of change in land cover and vegetation; the local extirpation and immigration of animal and plant species; and the major shifts in biogeochemical, hydrologic, and biosphere-atmosphere exchange processes all make this brief historical period the most tumultuous since the glacial ice sheets retreated 13,000 years ago.

In exploring the ecological changes wrought since European colonization, our studies seek to assess the direct consequences of human activity as well as the major social underpinnings of these dynamics. Of course, the range of land-use effects extend far beyond land conversion to agricultural, residential, or commercial land cover, or forest succession on abandoned farmland. Therefore, we also explore subtle and indirect human effects such as the introduction of pests and pathogens, changes in wildlife composition, and altered chemistry of Earth's atmosphere. One inevitable conclusion from consideration of these diverse dimensions of human activity is that modern New England is a cultural landscape, shaped extensively in pattern, structure, and process by the far-reaching hand of human history.

Establishment and Expansion of the New England Colonies

With the establishment of Plymouth in 1620, European expansion and landscape transformation proceeded across Massachusetts and the rest of New England at an uneven pace (Figure 4.15). By 1675, small coastal settlements were established from Long Island to southern Maine, and farther inland towns extended from New Haven north up the Connecticut River Valley to the frontier town of Deerfield in western Massachusetts. Small population size, limited transportation, and ongoing hostilities with native groups over the next half century constrained expansion to coastal and lowland regions as populations and commercial activity increased. As the Indian wars drew to a close and European immigration increased in the mid-eighteenth century, colonists pushed rapidly inland and northward. By the early 1800s, New England was settled in established townships, save for northern Maine and the rugged elevations of the Green and White Mountains. An effective, albeit rudimentary, road system connected villages with the coast and distant trading partners. Consequently, although European knowledge of New England extends back to Verrazano's sightings in 1524, vast areas, including the region around Petersham, were not settled for another two centuries or more.

Forest Area and Human Population Dynamics

Population growth, social change, and economic transformation initiated a sequence of novel disturbances on New England ecosystems that continue to shape their structure and function today. The most important of these in terms of understanding the modern forest landscape

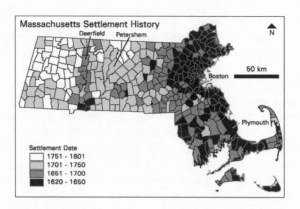

Figure 4.15. European settlement pattern across Massachusetts showing a general pattern of spread from the eastern coast and the Connecticut River Valley into interior upland areas. Dates indicate the actual timing of initial settlement and land clearance for each town, as opposed to the official date of town incorporation, which is oftentimes much later. Modified from O'Keefe and Foster 1998b.

is land-cover change. Although agricultural practices, woodland use, and local industry varied in detail from Connecticut to southern Maine, the overall similarities in land-use history and patterns of land-cover change are striking. These similarities, which are highlighted by the parallel trajectories in forest and open land among the six states are remarkable given the regional variation in physical and biological characteristics (Figure 1.6). Across New England, forest conversion to agriculture progressed rapidly through the mid-1800s, a time when open land peaked at 50 to 75 percent and exceeded 90 percent in individual townships. After a brief period of fairly stable landscape pattern, farmland was abandoned from active use and allowed to reforest naturally and haphazardly. Currently, forests cover from 60 percent to more than 90 percent of the New England upland, making it one of the most heavily forested regions in the United States. Similarity in this history across a heterogeneous region suggests that change was driven largely by broad social, demographic, and economic factors rather than by local changes in the quality of the land. This congruency in human history was particularly strong for the extensive uplands that support the major forest areas that are the focus of our interest.

Township Settlement, Forest Clearance, and Early Agriculture (1650–1790)

Early on, a unique form of land ownership and political organization emerged in New England based on dispersed settlement, private ownership, and a political hierarchy of town, county, and state. Towns (often approximately 100 square kilometers; see Figure 4.15) were granted to groups of individuals, and subsequent property divisions were made until most of the town area was privately owned. Initial land clearance proceeded slowly, land speculation was widespread, and agriculture served a small, reasonably self-sufficient population (Figure 4.16). Forest clearance ranged from less than 1 to more than 3 acres per year as farmers undertook the laborious process of cutting and burning the forest (Figures 4.17a and 4.17b). However, with continued immigration, improved transportation, and the emergence of markets across the region, throughout the Colonies, and abroad to the West Indies and Europe, commercial agriculture flourished and deforestation accelerated. In inland areas agriculture was based on commodities such as beef cattle driven to coastal and river ports, potash from the ashes of the forests, and timber floated to mills. All towns supported mixed agriculture, small industry, and commerce, and most individuals balanced farm work with household manufacture of such items as boots, shoes, hats, and brooms. Throughout New England's interior abundant wood resources provided fuel, building materials, household and farm products, and essential material for local sawmills, tanneries, coopers and other industries.

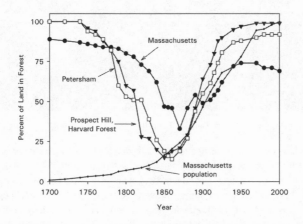

Figure 4.16. Changes in forest area in Massachusetts, Petersham, and the Prospect Hill tract of the Harvard Forest and state population. Population growth increased with industrialization in the mid-nineteenth century as people abandoned rural farms and concentrated in mill villages and cities. Information about land cover is derived from Dickson and McAfee 1988, MacConnell 1975, Rane 1908, and Baldwin 1942 for Massachusetts; from Raup and Carlson 1941, MacConnell and Niedzwiedz 1974, Cook 1917, and Rane 1908 for Petersham; and from Foster 1992 and Spurr 1950 for Harvard Forest. Population data compiled from U.S. censuses.

a) Pre-settlement Forest

b) Initial Clearing of the Land, A.D. 1740

Figure 4.17. Changes in the New England landscape depicted in the Harvard Forest dioramas and modeled after the landscape history of Petersham, Massachusetts. The same site and scene are shown through various stages, from the presettlement forest (a) to initial clearing (b), the height of agriculture (c), farm abandonment and establishment of white pine (d), logging of white pine (e), growth of hardwoods after logging (f), and finally the modern forest (g). Photographs (a through f) by J. Green, from Foster and O'Keefe 2000. The final scene (g) shows a stream and stone wall running through a modern forest in Petersham that developed through the same general history depicted in the dioramas. Photograph by D. R. Foster.

c) The Height of Agriculture, A.D. 1830

d) Farm Abandonment and White Pine Establishment, A.D. 1850

Figure 4.17. Continued

Commercial Agriculture and Local Industry (1790–1860)

Under intensive agriculture and expanded local industry, the landscape was transformed into the quintessential image of agrarian splendor, with many stately homes, white steepled churches, broad village greens ("commons"), and extensive fields and farms (Figure 4.17c). A well-developed system of local roads connected to regional turnpikes and regional travel and trade were surprisingly active. Coastal ports such as Salem and Boston and fishing and whaling centers such as Gloucester, New Bedford, and Nantucket were engaged in wide-ranging

e) Logging of Old-field White Pine, A.D. 1910

f) Growth of Sprout Hardwoods, A.D. 1930

Figure 4.17. Continued

international trade. This was an era of a relatively homogenous population distribution and agricultural landscape pattern that represents the extreme point in the physical transformation of the New England landscape (Figure 4.18). People lived close to the land and used it and its products actively.

Understanding the landscape and land-use patterns of this period is essential for interpreting modern landscapes. For example, areas that persisted as woodland in the mid-nineteenth century make up the majority of our *primary forests*: areas that may have been cut over or grazed but were never cleared of their native plant species. As a consequence of

g) The Modern Forest, A.D. 2002

Figure 4.17. Continued

this history, primary forests support many of their original species and are in general the least altered forest ecosystems. In parallel fashion, the practices of land clearance, plowing, pasturing, or manuring each exerted a distinct influence on the native plants, soil environment, and local hydrology. As we will see, these historical impacts persist, even on sites that were abandoned for agriculture and have been reforested for more than a century. In many ways, the present landscape is still recovering from the period of agricultural prosperity and intense impact that occurred 150 years ago. Therein lies one of the essential uses of history to the ecologist.

An analysis of farm census data for the mid-nineteenth century indicates three dominant land uses that we can arrange in order of increasing intensity of disturbance to the native vegetation and land: woodlot, pasture, and tilled land. Pastureland predominated across the region, as befits a generally hilly landscape of rocky, poor soils where the transportable crops were cattle, sheep, and other livestock that could be driven to market. Pastures were diverse in appearance and ecology, from sedge and tussock–dominated swales, to rocky hill slopes that retained some native species, to improved grasslands where the removal of larger stones, shallow plowing, and seeding provided good fodder. Plowed land occupied only about 5 percent of cleared area and produced grains for animal and human consumption such as corn, wheat, oats, rye, and barley and English hay, which was rotated with other crops or pasture. Limited transportation made local production of grain and hay critical.

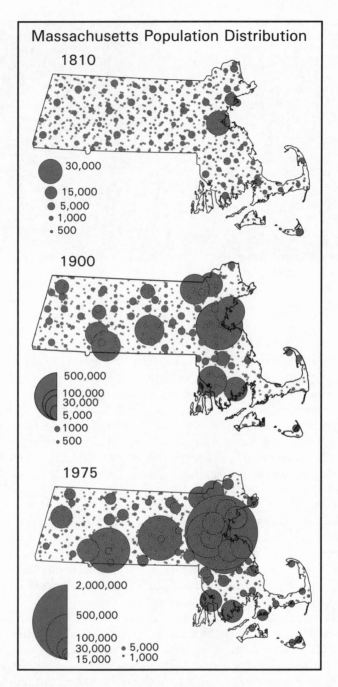

Figure 4.18. Changes in population size and distribution in Massachusetts over the past two centuries. The homogeneous distribution of people in small agricultural villages of 500 to 2,000 inhabitants changed abruptly with industrialization in the mid-nineteenth century as the population concentrated into urban centers and large mill towns. Data compiled from U.S. censuses, with maps modified from Wilkie and Tager 1991.

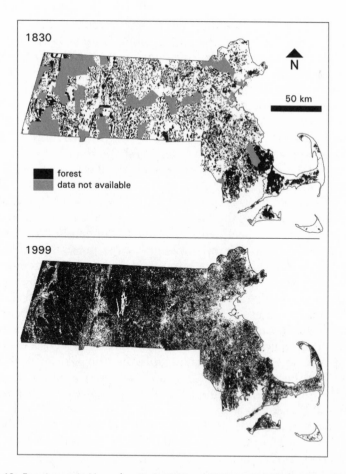

Figure 4.19. Forest cover in Massachusetts in 1830 and 1999. Much of the open area (white) in 1830 was farmland, whereas town centers and developed land contribute more to modern nonforested areas. Considerable variation in forest patterns exists across the state in both maps. In particular, large sandy outwash areas in southeastern Massachusetts and Cape Cod supported the largest forest blocks in the nineteenth century, whereas the hilly regions of north-central and western Massachusetts are most heavily forested today. Modified from Hall et al. 2002, with permission from Blackwell Science Ltd.; forest cover data for 1999 from MassGIS 2002. The original 1830 maps are stored at the State Archives in Boston, Massachusetts.

In order to develop a regional understanding of the broad patterns and variation in primary woodlands and farmland, we compiled statewide maps from 1830 that depict forest and open land cover. Across the state a general picture of deforestation, agricultural dominance, and fragmented woodlots emerges (Figure 4.19). Important variation is also notable: across the north-central region around the Harvard Forest, open areas were expansive and woodlots were small, numerous, and isolated;

to the west, woodlot size and total area increased on larger, rougher hills, and, quite strikingly, the largest block of forest occurs in southeastern Massachusetts on extensive sand plains at the base of Cape Cod. It is important to recognize that these nineteenth-century woodlands were seldom the tall and maturing forests of today but were primarily young, sprout woodlands of oak, chestnut, maple, and birch, disturbed by cutting, grazing, and fire. Woodlots were preferentially restricted to the poorer agricultural sites like wetlands, steep slopes, or remote rocky areas. Consequently, across the region, the shape, distribution, and composition of these woodland areas varied considerably with local physiography and soils.

Specific land-use patterns varied with soils and land ownership. On the Prospect Hill tract of the Harvard Forest, tilled land made up 15 percent of the area and occupied small fields bounded by massive stone walls, on well-drained loamy soils, and in close proximity to barns and farmhouses. Using hand tools, oxen, and horses farmers could work small areas, fitting agriculture very closely to slight variations in soil drainage, moisture, or rockiness. Depressions, swales, and drainageways became pastures, as did extensive areas at great distance from the buildings. On the Sanderson Farm on Prospect Hill, the one remaining woodlot occupied wetlands and poorly drained soils toward the back of the property. However, even swamp forest on deep peat can be logged in winter when the frozen ground and snow ease transportation. Historical reconstructions indicate that most of these stands were cut repeatedly from the late 1700s onward.

As documented in Hugh Raup's delightful article "The View from John Sanderson's Farm," the mid-1800s represented a time of prosperity and apparent stability in the New England countryside. Census data confirm that farmland productivity rose throughout this period. The development of rural town centers and the construction of expansive colonial homes through mid-century suggest that the inhabitants anticipated a long future working the land. Contrary to historical interpretation and popular assumption, the New England farmers did not scrape a hardscrabble existence from stony land. Nor did they exhaust the nutrient capital of the soils or their ability to produce crops. Instead, cultural changes quite external to the land brought the period of agricultural prosperity to an unanticipated close.

Industrialization and Reforestation (1860–1940)

Social and economic change across the eastern United States in the mid-nineteenth century was generated by at least four factors: the opening of the West following the Louisiana Purchase, improved transportation from the midwestern farmlands to the East and abroad, industrial development, and the discovery of gold on the West Coast. These forces brought competition to New England farmers and provided the lure of jobs and new lifestyles to rural families. In combination and with the added effects of the Civil War, they promoted regional exodus and massive redistribution of the eastern population from rural villages to developing industrial centers where they were joined by foreign immigrants (Figure 4.18). In upland regions, many town centers literally moved downhill as the construction of industries, shops, and houses shifted the focus from the agriculturally productive soils of the ridges to the valleys, where waterpower for industrialization coincided with transportation provided by the expanding railroad network (Figure 4.20). On larger rivers, planned industrial cities such as Lowell and Holyoke in Massachusetts, Manchester in New Hampshire, and Willimantic in Connecticut were developed to exploit water around a system of factories and canals. In a wholesale landscape reorganization the growing population focused into urban and eventually suburban areas, and reliance on the land declined. Increasingly, materials for local production and consumption were imported, and the need for local agriculture and natural resources declined. As the population moved away from the land, forests gradually returned (Figure 4.17d).

The timing, rate, and pattern of land abandonment and reforestation varied somewhat across New England, often controlled by distance from urban markets and regional patterns of soils and physiography. Southern New England reforested first, followed by New Hampshire, and gradually by Vermont, with its richer soils. Rough pastures were abandoned early and returned to forest most rapidly. Productive croplands in the major valleys and areas in reach of urban markets retained more agriculture. Farming narrowed from a diverse base that enabled regional self-sufficiency to products that could not be transported easily from outside the region, including perishable items such as milk, fruits, and vegetables and bulky materials such as hay and fuelwood. As a consequence, New England agriculture became identified with dairy farms, orchards, truck crops, or specialty items like shade tobacco for cigar wrappers.

Reforestation largely reversed patterns that were followed in initial clearing. In upland towns like Petersham, coarse pastures adjacent to woodlots on wetlands and steep slopes were abandoned earliest, whereas in the broader river valleys pastures on nearby hills were allowed to revert to forest first. In both cases, remaining farmland became increasingly restricted to the productive soils and areas closest to the farms. Regionally, this process led to the development of striking patterns of forest extent and shape (Figure 4.19). In the central Massachusetts uplands, forest currently makes up approximately 90 percent of the land area. Agricultural land is largely confined to narrow strips running north-south on hilltops and ridges. In the Connecticut River Valley or the Eastern Lowlands toward Boston, the productive soils continue to support

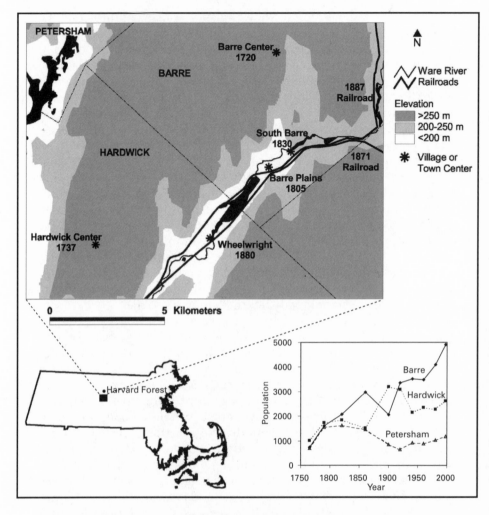

Figure 4.20. Development of the towns of Barre and Hardwick, Massachusetts, showing the original agricultural town centers on high elevations in the early eighteenth century and subsequent creation of nineteenth-century industrial village centers in the valley bottoms. Since the best agricultural lands are on the broad upland ridges, and the rivers provided power for industry, the towns' population centers literally moved downhill through time. The presence of waterpower and railroads transformed towns in the nineteenth century. Whereas the rural hill town of Petersham shows a population decline as agriculture waned, the industrial development of Barre and Hardwick allowed their populations to increase through the early 1900s. Population data from U.S. censuses.

extensive farmland, along with suburban and urban development. The bedrock ridges and larger swamps that were forested in the mid-nineteenth century continue to be so today, and the extensive areas of sandy outwash formerly dominated by pitch pine and scrub oak communities have been largely converted to urban uses, airfields, or industry.

Modern forest patterns provide many clues to landscape history (Figure 4.21). As fields were abandoned, opportunistic, light-seeded species such as white pine, gray and paper birch, and red maple and bird-dispersed species such as red cedar and black cherry were established from fencerows and surrounding woodlands. Pine and cedar, which are less palatable than the hardwoods, were also favored by cattle grazing, which frequently continued as lands were abandoned, and by their ability to compete with grasses in many fields. By far the most important vegetation to develop on abandoned fields from Connecticut to northern Vermont were forests of "old-field white pine." The extent of white pine

Figure 4.21. A typical landscape and topographical arrangement of vegetation corresponding to land-use history in central New England: open fields (valley bottom), white pine (lower to mid-slopes), hardwood forest (mid- to upper slopes), and hemlock (upper slopes). Forest was cleared for agriculture in the nineteenth century up to the lower boundary of the hemlock forest, which remained forested, though was cut. Abandonment of pastures on the upper slope gave rise to white pine stands that were subsequently logged and then regrew to hardwoods (see Figure 4.17). Later abandonment of the lower slope pastures and hayfields gave rise to the existing band of white pine forest. Current farming activities concentrate on the most productive and accessible lands in the valley bottom. Photograph from Winchester, New Hampshire, by D. R. Foster.

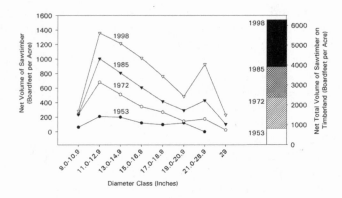

Figure 4.22. Change in forest size and maturity in Massachusetts over the past half-century as represented by sawtimber volume. The amount of wood increased continually in all size classes of tree (left) and in total volume (right). Modified from Berlik et al. 2002a, 2002b, with permission from Blackwell Science Ltd.

forest in the early twentieth century was sufficient to earn central New England the designation of "White Pine region" on forest maps.

As noted by Henry Thoreau in *Succession of Forest Trees* and as eventually relearned by foresters and ecologists across the region, the cutting of white or pitch pine on upland sites (Figure 4.17e) frequently gives rise to an ensuing forest of hardwoods such as maple, oak, birch, chestnut, and ash that previously grew alongside or beneath the pines (Figure 4.17f). The ability of the hardwoods to sprout and their greater tolerance of shade provide them with a competitive advantage over the pines. Consequently, we often see a landscape pattern in which primary woodlands support hardwood-hemlock forest, sites abandoned in the nineteenth century support hardwoods that succeeded cut pines, and white pines remain on the more recently abandoned sites and are adjoined by open fields.

Overall, the rate of reforestation in New England slowed through the twentieth century, and new development for housing, commercial activity, and highways initiated a decline in forest area in suburban areas. Despite the ongoing utilization of forests for a range of products, forest growth continues to exceed logging harvests. Having expanded greatly in extent over more than the past century, the forests of New England are continuing to grow and mature (Figures 4.17g and 4.22).

Historical Changes in Forest Use

Through history the remaining woodlands were subjected to a range of changing uses. Early on, wood availability exceeded need, and thus most cut trees were piled and burned in the process of land conversion. Transportation limited the contribution of wood products to the region's commercial base although oak and especially pine were selectively cut along major streams and were important exports from parts of New England. Only potash and pearl ash, derived from wood ash and useful as fertilizers and industrial chemicals, represented concentrated and valuable wood products that warranted transportation. These were widely produced across the region. Remote forests distant from waterways or farmland escaped initial impacts.

As forest area declined in the late eighteenth century and wood demands increased, the extent and intensity of cutting increased. Individual homes consumed 10 or more cords per year and accounted for up to 75 percent of wood consumption. Industrial demand for charcoal and fuel created regional patterns of concentrated activity. For instance, around the brass, lime, and iron industries focused on the Naugatuck Valley of Connecticut, thousands of acres of forest were clear-cut annually for charcoal. As technology enabled the production of paper from wood fiber as opposed to rag in the 1870s, pulpwood cutting spread across remaining forestlands. Wood-demanding industries were concentrated along major waterways such as the Connecticut River, and as river cities such as Holyoke, Massachusetts, became centers of the pulp and paper industry, far-flung regions in Vermont and New Hampshire became the source of fiber. The railroads augmented demand as they consumed small-diameter trees for fuel and ties and provided a regional transportation network for logs and wood products. In 1845, more than 54,000 cords of wood fueled trains on 560 miles of track in Massachusetts; by the early 1900s track length had increased fivefold.

The effect of rapacious cutting did not go unnoticed. In 1855 Thoreau wrote in his *Journals:* "Our woods are now so reduced that the chopping of this winter has been a cutting to the quick. At least we walkers feel it as such. There is hardly a woodlot of any consequence left but the chopper's axe has been heard in it this season." And George Emerson noted in 1846 *(Trees of Massachusetts)*: "A few generations ago, an almost unbroken forest covered the continent. . . . Now, these old woods are every where falling. The axe has made, and is making, wanton and terrible havoc. . . . The new settler clears in a year more acres than he can cultivate in ten, and destroys at a single burning many a winter's fuel, which would better be kept in reserve for his grandchildren."

Two critical concerns regarding the forest resources of New England were apparent: the limited size and timber volume of the remaining woodlots and frightful mismanagement consisting of overcutting, burning, and livestock grazing. Both authors warned of impending wood shortages.

Our studies of census data from 1885 substantiate this concern regarding the effect of frequent cutting on forest structure and condition (Figure 4.23). Across central Massachusetts, more than half of the log-

Forest Harvesting 1885

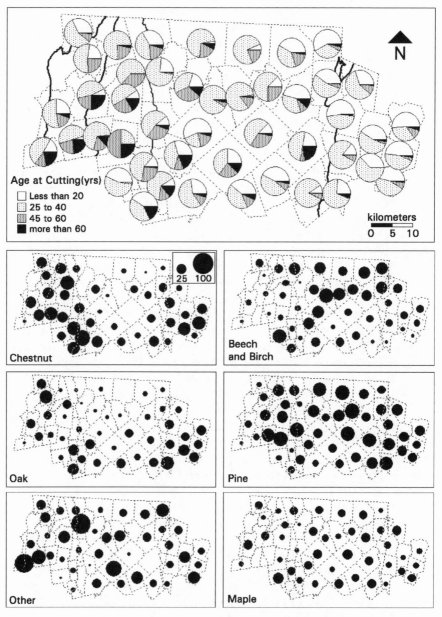

Figure 4.23. The age of forests cut in 1885 in central Massachusetts (top panel) and the relative amount of wood cut by tree species (lower panels). In the late nineteenth century average forest age was young, with many forests cut at less than twenty years old. The species harvested varied with regional abundance and wood value. Modified with permission from Foster, Motzkin, and Slater 1998, 107: fig. 9, copyright Springer-Verlag.

ging focused on stands twenty-five to forty years in age, suggesting both a dearth of older forests and heavy reliance on small-sized wood for fence posts, fuel, railroad ties, and containers. In the uplands, where woodlots were especially small, nearly half of the cutting occurred in woodlots fewer than twenty years in age. The result was a landscape of scattered, young sproutlands. At the same time, new stands of old-field white pine provided an abundant source of wood for timber as well as containers such as barrels, boxes, and crates. Since these essential products of an era before cardboard and plastics could be manufactured from small, short lengths of wood, stands could be harvested at a young age (thirty years and older) and with little regard for timber quality. The development of the portable sawmill in the late 1800s enabled logging in these widespread pine forest stands as well as remote forests that had been previously inaccessible.

The result has been called a "period of forest devastation" from 1880 to 1920, with a peak in 1909–10 when more than 2.5 billion board feet of timber were harvested across New England (Figure 4.24). In contrast to booms based on old-growth and pristine forest, this activity was rooted in agricultural decline and second growth. The fields had become white pine forests, the rural landscape provided a population in need of employment, and the old farmland paths provided efficient forest access. But loggers also reached out to remaining old-growth forest, including the last stands in Connecticut and important sites across the north. In southern New Hampshire they came within feet of one magnificent stand of immense 300-year-old pines and hemlock. This forest was saved only by the efforts of Richard Fisher (director of the Harvard Forest), head-

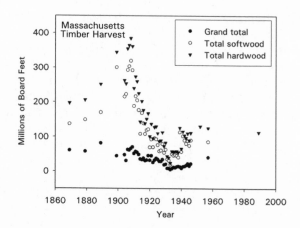

Figure 4.24. Historical changes in timber harvesting in Massachusetts. Figure modified from Berlik et al. 2002a, 2002b, with permission from Blackwell Science Ltd.

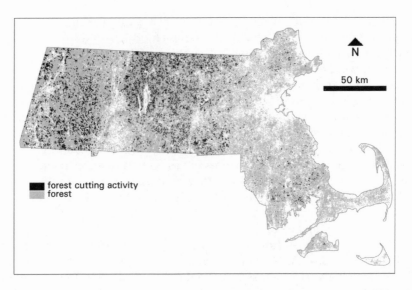

Figure 4.25. Forest harvesting patterns in Massachusetts, 1984–2003. Logging, ranging from selective removal of quality timber to intensive harvesting of most growing stock, is widespread and random in distribution across broad areas of central and western Massachusetts. Note: This map shows only forestry operations in areas that remain forested and does not include cutting for development. Updated from Foster 2002c and Kittredge et al. 2003; forest cover data from MassGIS 2002.

lines in the *New York Times,* and donations from across the country. Today it forms the Harvard Forest's Pisgah tract, a virgin forest centered in a 5,000-acre state park. Across the larger region, however, after 200 years of clearing, widespread old-field succession, grazing, burning, and repeated cutting, the forests could be characterized as "only a pitiful remnant of the once extensive forests of the colonists."

Although the current intensity of logging may have declined to less than half of peak values in the early 1900s, harvesting still exerts a pronounced impact on forest structure and function. Detailed and spatially explicit information on logging for the eastern United States is difficult to obtain because much of the land is owned by thousands of private landowners and the predominant form of logging—selection cutting of individual or groups of trees—is difficult to quantify from aerial photographs or satellite imagery. To overcome this problem we used a unique Massachusetts database—Forest Cutting Plans, which are required for all commercial logging activities—and documented a surprising pattern (Figure 4.25). Over the eighteen years of data collection, harvesting occurred at an annual rate of 1.5 percent of the land area and was selective for larger and higher quality material and species (for instance, red oak and white pine). Remarkably, the pattern of logging was spatially random with regards to topography, distance from roads, land-use history, and broad cover type. In conjunction with other information, these results suggest that logging is probably acting to homogenize forest composition and to selectively promote lower quality trees and less valuable species. Thus, despite the fact that regional forest growth greatly exceeds removals, the pattern of cutting is undoubtedly exerting a profound, albeit easily overlooked, effect.

Ongoing Land Conversion, Modification, and Protection

Although the broad forest history is one of recovery and de-creasing human use over the past century, other activities have substantially altered the physical setting of the region. These effects range from land conversion and hydrologic changes to the introduction of exotic pests, species, and chemicals (Figure 4.26). Many natural areas have been converted to permanent or semipermanent human uses. The selective distribution of highly modified lands, for example, varying with soil type, physiographic unit, or distance from the coast, results in a long-term loss of particular habitats. For example, despite the general decline in agriculture across New England, broad expanses do remain in open fields and farm use, preferentially concentrated on rich lowlands such as the Connecticut River and Champlain Valleys. Industrial and urban areas are situated preferentially on areas viewed as "wastelands" such as sandy outwash plains, wetlands, and coastal marshes. As a consequence, exemplary floodplain forests, sand plain ecosystems, and un-altered wetlands are uncommon.

Modification of many areas has resulted in seminatural ecosystems, substantially different from those that existed at the time of European set-tlement. Changes in local hydrology and watershed conditions have transformed the composition of freshwater and coastal wetlands. Similarly, dam construction and stream channelization have altered sediment loads, the hydrologic regime, and migrational pathways for fish and other organisms in many waterways. Alteration of hydrologic function occurs throughout many watersheds from high in the headwaters where remnant dams and road crossings trap sediment, to middle reaches where flood-control structures and reservoirs occur, to coastal areas where tide flows are regulated and marshes are ditched.

Dam construction for reservoirs and industrial power has also greatly increased the extent of surface water in lakes and reservoirs. Across many upland areas, natural water bodies are uncommon and historic water bodies were created, to the detriment of wetlands and valley communities. One extreme example of this process lies adjacent to the Harvard Forest in the Quabbin Reservoir, which provides drinking water for the greater Boston metropolitan area. Created in 1938, this 10,000-hectare reservoir is one of the largest water bodies in New En-

Percent Developed Land 1951 - 1999

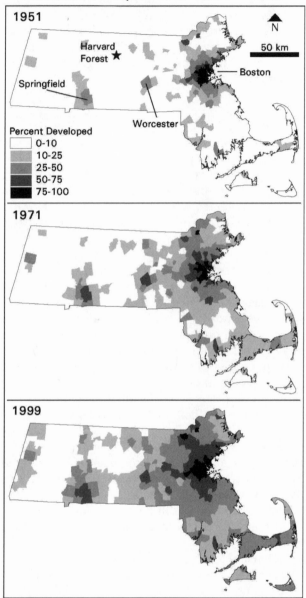

Figure 4.26. The suburbanization and development of Massachusetts. Over the past five decades, areas of dense population have expanded outward from Boston and other metropolitan areas and are contributing to a gradual decline in and perforation of extensive forest areas. Modified from Hall et al. 2002, with permission from Blackwell Science Ltd.; data from Mac-Connell 1973 and MassGIS 2002.

gland. Inadvertently, this artificial water body and its surrounding protected lands also provide the largest conservation property in southern New England and a focal point for wildlife conservation and natural area management.

Introduced Pests and Pathogens

The region's forests have been inadvertently exposed to a number of introduced organisms—including the chestnut blight, white pine blister rust, gypsy moth, Dutch elm disease, beech bark disease, and the hemlock woolly adelgid—that have had a dramatic effect on their composition and structure. All but the gypsy moths have specific hosts. From 1910 to 1920 the chestnut blight killed the aboveground stems of all chestnut trees across the region, removing a dominant species from the canopy of many forests. Although millions of chestnuts survive across the region, they become reinfected with the fungus and die by the time they reach a large sapling size. As a consequence, the ecological role of this species has been transformed from canopy dominant to understory tree. The loss of large chestnuts with their rapid growth, strong, decay-resistant wood, and edible nuts has significantly affected the forests, many wildlife species, and human activity.

Gypsy moths, introduced to Massachusetts in 1869 as potential silk-producing worms, have subsequently spread throughout the Northeast and into the Midwest, where they periodically defoliate hundreds of thousands of hectares and cause significant mortality among oaks and other favored species. White pine blister rust, a fungus with an alternate host on the shrub genus *Ribes,* and beech bark disease, a complex pathogen consisting of a fungus and a scale insect, have altered forest composition and caused significant mortality in these important tree species. Beech bark disease has exerted an uneven effect regionally but has converted broad areas of northern forest from open, older growth stands to dense thickets of beech sprouts. In many maturing forests, beech is exhibiting a striking contrast to the general trend of major tree species by undergoing a decrease in basal area and an increase in density of small stems. Mast from beech nuts is an important food source for bears and smaller mammals in northern forests, and so the disease is affecting the entire ecosystem.

Dutch elm disease, a wilt fungus transported by a bark beetle, transformed the appearance of most towns in the Northeast during the 1950s and 1960s by killing the spreading elms that lined and shaded the main streets, commons, and parks. Devastation of the elms was greatly facilitated by the creation of virtual monocultures in our urban forests. However, elm played a minor role in the natural landscape of the Northeast, and so this loss was less significant than those caused by the other pathogens.

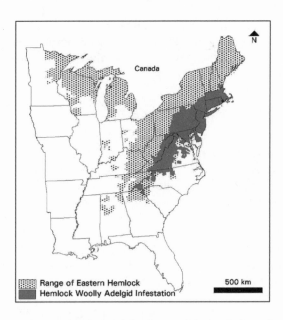

Figure 4.27. Range of eastern hemlock and distribution of the hemlock woolly adelgid in 2002. Since its arrival in Connecticut in 1985, the adelgid has moved continuously northward. Modified from U.S. Department of Agriculture 2002.

Today, the hemlock woolly adelgid is poised to exert a major impact on our forests as it slowly kills large areas of hemlock in southern New England and elsewhere in the Northeast (Figure 4.27). The loss of hemlocks is especially disturbing because this species is one of our longest-lived trees and is abundant in the least disturbed habitats, such as steep slopes, swamp forests, and stream banks, where it creates a unique cool, dark microhabitat. The unfortunate arrival of hemlock woolly adelgid provides us with an unusual opportunity to develop a comprehensive study at the ecosystem, landscape, and regional scales of an important process in New England's history and increasingly in forests globally: the spread and effect of an introduced pest.

The Unfolding Effect of an Exotic Pest: The Hemlock Woolly Adelgid

For many reasons the spread of the hemlock woolly adelgid (HWA) across the eastern United States warrants concern and study by ecologists, conservationists, and landowners. Hemlock holds a unique position among temperate forest trees, and its loss or decline will affect forests, wildlife, and landscape appearance profoundly and irreversibly. Hemlock is among the longest-lived and most shade-tolerant temperate trees, matched most closely only by beech and sugar maple, two of its hardwood associates. It grows slowly and has limited ability to disperse, which means it tends to spread slowly across the landscape and is among our last tree species to recolonize sites like old fields or areas that have burned intensively. As a consequence, large old hemlock trees tend to be restricted to primary forest sites that have received comparably less intense human activity and where hemlock has been present for a long time. It is a sad irony that as the general landscape is becoming more heavily forested and wilder and as old-growth forests, many of which are dominated by hemlock, are emerging as a conservation priority, this quintessential old-growth species is being threatened by an exotic insect.

Hemlock's predilection for growing in forested wetlands, along stream courses, and on lake shores and its habit of casting dense shade that often excludes other plants raise many questions regarding the consequences that a rapid hemlock decline will have for aquatic ecosystems and water quality. Hemlock canopies intercept snow and shelter wildlife during winter months. Meanwhile, the shady, acidic, and relatively cool and stable environments in hemlock forests provide unusual terrestrial and aquatic habitats in a landscape dominated by hardwoods. However, to most New Englanders, the immediate and perhaps greatest effect of the HWA will be aesthetic. Hemlock's evergreen and shade-tolerant nature makes it a distinctive background in nature reserves and an ideal tree for hedgerows and property boundaries. As our forests and backyards lose their hemlocks, we will discover that there is no substitute or ecological replacement for this unusual species.

Despite growing information on HWA biology, we know relatively little about the dynamics of HWA infestation and its effect on forest patterns and processes. Unanswered questions include the following: What factors determine the rate and pattern of spread of this wingless insect within a stand, from forest to forest, and across the region? Are there climatic or biological factors that may eventually limit the insect's spread? Once infested, how rapidly do trees decline and die and what are the physiological mechanisms involved? How does the decline and mortality of hemlock affect the forest environment and the availability and cycling of important nutrients? What species and forests will replace hemlock? And how will these major forest changes influence wildlife, streams, and ponds?

To address such questions, we developed a major research effort to examine forest response to HWA outbreaks in southern New England. The research focuses on a 100-kilometer-wide region that stretches from the Connecticut coast northward to southern Vermont. This transect encompasses a gradient of HWA infestation, ranging from the most heavily affected forests in New England to areas beyond the current extent of the insect, and embraces a wide range of different landscapes and forest conditions. Although the HWA has only just reached the Harvard Forest, it looms as a major ecological factor that we need to anticipate, study intensively, and understand comprehensively.

Forest Dynamics in Response to the Hemlock Woolly Adelgid

The HWA is a small, aphidlike insect introduced into a number of East and West Coast locations on nursery stock from Japan. It damages the trees by inserting a long stylet into the parenchyma tissue of the needle and inner twig, removing carbohydrates and nutrients and potentially injecting toxins and damaging cellular function. Since the HWA entered Connecticut across Long Island Sound around 1985, it has spread especially northward across southern New England, to reach more than half of the towns in Massachusetts (Figure 4.27). Across Connecticut we estimate that the abundance of hemlock has decreased by one-third because of mortality from the HWA and salvage or preemptive logging. The long-term prognosis for the species is bleak (Figure 4.28). In

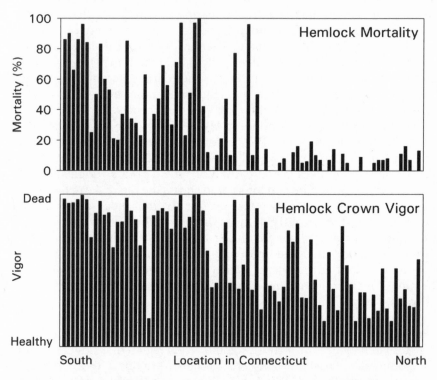

Figure 4.28. Effect of the hemlock woolly adelgid on forest vigor and mortality in a broad transect extending from Long Island Sound (south) to the Connecticut-Massachusetts border (north). The gradient in damage relates to the northward expansion of the insect and the duration of infestation.

Figure 4.29. Death of the hemlock overstory results in dramatic changes in environmental conditions and rapid establishment and growth of a new hardwood forest. Photograph by D. Orwig.

the 150 forests that we have examined, mortality exceeds 60 percent in half of the stands, and trees are dying at a rate of 5 to 15 percent per year. The remaining trees are deteriorating in all infested stands; most remain as thin and graying shadowy ghosts that retain less than 25 percent of their original foliage. Absolutely no sign of tree or forest recovery can be observed after heavy infestation, and we anticipate that the remaining trees will die within five to ten years of infestation.

As the dense hemlock canopies thin and begin to break up, light levels near the ground increase dramatically, and the forests are rapidly colonized with seedlings of black birch, red maple, and oak as well as weedy herbs and shrubs including pokeweed, fireweed, raspberry, and ferns. Seedling densities increase in proportion to the increasing light as hemlocks continue to decline and die. As a result, vegetation cover remains high, erosion is minimal even on steep slopes, and succession proceeds rapidly (Figure 4.29). In stands with less hemlock cover and correspondingly more hardwood canopy and residual shade, seedling establishment is slower, and hemlock replacement occurs gradually through in-growth from adjacent trees and up-growth of existing saplings or new seedlings.

To interpret broadscale patterns of HWA spread, we mapped all hemlock forests in the Connecticut portion of our transect from aerial photographs and then visited and sampled 120 stands in detail. The HWA occurred in nearly 90 percent of stands, but hemlock mortality varied regionally, from 20 to 100 percent in the south compared with 0 to 15 percent in the north. Trees exhibited no apparent resistance although there is considerable within-stand variation in infestation rate and decline. The broadscale south-to-north trend in decline parallels the migration path of the adelgid and the duration of infestation. Birds and other animals, including people; wind; and transportation of nursery stock and forest products are the main facilitators of HWA movement. The rapid spread to the north may be a consequence of the migratory route of many bird species.

Across southern New England and increasingly to the north and beyond the current range of the adelgid, landowners and public agencies are logging hemlock forests at an unprecedented rate. In southern Connecticut alone more than 25 percent of hemlock forests have been logged, and many more will be cut before they die from HWA. The decision to harvest these forests is largely driven by economic, safety, and aesthetic concerns, and therefore, many of the major impacts of the HWA may be indirect: widespread salvage or preemptive logging, road construction, and associated effects. Initial studies suggest that these land-use side effects of the HWA initiate more pronounced ecosystem responses than the HWA itself. In logged stands, canopy removal is more complete and immediate, soil disturbance and changes in light and soil microenvironments are greater, and the transition to hardwood forest occurs much more rapidly than in intact forests where the trees decline gradually from HWA. Recognition of these differences between logging and HWA impacts should figure into landowner decisions and land policy development regarding responses to the arrival of this exotic pest.

Outlook for Hemlock Forests

The future of hemlock and hemlock forests in New England is dim. The adelgid continues to migrate northward, and its progress seems unimpeded by climate or other factors. Although severe winters initiate heavy mortality, they only check the insect's population growth temporarily because of its parthenogenetic (self-fertilizing) reproduction and the occurrence of two generations per year. In Japan, where the HWA is native, it grows under winter regimes that are as severe as any in the eastern hemlock's range. The complete absence of recovery in any infested trees or stands suggests that although trees may linger for long periods after initial infestation, they continue a progressive decline. Biological controls based on native predators on the HWA have been tested but appear unable to prevent or control the spread of the insect. The same is also true of chemical insecticide. Consequently, hemlock may be drastically reduced or eliminated across broad portions of New England and the East in a few decades. Our ongoing studies will evaluate the many dimensions and consequences of this new ecological force in our landscape. They will provide new perspectives to the history of earlier pathogens, a better understanding of where our forests are heading, broader insights into the important effects of introduced organisms on natural forest ecosystems, and information that landowners, managers and planners can use as they seek to shape our landscape into the future.

The Future and Remnants of the Past

As human population increases and disperses from urban centers, suburbanization is increasingly affecting our forests and open spaces (see Figure 4.26). Over the past fifty years both the average duration of ownership, now less than ten years, and the average size of forest properties, now less than 10 hectares, have decreased significantly as our population has become less agrarian and more mobile and as our suburbs encroach on rural areas. These social and geographical trends will strongly influence our perception and use of the forests in the coming century.

Despite the long history of agriculture, clearing, and intensive harvesting in the New England landscape, recent environmental interest has led to the discovery of remnant patches of old-growth forest, once assumed to be entirely eliminated. Although definitions of "old-growth" vary considerably, these forests typically show minimal evidence of human disturbance and include dominant trees well over 200 years old. At present, more than 400 hectares of old-growth forest are recognized in Massachusetts, and the number continues to rise slowly as additional areas are investigated by scientists with a constantly improving understanding of what these ancient trees look like. The largest stand, more than 50 hectares in extent, was recently identified on steep slopes below the summit of Wachusett Mountain in an area heavily used for recreation for 150 years, and less than 40 kilometers from Boston. Although these old-growth stands are among our most natural forests, even these sites are not protected from subtle but pervasive human disturbances such as atmospheric pollution, rising CO_2 levels, and associated climate change. Nonetheless, the ability to identify and protect old-growth stands within an extensive matrix of forest cover across the modern New England landscape is a remarkable outcome of 400 years of landscape change. The extent to which these modern forests resemble their pre-European predecessors is a major question with important ecological and conservation ramifications.

CHAPTER 5

Broadscale Forest Response to Land Use and Climate Change

J. FULLER, D. FOSTER, G. MOTZKIN, J. McLACHLAN, and S. BARRY

Questions Concerning the Regional Patterns of the Forest in New England

In New England, where the landscape has been shaped by four centuries of intensive land use following European settlement, a history of environmental, human, and vegetational change over the past millennium or more has great relevance to the study, conservation, and management of modern ecosystems. As ecologists investigate current relationships between forest ecosystems and the environment in order to anticipate future changes, it is important to understand the factors that control current landscape conditions and the rate at which these are changing. Similarly, as conservationists or forest managers work to preserve, restore, or manage species and ecosystems, they frequently seek to base their management decisions on a knowledge of the conditions and processes that have existed in the past. Therefore, in our desire to understand the modern landscape of central New England, we used approaches from paleoecology, archaeology, and history, in addition to modern vegetation and environmental research, to investigate the long-term consequences of the environmental, cultural, and disturbance processes described in the previous chapter. In our attempt to evaluate the factors that control the rates and patterns of change in ecological systems and to provide practical information concerning past conditions and the development of the modern landscape, we centered on several fundamental questions:

- What were the regional patterns of forest composition and dynamics before European settlement, and how did these relate to broadscale variation in climate, soils, and human activity?
- What was the vegetation response to broadscale human disturbance, particularly deforestation, agriculture, and intensive woodlot management?
- How have forest patterns changed since European settlement? To what extent are their regional characteristics determined by the modern environment versus historical factors?

Although our studies focus on characteristics of the New England landscape and the peculiarities of its land-use history, the questions addressed and our framework for integrated historical and ecological study are relevant to other parts of the world. Much of the eastern United States, along with portions of the Caribbean, Central America, northwestern Europe, and even parts of Southeast Asia, share a history of recent declines in human activity following intensive land use. Moreover, studies from areas with such a history of changing land use have application to regions that are currently subject to intensive forestry and agriculture. How do natural systems respond, on a broad scale, to new disturbance? And if we choose to restore disturbed landscapes to more "natural" conditions by ceasing these activities, how will the regional patterns of vegetation respond and recover? As we set about examining modern patterns of vegetation across broad areas, information about the history and trajectory of forest change as well as the factors that have controlled these will provide fundamental information for planning research and interpreting results.

Research Approaches and Considerations

In order to evaluate broadscale vegetation patterns and dynamics, we focused on the central Massachusetts subregion surrounding the town of Petersham, integrating historical approaches with the analysis of modern conditions (Figure 5.1). We designed our study so we could examine vegetation and human history across natural gradients in climate and physiography, moving from the Connecticut River Valley to the Central Uplands where the Harvard Forest is located, and down onto the gentle Eastern Lowlands (see Chapter 2).

Paleoecology: Extending Our Long-Term Studies

To extend the record of vegetation and environmental change back through the period preceding European settlement, we used a wide range of paleoecological methods, including pollen, chemical, and physical analyses of lake sediments. We focused especially on the past 1,000 years in order to have a lengthy record of the conditions before European arrival as a background for evaluating subsequent changes. In particular, we were interested in documenting the pre-European fire and disturbance regimes and the long-term rates of vegetation change. Because the pollen records from small, closed lakes (that is, water bodies with no inlet or outlet streams) sample the vegetation from approximately a 10-kilometer distance, we selected ten such lakes for comparative study.

Paleoecological studies provide some unique advantages over other historical methods as they yield continuous, long-term records in which

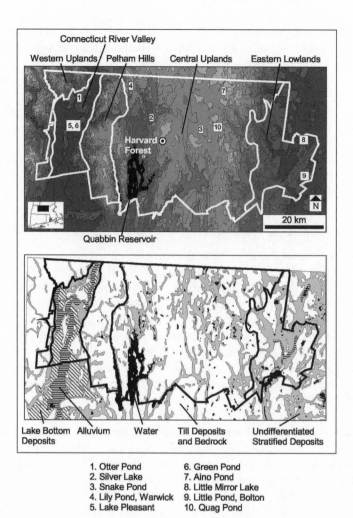

Connecticut River Valley

Western Uplands Pelham Hills Central Uplands Eastern Lowlands

Harvard Forest

Quabbin Reservoir

20 km

Lake Bottom Deposits · Alluvium · Water · Till Deposits and Bedrock · Undifferentiated Stratified Deposits

1. Otter Pond
2. Silver Lake
3. Snake Pond
4. Lily Pond, Warwick
5. Lake Pleasant
6. Green Pond
7. Aino Pond
8. Little Mirror Lake
9. Little Pond, Bolton
10. Quag Pond

Figure 5.1. The central Massachusetts subregion showing topography, major physiographic regions, and the location of the small lakes sampled in the paleoecological study (top) and surficial geology (bottom). See inset and Figure 1.1 for location information. Compiled from Foster, Motzkin, and Slater 1998, 98: fig. 2 (surficial geology data from Heeley and Motts 1973), and Fuller et al. 1998, 78: fig. 1, with permission from Springer-Verlag (copyright 1998) (elevation data from U.S. Geological Survey 1993).

the resolution of vegetation pattern and composition remains relatively constant through time. In contrast, most of our historical vegetation records are restricted to the past 100 years, and the methods, details, and resolution of each historical survey often vary considerably. Nonetheless, paleoecological approaches have inherent limitations. Each of the different types of fossil analyses may actually record phenomena at somewhat different spatial scales. For example, pollen may be derived largely from vegetation within 5 to 10 kilometers of the lake, most large charcoal fragments come from within the watershed, and organic matter may originate from organisms in the lake (for example, plankton, zooplankton, and macrophytic plants) as well as vegetation along the lake shore and through the watershed. The taxonomic resolution and abundance of the different fossil materials also vary; with some notable exceptions, pollen is primarily identifiable at the level of genus, such as birch and oak. Different growth forms of plants, such as herbs, shrubs, and trees, and different species also vary in the amount of pollen that they produce and its dispersibility. Necessarily, all of our interpretations must attempt to take these factors into account.

Historical Sources: A Range of Sources and Insights

The analysis of historical land use generally entails utilizing many sources. These include state and federal censuses collected according to politically defined sampling units (often towns, roughly 10 by 10 kilometers) at regular intervals (generally ten years) for primarily economic purposes, and according to guidelines and regulatory needs that may vary through time. We then combine these long time series with information derived from town histories and sporadic additional surveys and maps in order to interpret European settlement patterns, land-use history, and land-cover changes. Despite some obvious limitations, these data provide many remarkable insights.

Most useful are decadal census data on population, livestock, crops, and land cover and infrequent tallies of forest cutting activities and products. In addition, and most fortuitous for our investigations, there is the statewide map series from 1830 that depicts forest and landscape patterns at approximately the peak of deforestation and intensive agriculture (Figure 5.2). These maps underscore the opportunistic nature of all historical research and the important role that serendipity plays in scientific endeavors (see Box 5.1). Commissioned by an act of the state legislature that required each town to produce a map showing cultural features (such as roads, houses, churches, and mills), physiography, and water bodies, these maps also depict forests, open land, and lowland meadows that were an important source of hay for colonial livestock. By transferring these nineteenth-century maps to modern base maps and then digitizing them onto a geographic information system (GIS), it is possible to piece together regional maps that are suitable for qualitative and (with caution) quantitative analysis. The resulting maps provide a perspective on the patterns of vegetation and human activity, as well as insights into the decision-making processes of the inhabitants, at the period of the most intensive land use in the region's history. The 1830 series therefore provides one of our most important regional perspectives on land cover before the twentieth century.

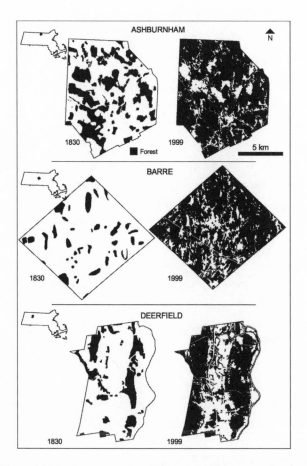

Figure 5.2. The availability of town maps from the 1830s enables us to explore variation in the patterns of vegetation and land-use history among towns across the state. These three central Massachusetts towns exhibit major differences in the amount and distribution of forest in the mid-nineteenth century and equally striking variation in patterns of reforestation. Ashburnham is a rural hill town with rugged topography; Barre occupies more gentle terrain and progressed from strongly agricultural to industrial (see Figure 4.20); and Deerfield lies largely in the Connecticut River Valley where rich, level terrain supported early settlement and extensive agriculture. See inset maps for locations in Massachusetts. Modified from O'Keefe and Foster 1998b; forest cover data for 1999 from MassGIS 2002.

Data on the Modern Landscape: Surprisingly Weak

One of the consequences of the twentieth-century growth of New England's industrial, commercial, and technological economy and the decreased dependence on local agriculture and natural resources is that the data characterizing modern vegetation and land cover are of lower quality than the best historical data. Similarly, information on soils, surficial geology, and many environmental variables are less reli-

Box 5.1.
The Art and Serendipity of Historical Studies

Much of the challenge and excitement in designing long-term studies lie in the development of strategies for identifying, collecting, and integrating diverse types of modern and historical data. Inevitably, this process involves ingenuity and compromise as we attempt to compare sources of information that vary considerably in their spatial, temporal, and taxonomic resolutions and as we seek to adapt data to our specific ecological interests that were originally collected for a wide range of mostly economic or regulatory purposes. In our regional study of forest dynamics, one good illustration of this process of data mining and adaptation comes from the piecing together of very different historical sources to interpret changes in vegetation and land cover.

At the time of the establishment of most towns across the Commonwealth of Massachusetts (as well as other New England states), the original settlers, called "proprietors," commissioned surveyors to delimit the town boundaries and major property divisions through so-called "lotting" surveys and to lay out roadways. Although these surveys were based on geography rather than on vegetation, the surveyors blazed "witness" trees and recorded the tree species at the intersecting corner of each property line. If we assume that these trees were selected objectively and without bias (that is, for species or size) and that the survey lines are numerous enough to sample the towns representatively (large and oftentimes rather uncertain assumptions), then a compilation of the recorded trees should offer an extraordinary opportunity for us to assess compositional variation of forests across the region at the time of European settlement.

Were the surveys biased with regards to particular species, tree sizes, or forest types? Were the surveyors accurate in their tree identifications? Are the samples of trees large enough and sufficiently independent to yield accurate assessments? And can samples defined by political and ownership boundaries provide information that can be reconciled with other forestry data and with ecological scales of variation? To answer these questions, we initially collected proprietors' data from more than fifty towns and examined the resulting patterns with these questions in mind (see Figure 5.3). To our satisfaction, the data formed a remarkably consistent geographical pattern that closely parallels underlying environmental gradients and the vegetation records from pollen in lake sediments across the region. A similar conclusion was reached in a larger companion study that we undertook in collaboration with Charlie Cogbill from the Hubbard Brook LTER site. In this subsequent effort we assembled proprietors' data for nearly the entire New England region. Having confirmed the general accuracy of these data, we gathered other later sources of vegetation, land-use, and environmental information that could be collected or aggregated to a town scale. In each case, we went through a similar process of searching sources, developing methodologies, and then checking results. In the end, by utilizing diverse sources of historical information in new ways, we emerged with a valuable perspective on centuries of change in New England.

able and detailed in New England and many eastern states than in other parts of the country, where natural resources are an important part of the economy. For forests, the most comprehensive statewide data are for 1985, and these provide only broad height and cover classes, with no information on forest composition. In fact, there are no reliable maps of forest types (that is, based on forest composition) in the state for the past half century. Consequently, and ironically, a major hurdle in our studies was to develop modern vegetation data that were commensurate with the species-specific data that are available from historical sources such as proprietors' survey records (see Box 5.1), which date from the seventeenth and eighteenth centuries. We accomplished this by sampling more than 450 plots distributed randomly in forests across central Massachusetts. The paucity of biological and environmental information for one of the wealthiest states in the United States unfortunately leaves us with a limited ability to assess current ecological conditions, let alone to predict or shape future ones.

Data: Strengths and Weaknesses

In comparing the historical and modern vegetation data with the paleoecological record, we can see some complementary strengths and weaknesses. Whereas we often have a good idea of the spatial and temporal precision of each data set, the resolution of data from different sources is often quite different. In fact, the scale, methods, and objectives vary among almost all historical and modern sources. Unlike sediment analyses in which we can control (with some limitation) the temporal frequency and resolution of the sampling, with historical data we generally have little control over these factors and need to evaluate whatever data are available. As a result, our reconstructions of past landscapes are based on irregular sampling intervals and offer ample opportunity for ongoing discussion and refinement.

Presettlement Vegetation and Landscape Dynamics

Historical and paleoecological sources provide consistent and complementary pictures of the landscape that the earliest European explorers and settlers encountered. From both perspectives, we see strong variation in forest composition that corresponds to physiographic and elevational variation across the region (Figures 5.3 and 5.4). The cooler upland areas in the north-central part of the region were dominated by hemlock, white pine, and such northern hardwood species as beech, yellow birch, and lesser amounts of red oak, whereas the lower elevation and warmer Connecticut River Valley and Eastern Lowlands had considerably more oak (presumably all three major species—red, black, and white oak) and hickory. White pine, ash, and chestnut were rather evenly distributed across the landscape. Red maple, paper birch, and gray birch, which are important today, were not common.

Identifying the factors that were responsible for controlling this pronounced geographic pattern across a relatively small geographic area warrants discussion, however. The strong similarity in the dominant tree species in the two lowland portions of the study area, which vary considerably in soils, physiography, and geology, suggests that the primary factor controlling vegetation variation was climate, which is roughly similar in the two areas. In support of this interpretation, analy-

---

Figure 5.3. Witness tree surveys from early European colonial history were compiled to examine changes in the distribution and abundance of tree species over the past three centuries. Regional variation was striking in the early eighteenth century and corresponded closely to elevational and climatic gradients (growing degree days). However, after more than 200 years of land use and reforestation, tree species abundances show little variation in the modern landscape. Modified from Foster, Motzkin, and Slater 1998, 110: fig. 11, with permission from Springer-Verlag (copyright 1998).

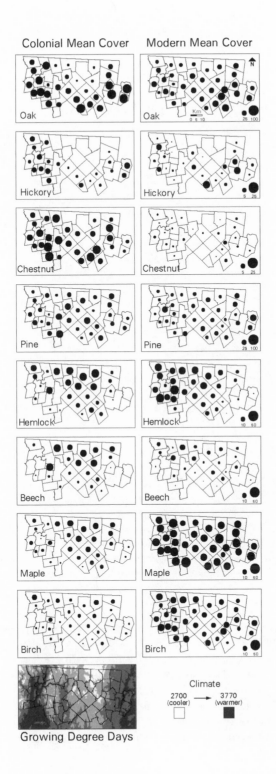

Colonial Mean Cover   Modern Mean Cover

Oak

Oak

Hickory

Hickory

Chestnut

Chestnut

Pine

Pine

Hemlock

Hemlock

Beech

Beech

Maple

Maple

Birch

Birch

Growing Degree Days

Climate

2700 ⟶ 3770
(cooler)   (warmer)

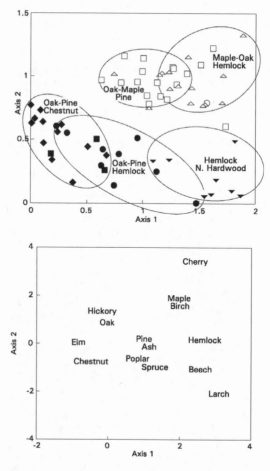

Figure 5.4. Detrended correspondence analysis (DCA) of early settlement (closed symbols) and modern (open symbols) vegetation for the central Massachusetts region (top); and the DCA loadings for major tree species (bottom). At the time of settlement, the vegetation varied along a compositional gradient corresponding to elevation and climate, whereas the modern vegetation is distinctly different and much more homogeneous in composition. Modified from Foster, Motzkin, and Slater 1998, 111: fig. 12, with permission from Springer-Verlag (copyright 1998).

sis of the relationship between vegetation and climate based on both historical and paleoecological data demonstrates that the abundance of each of the major tree species is strongly correlated with the modern variation in the length of the growing season across the region (Table 5.1). Thus, there appears to be a strong correlation, if not a direct cause-and-effect relationship, between variation in forest composition and climatic variables known to control broad biogeographic patterns.

However, the interpretation is not completely straightforward, as other factors, including disturbance regimes and cultural activity, covary with these physiographic and climatic gradients. In the uplands, the abundance of long-lived and shade-tolerant species, which are characteristic of mature forests, implies that disturbance, especially fire, was probably infrequent. Hemlock, in particular, is quite intolerant of burning, and our studies of individual woodlands show that it may take many centuries for hemlock to recover its former abundance after intense surface fires. During such prolonged postfire periods, hemlock is generally replaced by sprouting and early- to mid-successional hardwood species, including birch, oak, chestnut, and maple, or by the fire-tolerant white pine. Therefore, the consistently high pollen and witness tree values for hemlock and low values for oak and chestnut from across the uplands suggests a low incidence of fire at the time of European settlement. Indeed, charcoal levels in lake sediments at all uplands sites are low throughout the pre-European period, bolstering this interpretation. The low abundance of early successional species such as paper birch and red maple suggests that other disturbances, including windstorms, pathogens, or human activity, were either infrequent or local in extent.

In contrast, oak and hickory were more abundant in the lowlands,

Table 5.1. Results of Simple Linear Regression of Tree Abundance (Percent Occurrence) and Climate (Growing Degree Days) for the Colonial Period

	Colonial Forest		Modern Forest	
	r^2	Slope	r^2	Slope
Oak	0.53	0.121‡	0.00	0.004
Hickory	0.44	0.013‡	0.01	0.008*
Hemlock	0.57	−0.052‡	0.06	−0.017
Beech	0.53	−0.043‡	0.16	−0.011*
Maple	0.47	−0.028‡	0.04	0.010
Birch	0.18	−0.010*	0.00	0.001
Ash	0.03	−0.004	0.00	0.001
Chestnut	0.02	0.005	0.01	0.001
Pine	0.00	−0.002	0.05	0.012
Cherry	0.06	−0.002	0.12	−0.010*
Larch	0.06	−0.002	?	?
Poplar	0.02	−0.002	0.05	0.002
Spruce	0.03	−0.002	0.23	−0.004
Elm	0.06	0.003	0.19	0.001*
Oak-hickory	0.55	0.133‡	0.02	0.012
Northern hardwood-hemlock	0.61	−0.122‡	0.05	−0.018

Source: Reprinted from Foster, Motzkin, and Slater 1998, table 3, p. 110, with permission from Springer-Verlag (copyright 1998).

Note: Based on proprietors' survey data (approximately 1700–1800) and modern forests (1993–96) for each township in north-central Massachusetts. Northern hardwoods include birch, maple, and beech. Climate was based on the parameter that provided the best regression fit for each period, which was maximum degree growing days in a township for the colonial period and average degree growing days for the modern period. Significance levels: *p < .05, ‡p < .001

where it appears that more fire accompanied the warmer temperatures and longer growing seasons. Charcoal levels in lake sediments are generally higher in the Connecticut River Valley and Eastern Lowlands than in the uplands. In fact, charcoal abundances from lakes on the flat, dry Montague sand plain in the Connecticut River Valley approach the very high values observed from sandy outwash plains along the southeastern coast of Massachusetts and on Cape Cod. Archaeological studies reveal a consistent Native American presence throughout the Connecticut River Valley and Eastern Lowlands, and these landscapes have always been interpreted as providing diverse resources for a fairly high human population. Although the uplands are not as well-studied, the available evidence indicates many fewer Native Americans, primarily bands from the lowlands engaged in seasonal activity, and very scattered settlement. Thus, although the climatic gradient may have been a primary factor controlling the vegetation pattern, this climatic influence was apparently reinforced by greater fire frequency. Certainly, the absence of hemlock and northern hardwoods and the abundance of fire-tolerant and sprouting oak and hickory in the lowlands are consistent with regional patterns of climate, fire, and human activity.

In this regard, it is important to underscore that the inertia inherent in forest ecosystems may enable even slight variation in regional disturbance patterns to exert a rather pronounced effect on vegetation patterns. Specifically, if fire was extremely uncommon across much of the uplands, occurring a few times a millennium or so as indicated by our admittedly limited studies in Petersham, or even less frequently as suggested by studies in the Berkshire Uplands, then it may not require a particularly active fire regime in the lowlands to favor species like oak and other hardwoods over hemlock. Because hardwood species are capable of sprouting and hemlock cannot tolerate fire or invade new sites rapidly, surface fires every century may be all that is required to maintain the oak-hickory forest. This is an important consideration, as much of the discussion on fire in New England has focused on reconciling references from early settlers suggesting that frequent fires maintained open forest conditions with historical and fire ecology studies suggesting a low incidence of fire. Clearly, we would be well-advised to consider the potential importance of infrequent disturbances.

Patterns of Forest Response to European Activity

Before considering the role of pre-European fire and Indian activity in more detail, we may come forward to the present and ask how 300 years of land use, including extensive deforestation, agriculture, repeated cutting of the remaining forests, and widespread reforestation, have changed the pattern of vegetation that occurred in the seventeenth century. The answer is twofold: forests per se have proved to be ex-

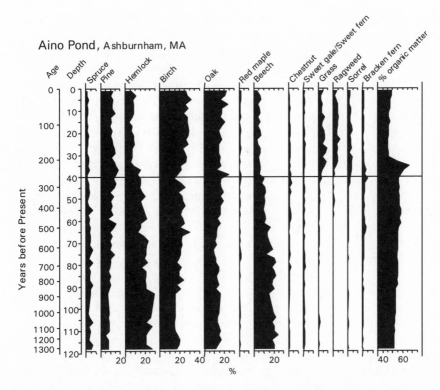

Aino Pond, Ashburnham, MA

Age · Depth · Spruce · Pine · Hemlock · Birch · Oak · Red maple · Beech · Chestnut · Sweet gale/Sweet fern · Grass · Ragweed · Sorrel · Bracken fern · % organic matter

Years before Present

%

Figure 5.5. Pollen diagram from Aino Pond in north-central Massachusetts. Vegetation change has occurred continuously but was most rapid just after European settlement. The pre-European declines in hemlock and beech during the Little Ice Age (beginning approximately 500 years B.P.) are quite apparent. Pollen diagram, by N. Drake, modified from Fuller et al. 1998, 82: fig. 2a, with permission from Springer-Verlag (copyright 1998).

tremely resilient to intensive human activity and have reestablished widely throughout the region after agriculture and other intensive land-use practices ceased. However, the abundance and distributional patterns of major tree species, as well as their relationship to underlying environmental factors, have been substantially altered over the past 300 years (Figures 5.3, 5.4, 5.5). In fact, across north-central Massachusetts today the abundance of tree species varies remarkably little as one moves from the Connecticut River Valley up through the hilly uplands and down into the Eastern Lowlands (Figure 5.3). From the pre-European pattern in which hemlock and northern hardwood species dominated in the uplands and oaks and hickory dominated the lowlands, the landscape has changed to one in which oak, red maple, and black birch are ubiquitous across the region, with lesser amounts of white pine and hemlock and very little beech.

The relatively uniform composition of forests across the region shows no statistical relationship to the regional climate (for example, growing degree days or average temperature) nor to the variation in specific land-use practices that exists among the three physiographic areas (Table 5.1). Instead, it appears that current forest composition and structure at this scale are largely the products of the relatively brief, intense, and novel disturbance regime that originated with the arrival of Europeans. The detailed patterns of land use and land cover do exhibit a degree of regional variation that is quite striking: as we drive through the Connecticut River Valley we view large, open expanses with tobacco sheds, cornfields, and vegetable stands that contrast with the few remaining dairy farms in the more forested uplands and with the apple orchards, market gardens, and suburbs of the Eastern Lowland. Nonetheless, the overwhelming similarity in the history of land clearance, intense logging, and old-field succession across the entire region evidently outweighs the influence of environmental factors, fire, or modest land-use differences that controlled the vegetation in the pre-European period. The consequence of the massive land-use transformation of New England over the past 300 years is the fairly uniform distribution of mid-successional and sprouting species that dominate the landscape today.

The major changes in composition that occurred in the past 300 years lend support to the interpretation of broadscale land-use history as a major determinant of modern forest geography. The species that were most important at the time of European settlement and that declined most abruptly after European settlement are beech and hemlock, both of which are susceptible to disturbances such as fire and compaction of the soil surface by grazing animals. Similarly, field studies and demographic modeling of beech and hemlock highlight the fact that these slow-growing and shade-tolerant species are slow to reinvade sites from which they have been removed by agricultural activity or fire. Relative to other species, both are also slow to recover and dominate sites where they have been heavily cut. In contrast, the species that have increased across the region, including red maple, gray and paper birch, and the oaks, exhibit distinctly different behavior. Each of these responds well to disturbance by fire, cutting, and field abandonment as they sprout prolifically, grow relatively rapidly, and disperse and establish relatively easily in open sites. Consequently, and perhaps not surprisingly, the imposition of a regime of frequent and high-intensity disturbance in the uplands, which previously were dominated by slow-growing, shade-tolerant tree species, resulted in a shift to rapid-growing, moderately tolerant to shade-intolerant species that are more typical of young and early-successional forests. Thus, the composition of the upland areas became more compositionally similar to that of the lowlands over the past few centuries.

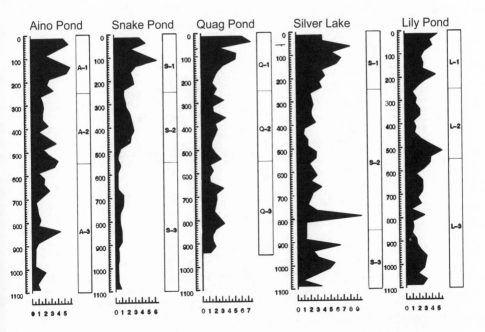

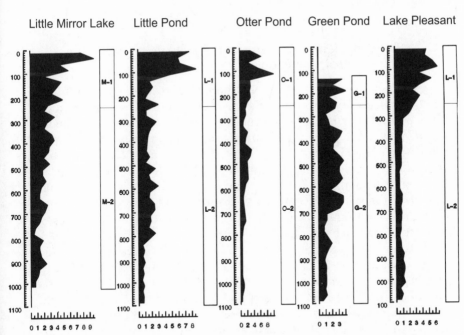

Figure 5.6. Rates of vegetation change around ten lakes in central Massachusetts based on multivariate analyses of pollen data. All sites exhibit continual though varying rates of change, underscoring the inherent dynamics in vegetation. However, most sites exhibit the highest rates of change during the period of European settlement (labeled with top rectangle showing zone 1,250–0 years B.P.). Modified from Fuller et al. 1998, 86: fig. 4, with permission from Springer-Verlag (copyright 1998).

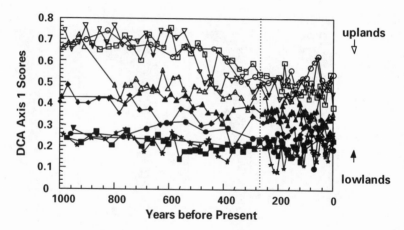

Figure 5.7. Long-term vegetation trends in central Massachusetts over the past 1,000 years based on detrended correspondence analysis (DCA) of pollen data. Samples that have similar scores are similar in vegetation composition. Although vegetation differences were great 1,000 years ago, they have lessened over time as a consequence of climate change and European land use. Currently, sites from the hilly uplands of central Massachusetts (see Figure 5.1) are quite similar to those in the Connecticut River Valley. Modified from Fuller et al. 1998, 88: fig. 7, with permission from Springer-Verlag (copyright 1998).

The Long-Term Perspective on Change in the Vegetation, Environment, and People

Knowing the extent of the differences between modern forests and those that the first European settlers encountered, a series of questions arises that can only be answered by a longer and more continuous record. How do the extent and rate of these changes compare with presettlement dynamics? What was the actual timing of shifts in major tree species? Is the modern forest composition gradually reverting to its earlier, presettlement composition? And are all of the changes that occurred over the past few centuries the result of human activity? For such a long and continuous perspective on landscape dynamics, we necessarily turn to the paleoecological record.

The story that emerges from our network of ten lake sites provides some interesting insights and a few unexpected observations on long-term vegetation dynamics. Not surprisingly, our results also raise new questions that will challenge and motivate our research through the next decades. An analysis of the overall rate of vegetation change over the past 1,000 years, which is largely driven by variation in the abundance of pollen from trees, indicates that vegetation has changed continuously, though at highly variable rates (Figure 5.6). Forest ecosystems, whether analyzed at the landscape level from an individual pollen diagram or at a regional scale using many lake sites, are characterized by constant change. Nevertheless, the rate of change increased greatly at essentially all sites after European settlement and has been maintained at enhanced levels to the present. Thus, the background levels of vegetation change generated by change in climate and possibly by Native American activity have been greatly exceeded by the effects of European land use.

Our pollen stratigraphies also indicate the rather unexpected result that a considerable amount of the regional change and convergence in tree composition occurred before and very shortly after European settlement (Figures 5.5 and 5.7). During this time, beech and hemlock declined and birch, oak, and red maple (which is insect pollinated and therefore underrepresented in the pollen rain) increased. The rapidity and timing of this shift in tree composition are notable, for at many sites it precedes the intense and broadscale disturbance associated with the peak of agriculture in the mid-nineteenth century. Subsequent to this early change, agricultural activity followed by natural succession, reforestation, widespread logging of white pine, the chestnut blight, the 1938 hurricane, and ongoing forest cutting have kept the regional vegetation in a mode of continual readjustment that has sustained high rates of forest change. This history of ongoing, albeit varied, disturbance is undoubtedly a major factor promoting the persistent dominance of mid-successional species and inhibiting the return of forest composition to presettlement conditions.

An examination of our pollen records, which extend back nearly a millennium before the establishment of Plymouth Plantation in 1620, implicates another environmental factor contributing to the compositional change that occurred with European settlement. Forest composition was actually changing and becoming more homogeneous on a subregional scale before European settlement (Figure 5.7). Notably, in the uplands there are two distinct pre-European periods that are differentiated at approximately A.D. 1400. At this time, bracken fern *(Pteridium aquilinum)* and grasses increase, whereas beech and hemlock begin a marked, though gradual, decline (Figure 5.5). Thus the subsequent decline in these two prominent late-successional tree species at the time of European settlement was preceded by a period of more than two centuries during which they were already decreasing in abundance. Explanations for this early change are based largely on circumstantial evidence and rely on comparisons with studies and data from other regions. However, these results fit an emerging pattern of forest dynamics seen from Cape Cod and the coastal islands to eastern Canada and beyond.

Interpretations of pre-European vegetation dynamics, especially the increase in bracken and decline in beech and hemlock, may be split into environmental and cultural arguments. However, as we have seen previously these two factors often act together and are not mutually exclusive. The *environmental argument* for the fifteenth-century change in vegetation focuses on the so-called Little Ice Age, a subtle but broadscale change in climate that has been recorded at sites around the globe (see Figure 1.2). This climatic anomaly occurred from the fifteenth century or earlier until the mid-nineteenth century and was responsible for the advance of mountain glaciers in the Alps, the Andes, and southeastern Alaska; the annual freezing of the Thames River in London during the Elizabethan era; and extreme weather conditions in the North Atlantic throughout the period of European exploration of North America. This approximately 400-year period was characterized by colder temperatures and more highly variable temperature, precipitation, and length of growing season than occur today. However, the exact mechanism linking Little Ice Age climate conditions with the decline in hemlock and beech and open forest conditions favoring bracken fern is unclear. In fact, a decrease of northern species with the onset of apparently cooler conditions appears to be counterintuitive. Nonetheless, if greater variation in annual and growing-season precipitation accompanied these changes, then a decline in drought-sensitive hemlock and beech is more understandable. In fact, although there is little independent climatic evidence for the Little Ice Age in New England, historical records from agricultural diaries indicate a greater frequency of extreme winters and cool summers as well as more variable growing seasons in the seventeenth and eighteenth centuries. In any case, the temporal synchronicity between widespread vegetation changes in New England and Little Ice Age phenomena noted elsewhere is striking. A similar shift in forest composition appears at a number of sites throughout northeastern North America, again arguing for a regional driver like climate.

The *cultural argument* notes that important changes in Native American subsistence activities may have followed the arrival of maize agriculture in New England around A.D. 1100 and that these could have initiated regional vegetation changes. More intensive and broadscale human impacts including increased sedentarism, village expansion, and widespread forest burning could be associated with an increase in horticulture and population. In turn, these disturbances could have encouraged weedy species locally and bracken and oak regionally because they are more tolerant to fire than long-lived trees such as hemlock and beech. Under this scenario, upland areas would respond strongly because of the abundance of hemlock and northern hardwoods and the historically low frequency of fire. Forests in the lowlands would exhibit less change in response to a slight increase in fire frequency because of the ongoing presence of fire-tolerant species.

Evaluating the cultural argument requires examining both the archaeological evidence for change in human subsistence patterns and the paleoecological record for evidence of a corresponding change in disturbance regimes accompanying the shift in forest composition. Evidence for the arrival of maize into the Connecticut River Valley by the eleventh century A.D. appears to fit established chronologies and evidence. There is, however, meager support for the notion that this was accompanied by any major and broadscale shift in cultural patterns. Indeed, the interpretation currently forwarded by our collaborator Elizabeth Chilton at the University of Massachusetts is for a "mobile farmer," essentially a seasonally mobile lifestyle based on hunting and gathering in which limited horticulture played an important though subsidiary role. In the absence of archeological sites in the valley or elsewhere indicating large permanent settlements, it appears that although Indian activity may have had local effects on the vegetation, widespread clearing or depletion of wood resources associated with a sedentary lifestyle is unsubstantiated.

Likewise, there is little evidence in the charcoal and pollen records that fire activity changed at this time and became important on a broad scale in southern New England. Charcoal abundance does rise after A.D. 1100 at a few lowland sites and is notably high at the few Connecticut River Valley sites; however, at the majority of upland sites, charcoal values remain low through the pre-European period. Most important, the abundances of charcoal, beech, and hemlock do not exhibit any consistent relationships.

After we examine the evidence from across New England and results from central Massachusetts, it appears clear that the major driver of change during the pre-European period is broadscale climate. Although the exact nature and magnitude of Little Ice Age climate change remain poorly understood, the widespread and temporally consistent nature of changes occurring from the coastal region through central Massachusetts to the Berkshires and northern Vermont implicates regional drivers. As we look across this region, the extent of vegetation change and the species involved vary. But there are few obvious mechanisms, other than fire, by which a relatively small human population could trigger such broadscale dynamics. Indeed, although there is every reason to expect that the New England Indian populations did affect their local habitats as well as specific resources, including fish and wildlife populations, little evidence remains for their impact at the landscape to regional scale that pollen analysis can sense. Though this result differs from many other parts of the globe where indigenous populations had a controlling influence on vegetation patterns and dynamics, it fits well with current archaeological data and supports the interpretation of the early historical landscape as one largely dominated by mature forests composed of long-lived species.

On the other hand, these results identify major gaps in our knowledge about the nature, timing, and specific mechanisms of pre-European landscape and environmental change. The direction of future research is clearly defined. In order to identify the timing and spatial variation in the extent and nature of vegetation change, we need a dense network of detailed pollen diagrams. These may well show, for example, that species vary considerably in their dynamics and that even the amount of change in a single species like beech varies across its range. To tease apart the relative role of climate change, fire, or other human activity in initiating these forest dynamics, we need to work with paleolimnologists, dendrochronologists, and archaeologists to develop independent records of physical, biological, and cultural factors.

The Other Vascular Plants: Effects of Regional History on Modern Distributions

One striking characteristic of most long-term assessments of forest change, including much of the work discussed above, is that it focuses on the major tree species to the exclusion of most other biota. Thus, as ecologists debate fundamental processes, such as the integrity of plant communities, the rates of postglacial migration, or, in our case, the regional response of forests to changes in climate and land use, much of the discussion is based on only a meager subset of the species that occur in the landscape. The explanation for this is largely pragmatic: paleoecological and historical records are primarily restricted to trees. In addition, there is some ecological and practical argument for focusing on the dominant constituents of the forest, as they largely control many important ecosystem characteristics because of their great size and biomass. However, in order to assess many important ecological processes and to consider some of the dramatic transformations that are occurring on our landscape today, we need to broaden our considerations to other biota. For example, we are very interested in knowing whether the process of regional homogenization, which characterizes the history of the major tree species in central Massachusetts, has occurred in the shrubs, herbs, and ferns.

Limitations in historical and modern data restrict the scope of inquiry considerably. Almost no information exists on the prehistoric or historical changes in the distribution of most species of herbs, shrubs, or other plants, and therefore we are not able to follow changes for these species as we did for trees. However, for the central Massachusetts study region, our sample of vascular plants in more than 450 randomly placed plots allows us to evaluate broad patterns of current distribution and abundance of many plants and to contrast these with tree patterns. At a regional scale, do these species vary with gradients in physiography, climate, or land use?

As was expected, a large number of herb and shrub species occur infrequently, so that even with several hundred plots, it is difficult to discern distinct geographic patterns of distribution. Other species such as clubmoss *(Lycopodium obscurum)*, partridgeberry *(Mitchella repens)*, star flower *(Trientalis borealis)*, wild oats *(Uvularia sessilifolia)*, and blueberry *(Vaccinium angustifolium)* are distributed fairly uniformly

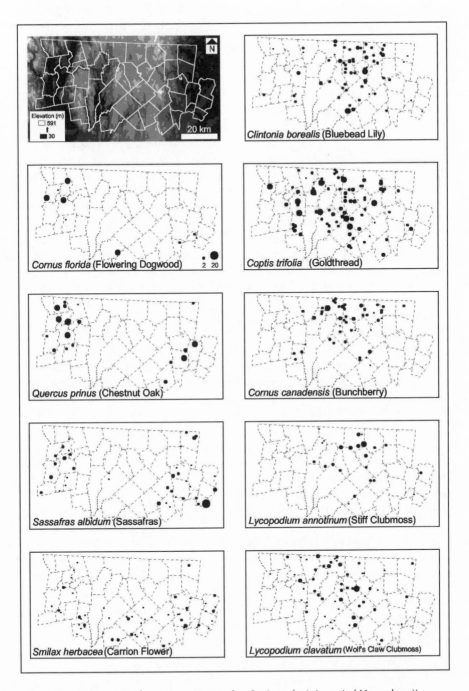

Figure 5.8. Distribution of nine minor trees and understory plants in central Massachusetts. Unlike the major tree species, which are rather homogeneous in their geographical distributions, these species show an apparent strong link to climatic and elevational variation. Species on the left preferentially occur at warmer, low elevations and southern sites, whereas those on the right are more common to the north and at higher elevations. Unpublished vegetation data from the Harvard Forest Archives; elevation data from U.S. Geological Survey 1993.

across the region and do not display strong patterns that are easily re-lated to the physiographic subregions or regional variation in distur-bance history. However, the modern distributions of many species do appear to be strongly related to physiographic and climatic variation across the region (Figure 5.8). Several species, including bunchberry *(Cornus canadensis)*, bluebead lily *(Clintonia borealis)*, goldthread *(Coptis trifolia)*, dewdrop *(Dalibarda repens)*, some clubmosses *(Lycopodium annotinum, L. clavatum)*, Canada yew *(Taxus canadensis)*, painted tril-lium *(Trillium undulatum)*, and southern wild raisin *(Viburnum nu-dum)*, are much more common in the northern portion of the Central Uplands than in the Connecticut River Valley, Eastern Lowlands, or southern portion of the Central Uplands. In contrast, species such as car-rion flower *(Smilax herbacea)*, flowering dogwood *(Cornus florida)*, sas-safras *(Sassafras albidum)*, and chestnut oak *(Quercus prinus)* are more characteristic of the lowlands or the southern portion of the Central Up-lands and are infrequent in the northern portion of the Central Uplands.

These results suggest that for at least some species, patterns of mod-ern distribution at the subregional scale are related to variation in phys-iography and climate. In many cases these regional distributions fit the broader patterns of the species' geographic ranges, as in the case of bunchberry and Canada yew, which extend well into the boreal forest; or dogwood, sassafras, and chestnut oak, which have a corresponding southern range into the Appalachians. It is quite likely that these pri-marily "field layer" species may therefore serve as more useful indica-tors of site conditions and environment than many tree species that are apparently tolerant of a wide range of physical conditions and whose modern distributions appear more closely linked with disturbance his-tory. In fact, this rationale is what has led researchers in Fennoscandia and elsewhere to base their vegetation classification and forest site eval-uations on a combination of understory and overstory species. Although locally the distribution and abundance of even these restricted field layer species have undoubtedly been influenced by historical land use and other disturbances (see Chapter 8), these local effects have not ob-scured the broader environmental controls over their distribution pat-terns. A few of the less common tree species also have distinct subre-gional distributions, as, for example, red spruce, which is much more abundant in the north-central uplands, and chestnut oak, which is more abundant in the lowlands.

Dynamics of Some Vascular Species in Response to Land-Use History

We can gain some insight into how some herb and shrub species have responded to regional land use for those species where herbarium specimens and other sources are available to document historical distri-

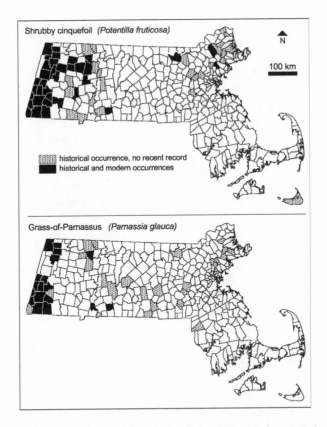

Figure 5.9. The approximate historical (late nineteenth to mid-twentieth centuries) and modern (since 1978) distributions of two uncommon plant species in Massachusetts. Both species declined in geographic extent and abundance during the twentieth century and are representative of a group of species that have become conservation concerns. Many plants and animals increased and thrived under open agricultural conditions during the nineteenth century but declined as grazing, mowing, and burning declined and the landscape reforested. Unpublished data from Glenn Motzkin and the Harvard Forest Archives.

butions. We searched for such cases with a particular focus on the historical and modern distributions of several species that are uncommon in Massachusetts and that occur today in habitats that are priorities for conservation. In particular, we examined a group that is thought to include good indicators of nutrient-rich, calcareous wetland conditions that frequently support unusual plant associations. A sequence of historical maps makes it clear that some species distributions have changed dramatically over the past several hundred years. A notable example is shrubby cinquefoil *(Potentilla fruticosa)*, a plant that is of considerable conservation interest and that today is largely restricted to Berkshire County and the hill towns west of the Connecticut River on circumneutral or calcareous soils (Figure 5.9). This species, however, formerly occurred fairly commonly in the Connecticut River Valley as well as in northeastern Massachusetts. Indeed, as recently as the early twentieth century, this shrub was widespread and seen frequently in old fields, where it was considered a "noxious weed" that warranted eradication by farmers. Referring to the distribution of this species in Vermont, L. R. Jones and F. V. Rand stated in 1909 that "shrubby cinquefoil was originally found only occasionally bordering swamps and in a few cool, rocky gorges or cliffs. During the last generation, however, it has been spreading persistently over pastures in certain sections until today it must be ranked as the most aggressive invader among the shrubby weeds, quite outclassing the hardhacks [i.e., *Spiraea* species]. Plowing and close pasturing have been used to successfully check its progress. . . . Reforestation of any kind, however, will soon suppress it."

Thus, we see that a species that is becoming uncommon in the modern landscape and is often used as an indicator of unusual soil conditions that may support rare species was formerly much more widespread and viewed as an undesirable weed. Several other uncommon species show similar dramatic shifts in their distributions during the past century, including grass-of-Parnassus *(Parnassia glauca)* (Figure 5.9) and yellow sedge *(Carex flava)*. Similar to shrubby cinquefoil, these species have disappeared from the wet meadows and pastures that they had colonized during the agricultural period but that have since reforested.

Thus, although data are frequently lacking, there is good reason to believe that the distribution and abundance of many taxa have changed substantially throughout the historical period in response to historical land-use activities. Because each species responds to such disturbances individualistically, the current landscape is characterized by species at differing points in their response to historical factors, and we may assume that substantial changes in species distributions and assemblages will continue to occur. These conclusions not only are interesting ecologically but have major implications for the development and implementation of conservation policies.

CHAPTER 6

Long-Term Forest and Landscape Dynamics

J. McLACHLAN, D. FOSTER, S. CLAYDEN,
and **S. BARRY**

Stand Dynamics in a Cultural Landscape

In the previous chapter, we saw that broad changes in forest composition accompanied the transition from presettlement forests to those dominated by European land use. These regional trends represent the aggregate consequences of the individual decisions of thousands of landowners and the effects of many separate disturbances, which by and large occur at the level of the forest stand. The stand is a coherent unit with a broadly uniform history and underlying environment. It is often the scale at which management takes place, whether in a field or a forest. It is also a scale at which forest dynamics can be mechanistically tied to biological processes such as competition for light and other resources.

The ecological dynamics of disturbance and succession discussed in this chapter drive the composition and structure of forest stands. Whether a disturbance is a natural event, like a windthrow, or a cultural intervention, like a logging operation, the distribution of biological resources (such as water, nutrients, and light) is rearranged, the influence of competitors is altered through selective damage or mortality (each species is more or less susceptible to different types of disturbance), and the physical substrate and environment may be rearranged. As time passes, resource availability continues to change because of the relative abilities of different species to establish or resprout and grow and then utilize these resources, resulting in a successional sere. Traditionally, ecologists and foresters characterize tree species as having an early (for example, paper birch), mid (red oak), or late (hemlock) successional role. In this chapter we will examine how detailed historical records of stand development can add insights to such simple characterizations and to the long-term developmental history of the broader New England landscape.

Early ecologists and foresters at Harvard Forest developed a strong understanding of how the effects of historical land use played out at the stand level. They characterized the most ubiquitous stand-level compositional trajectory on the landscape in the Fisher Museum's dioramas (Figure 4.17). Through the sequence of historical scenes, we see the pre-European forest transformed by forest clearance and increasingly intensive agriculture, and then we follow the process of forest succession on abandoned farmland leading from white pine to the hardwood forests that currently dominate much of central New England.

This sequence forms the backdrop for the work discussed here. The dominance of the postagricultural signal in regional vegetation history is confirmed by broad consistency between the regional pollen signal discussed in the previous chapter and the Fisher Museum's stand-level dioramas. But the ubiquity of agricultural abandonment in the region presents forest ecologists with a perplexing question: if the bulk of the landscape is shaped by a particular historical process, what kinds of changes occurred in the rest of the landscape, especially the less intensively disturbed areas that remained as forested woodlots? This question has important implications both for basic ecology and for conservation biology because fragments of primary forest that escaped agricultural clearing and have always supported native vegetation are considered to function as more "natural" systems both in the reforested landscape of New England and in parts of the world currently undergoing deforestation. They are thought to provide a link between the cultural present and the presettlement past.

Most of the examples in this chapter come from historical reconstructions of primary forests in and around the Harvard Forest. At the broadest level, we are interested in knowing whether the least-disturbed forests on the modern landscape maintained elements of the composition and dynamics of the presettlement landscape or whether they tracked the cultural transitions of the past 300 years in parallel to the postagricultural landscape in which they are embedded. This general framework leads to specific stand-level questions for presettlement and postsettlement times.

For the pre-European forests and their stand dynamics, we are interested in knowing the following: How common were late-successional communities? Were stand dynamics driven by disturbance or autogenic processes? Was stand composition stable, changing slowly, or in constant flux? When we look to the dynamics of the postsettlement primary forests we ask: To what extent do primary forests retain their presettlement character and species mixtures? Do sites with similar modern composition share similar histories?

Long-term stand-level reconstructions also help us address questions about the ecology of individual species because they tie the population dynamics of those species to local environmental change. In particular we would like to know: Does our understanding of the successional status of a species account for its response to the dynamics seen in the historical record? What was the successional role of species, such as chestnut, that no longer exist in the canopy of New England's forests?

The Forests and Approaches Used to Unravel Long-Term Forest History

We describe in detail the dynamics of four stands on three different sites (Figure 6.1). Hemlock Hollow is a small, seasonally flooded forest hollow (or vernal pool) in a protected lowland on the Prospect Hill tract that is currently surrounded by an extensive forest dominated by hemlock (Figure 6.2). The stand, which lies at the edge of the Black Gum Swamp, was an active woodlot owned by John Sanderson's family during the nineteenth century. Scattered dead chestnut poles leaning against the large hemlocks indicate that the site had a dynamic past despite the abundance of sizable late-successional trees (Figure 6.3). Chamberlain Swamp is a small wetland basin (40 by 70 meters) on a broad upland ridge adjacent to the Quabbin Reservoir that is completely dominated by oak forests (Figure 6.4). The "oak side" of the swamp, where one sediment core was retrieved, is a stand typical of the vegetation across the ridge, as it is dominated by red oak and other hardwoods. In contrast, there is a small hemlock stand on the "hemlock side" of the wetland basin, where another sediment core was taken. Sediment

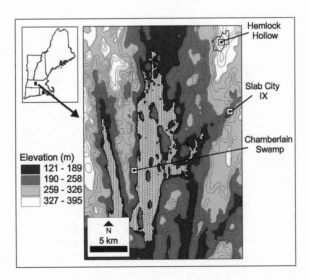

Figure 6.1. Location of Chamberlain Swamp on the Prescott Peninsula in the Quabbin Reservoir watershed, Hemlock Hollow at the Prospect Hill tract, and Slab City IX in the southern part of Petersham. Elevation data from U.S. Geological Survey 1993.

Figure 6.2. Small hollows, vernal pools, and organic soils can provide lengthy and detailed records of local vegetation and environmental change. The closed canopy and small size of the basin and pollen-collecting area mean that most pollen, charcoal, and other material come from the neighboring forest. Hemlock Hollow, pictured here, provided a continuous 9,000-year record of forest dynamics (Foster and Zebryk 1993). Photograph by J. Gipe.

records from these small hollows give long presettlement histories of two contrasting environmental settings: a moist lowland and an exposed ridge top.

In many hemlock-dominated stands, deep accumulations of mor humus preserve pollen as do the sedimentary settings used more typically in paleoecological studies. The final stand we describe, on the Slab City tract south of the center of Petersham, contains such a pollen record extending to presettlement times (Figure 6.5). The surrounding forest is dominated by large old hemlocks and has been the repeated subject of investigation by ecologists such as Steven Spurr and Hugh Raup because it has many of the physical and biological characteristics that forest ecologists associate with late-successional undisturbed forests.

In each study site, local pollen records provide the strongest evidence for long-term compositional change at the stand level. Forest hollows and humus profiles collect most of their pollen from a scale approximating the local stand (for example, an approximate 50-meter radius), in contrast to the lake sediments described in Chapter 5, which collect pollen produced at the landscape or subregional level. Over the past few hundred years, pollen records can be usefully augmented by tree-ring and archival records of forest history, which provide informa-

Figure 6.3. Hemlock stand in the center of the Prospect Hill tract of the Harvard Forest showing the leaning stems of chestnut that formerly dominated the stand but were killed by blight in 1913. Understory hemlock grew rapidly in the open environment to become the major canopy species. The remarkable decay resistance of chestnut wood has enabled these stems to remain standing over the ninety years since their demise. Photograph by D. R. Foster.

tion on disturbance history and structural changes that accompanied the compositional changes observed in the pollen record. Each of the stand histories described here was developed using a combination of reconstructive approaches. We illustrate the strength of this complementary approach in detail for the Slab City hemlock forest.

Presettlement Forest Dynamics

Forests at Hemlock Hollow have been dominated by hemlock since the species' arrival into the landscape about 8,000 years ago (see Figure 4.14). Because of the site's moist, sheltered environment, fire and other disturbances were infrequent, apparently only affecting the stand once every 500 to 2,000 years. For the first 5,000 years of hemlock dominance, succession after fire or other disturbance involved an initial decline in hemlock and increases in oak, pine, and birch. Hemlock eventually resumed its dominance in approximately 500 years after the

Figure 6.4. An extensive oak forest dominates the long ridge containing Chamberlain Swamp. The forest is composed of many multiple-stemmed trees due to the history of logging and fire. Photograph by D. R. Foster.

disturbance, presumably because of its shade tolerance and longevity. The last major change to affect the composition and dynamics of the stand in the presettlement period was the arrival of chestnut around 1,500 to 2,000 years ago (Figure 6.6). An interesting result of the immigration of this new species to the region was that it changed the stand's dynamics following disturbance. After its arrival, chestnut, rather than oak, pine, or birch, increased after fire, illustrating quite clearly that a change in the pool of available species may significantly alter the competitive balance and successional dynamic in the forest.

The ability of chestnut to dominate the stand after disturbance more effectively than other species did not affect hemlock's importance in the absence of disturbance, however. In fact, the continued low frequency of

Figure 6.5. Old hemlock forest on the Slab City tract of the Harvard Forest in the southern part of Petersham. On the basis of the size, age, and composition of the trees, Hugh Raup described this stand in the 1930s as the closest to a climax stand that he had seen. Photograph by D. R. Foster.

disturbance at this site allowed hemlock to dominate through much of the time preceding the arrival of European settlers 250 years ago. Overall, presettlement forest dynamics at Hemlock Hollow were fairly stable and relatively predictable. Over a 4,000-year period, only minor changes occurred in forest composition as a consequence of long-term climatic change. The low frequency of disturbance persistently favored hemlock as a late-successional species.

In contrast to Hemlock Hollow, uplands on both sides of Chamberlain Swamp were dominated by oak for thousands of years before the arrival of chestnut (Figure 6.6). There is strong evidence for aboriginal populations in the valleys surrounding this ridge, so the association between local aboriginal burning and oak dominance hypothesized more broadly for other parts of the Northeast may hold here. It is also likely that compositional differences between Chamberlain Swamp and Hem-

Hemlock Hollow, Harvard Forest

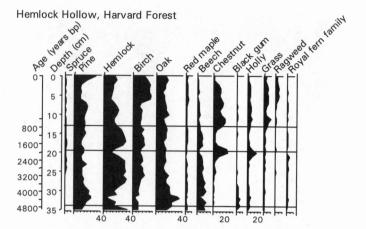

Chamberlain Swamp – Hemlock Side, Quabbin Reservation

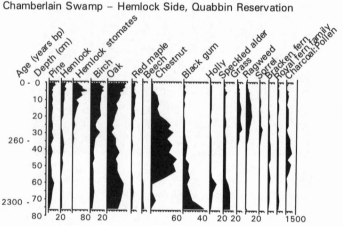

Chamberlain Swamp - Oak Side, Quabbin Reservation

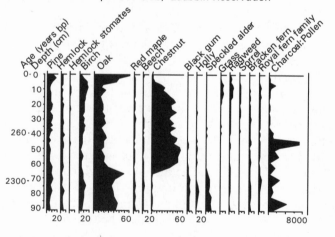

lock Hollow reflect differences in soils, hydrology, and exposure to disturbance, including damage from natural fire, wind, and ice. At Chamberlain Swamp, the arrival of chestnut had a very different effect on stand composition than at Hemlock Hollow. During the first 500 years after chestnut's arrival, it was a minor component of the forest. After an apparently abrupt change to a cooler and most likely moister climate approximately 1,500 years ago, chestnut rapidly became the dominant species; chestnut pollen percentages from that time to the twentieth century are among the highest recorded at any paleoecological site. It is interesting to note that at both Hemlock Hollow and Chamberlain Swamp chestnut seems to have taken over the role of oak in responding to disturbance. At the moister and less fire-prone hemlock-dominated site, chestnut had a transient increase after disturbance, whereas on the drier site with more frequent fire, it replaced oak as the dominant species.

The contrast between the two stands in different parts of the landscape allows us to better understand some of the landscape variation in forest pattern and process, including the dynamics of species invasion. Chestnut arrived in New England forests much later than most hardwoods with similar modern distributions. Whether this migration lag was due to dispersal limitation in this large-seeded species or to subtle climatic limitations on its range has been the subject of debate. Our studies show that, once it arrived, chestnut existed in low populations in central Massachusetts for hundreds of years, and its rise to prominence was triggered by changes in environmental factors. At Hemlock Hollow the long periods between disturbances caused its abundance to vary considerably over time, as chestnut functioned as a successional species. At Chamberlain Swamp, a subtle change in site water balance, indicated by changes in the abundance of wetland taxa here and at nearby Lily Pond, allowed chestnut to reach its compositional dominance in the stand, where it was presumably maintained by fire. Interestingly, following these environmental shifts chestnut maintained a consistent though different niche in the two forests.

Figure 6.6. Pollen diagrams from Hemlock Hollow (see Figure 6.2), a low-lying site dominated by hemlock in the Harvard Forest, and from the opposite sides of Chamberlain Swamp, which lies on an oak-dominated ridge near the Quabbin Reservoir (see Figure 6.4). Locally around Chamberlain Swamp there is a small hemlock stand on one side and broad hill of oak on the other. Because of the small size of the sites, each of the diagrams shows local vegetation dynamics and fire history within approximately 50 meters. At each site, there has been considerable vegetation dynamics involving chestnut, oak, and other species, including hemlock, as a result of climate change, fire, insect outbreak, and logging. Modified from Foster and Zebryk 1993, with permission from the Ecological Society of America, and Foster, Clayden et al. 2002.

The Effect of European Land Use

The influence of European land-use practices is first detected at most paleoecological sites by increases in agricultural weed pollen. These first pollen changes generally do not represent actual disturbance or change at the site; rather, they reflect regional changes, including agriculture and deforestation in the earliest settled part of the region, that alter regional pollen signals. Subsequent more-pronounced changes mark direct cutting and disturbance to the local forest. At Hemlock Hollow, this initial episode is marked by a vegetation response qualitatively similar to disturbance response in the pre-European forests. The early period of colonial activity involved intensive forest cutting locally and extensive clearing in surrounding areas. Active cutting of hemlock and pine in the stand around Hemlock Hollow as a source of timber and use of hemlock and chestnut for tanbark and chestnut poles in the nineteenth century resulted in the establishment of a new forest dominated by chestnut (a prolific sprouter) along with increases in birch and red maple (Figure 6.6). The elimination of chestnut from the forest canopy by the chestnut blight in the early twentieth century resulted in the development of the current hemlock stand at the site. It is interesting that after 250 years of anthropogenic fire, logging, and pathogen outbreaks, a stand has developed that is dominated by large hemlocks and is perhaps structurally similar to the presettlement forest at the site. Our knowledge of history and tree-ring records from the stand, however, suggest that an anthropogenic disturbance regime favoring hemlock played as much of a role in the increase of hemlock at the site as the natural process of autogenic succession.

At Chamberlain Swamp, forest cutting resulted in a decrease in chestnut, in contrast to Hemlock Hollow. Birch appears to have increased because of nineteenth-century disturbance (Figure 6.6), and the site is called a "sproutland" in archival records from the 1930s. Through the twentieth century, birch declined and oak increased, as intensive use of the site lessened and as chestnut was eliminated by the blight. Overall, the postsettlement dynamics of this side of Chamberlain Swamp indicate a transition from chestnut dominance to oak dominance as a result of disturbance. The way this transition qualitatively mirrors the displacement of oak by chestnut 3,000 years earlier might suggest that oak is simply regaining its competitive dominance at the site with the elimination of chestnut, but the compositional differences between the two sides of the swamp illustrate how complicated the long-term dynamics of forests can be.

Until European settlement, pollen records from both sides of the swamp showed the same trends, suggesting that no strong underlying environmental difference exists between the sites. The sites begin to strongly differ in composition only after the chestnut blight in the early twentieth century. At the hemlock side of the swamp, the increase in oak pollen with chestnut decline is also accompanied by increases in red maple and hemlock and the persistence of high birch abundances. Currently, the site is dominated by hemlock and is compositionally more similar to the forest at Hemlock Hollow than to the oak stand at Chamberlain Swamp. The postsettlement forest dynamics at Chamberlain Swamp suggest two important lessons. First, strong differences in the composition of adjacent stands can be predominantly the product of recent human impact; and second, this can be true even for stands dominated by large late-successional trees.

Convergence in Forests Following Different Histories

The contrast in the compositional history between Hemlock Hollow and the hemlock side of Chamberlain Swamp suggests that sites that appear structurally and compositionally similar today may actually be the product of very different histories. We investigated this possibility through detailed reconstructions of the history of Slab City IX and three other primary hemlock stands at the Harvard Forest. Currently, the stands are all quite similar, and their dominance by large late-successional hemlocks suggests that they might be the stable product of autogenic succession. However, the pollen records derived from the humus soils in each stand show that the forests were all different from each other at the time of European settlement, that they all experienced major compositional change in the eighteenth and nineteenth centuries due to forest cutting, and that they converged on their current hemlock-dominated assemblages in the twentieth century.

When we combine tree-ring data, a stand-level pollen record, and the history of known disturbances from archival records, we develop insights into the ecological mechanisms underlying the compositional history of one of these sites. In Figure 6.7A we see that the oldest trees currently in the stand are hemlocks that originated around the turn of the nineteenth century, but that most trees, including all the hardwoods, recruited in the few decades surrounding the turn of the twentieth century. In Figure 6.7B, three growth releases in the nineteenth century confirm archival data suggesting that the stand was repeatedly cut. The last of these cuttings initiated the origin of the modern stand. Although the chestnut blight is not strongly reflected in the tree-ring record, Figure 6.7C shows that the compositional response to nineteenth-century cutting was an abundance of sprout chestnut. As at Hemlock Hollow, the elimination of chestnut by blight after 1913 allowed hemlock, which already was established as advanced regeneration, to dominate the stand. Because of the specific disturbance history of this stand, including forest cutting, wind, and pathogens rather than agricultural clearance and

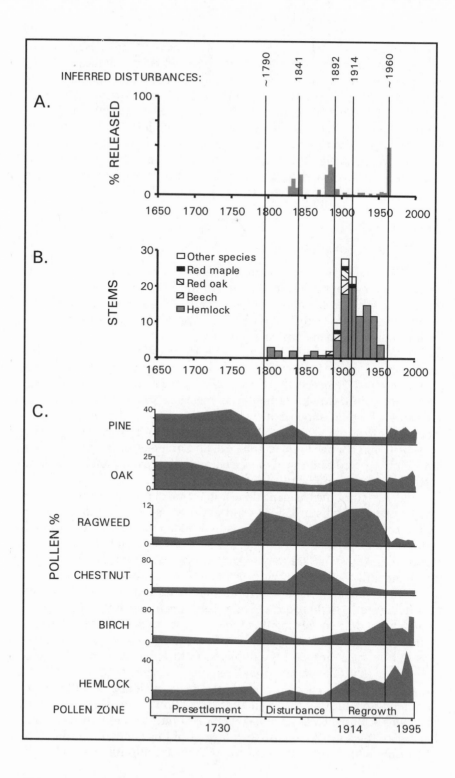

INFERRED DISTURBANCES:

A.

B.

C.

136

fire, the stand is now dominated by large old hemlocks. It is worth noting that before European settlement the stand was dominated by oak and pine rather than by the late-successional hemlock-hardwood assemblage that occurred at the Hemlock Hollow site. In fact, none of the three other hemlock sites we investigated had the same presettlement assemblages, and none of the trajectories of compositional change shows signs of returning to pre-European composition.

The Ecology of Hemlock and Chestnut

In addition to deepening our understanding of the ecology of forest communities, reconstructing the long-term dynamics of forest stands provides insight into the basic ecology of tree species, whose life-history characteristics often remain poorly understood because of their long generation times. It is a major goal of forest ecology to evaluate the potential response of forest species to accelerating changes in land use, climate, and atmospheric chemistry, but tools such as computer-based forest models must have accurate information about the ecological attributes of these species in order to evaluate their performance effectively. The sites discussed in this chapter provide specific information on the autecology and competitive dynamics of two important eastern tree species: eastern hemlock and American chestnut.

The common perception of the late-successional role of eastern hemlock is well-characterized by its presettlement dynamics at Hemlock Hollow. For thousands of years, hemlock dominated undisturbed late-successional assemblages. It was dramatically reduced after disturbance by fire and took hundreds of years to reassert its former dominance. Postsettlement dynamics of this species at Hemlock Hollow, as well as at the hemlock side of Chamberlain Swamp and at Slab City, present a

---

Figure 6.7. Stand dynamics at the Slab City IX site. Vertical lines signify disturbances to the site (historically documented logging operations in 1841 and 1892, the chestnut blight of 1913, and disturbances inferred from pollen and tree ring records at approximately 1790 and 1960). (A) Growth releases in hemlock tree rings. Hemlocks in the stand experienced growth releases surrounding the two cutting episodes and after an undocumented disturbance in 1960. (B) Year of recruitment of current stems. A few overstory hemlocks date to the early nineteenth century, but most hemlocks in the stand and all of the hardwoods recruited after the 1892 cutting. The lack of recruitment after 1950 is a methodological artifact (trees of less than 5 centimeters in diameter were not aged). Hemlocks have continued recruiting into the stand until the present. (C) Pollen percentages from soil humus. Presettlement forests were dominated by pine and oak. Regional increases in ragweed pollen indicate the period of regional historical deforestation. At Slab City IX chestnut sprouts followed local cutting. Blight eliminated chestnut in 1913. It was replaced by the hemlocks that now dominate the site. Modified from McLachlan et al. 1999, with permission from the Ecological Society of America.

more obscure aspect of its ecology but one that may be more important in the strongly human-influenced forests of today. At these sites, wind, cutting, and pathogens acted to promote hemlock dominance rather than to reduce it. Instead of characterizing hemlock as a disturbance-sensitive, late-successional species, in New England it is more accurate to consider the specific effects of different disturbance types, intensities, and patterns on a species that has a rather broad successional role.

Chestnut was an important species in the region before it was eliminated from the forest canopy by blight early in the twentieth century. Historical sources give important accounts of the character and distribution of this species, but the details of its role in forest succession before the blight are not well known. Because of this lack of information, evaluating the impact on current forests of the loss of this species is difficult. The records examined in this study highlight the complex response this species had to natural and anthropogenic disturbance. After its arrival in New England, chestnut replaced oak on some sites as the most successful species following fire. The Hemlock Hollow and Chamberlain Swamp records indicate that this was true on both relatively wet sites, where its successional role was only temporary, and on dry sites, where it dominated stand composition for millennia. As we found for hemlock, however, chestnut's predictable response to presettlement environments was not mirrored in its response to the anthropogenic disturbances of the postsettlement period (Figure 6.8). At Hemlock Hollow, chestnut sprouted vigorously after cutting of the original hemlock stand, but at Chamberlain Swamp and Slab City, chestnut's relative abundance actually decreased in response to similar eighteenth- and nineteenth-century disturbance.

A final and unfortunate comparison between hemlock and chestnut is that both species have suffered greatly from introduced organisms. Chestnut has been reduced to an understory shrub for nine decades and, despite ongoing efforts at reducing the effect of the blight or breeding resistance into chestnut, hopes for the tree's imminent recovery are dim. As we have seen in this chapter, the direct impact of the blight on stand-level dynamics is obscured by other twentieth-century phenomena such as land use. Nevertheless, the dynamics of eastern forests in the twentieth century were undoubtedly driven to a large extent by long-term compositional responses to this event. The decline of hemlock populations since the arrival of the hemlock woolly adelgid to southern New England in the 1980s suggests that forest dynamics in the twenty-first century might be characterized by a similar response to major decline of a dominant forest tree. Even if this grim scenario fails to occur, we can be certain that New England's forests face a period of intense and novel dynamics. The historical record provides an important tool for evaluating how they will respond.

Figure 6.8. Coppice forest of American chestnut in Massachusetts around 1900 showing the open structure of the forest and sprout form of the trees. Reprinted from Paillet 2002, with permission from Blackwell Science Ltd.

Conclusions

A broad examination of forest history underscores the contrasting patterns of forest development on different sites in New England's postagricultural landscape. Perhaps the most common, and certainly the best recognized, pattern is the sequence of changes in forest composition on former agricultural land, as depicted in the Fisher Museum's dioramas. Original transition hardwood–hemlock forests were converted to agricultural land that remained open for a century or more before being abandoned from active use. Today, most of these sites are forested with mid-successional hardwoods like red oak and red maple that assumed dominance after the initial postagricultural forests of white pine and other early-successional species were cut or blown down. Late-successional species such as hemlock and beech are becoming established only very slowly. On primary forest sites, including the two examined here, the long-term dynamics were nearly as dramatic as those of postagricultural stands. The original chestnut forests of Chamberlain Swamp and the old-growth white pine–oak community of Slab City were repeatedly cut over during the eighteenth and nineteenth centuries, resulting in assemblages of sprout hardwoods. Over the past 100 years, the effect of human activity has grown less apparent, and the stands now are among the oldest in New England.

These historical sequences illustrate some important generalities about the current forests of central New England. First, the regional homogenization of forest types shown in the previous chapter masks a good deal of variability at the stand level. Second, this stand-level variation may have more to do with site history than with the underlying environment. Finally, stands in our landscape that show characteristics of old natural forests are often anthropogenic in origin. This last conclusion does not diminish the value of these forests for conservation purposes, including landscape diversity and wildlife habitat. Indeed, they illustrate the potential of the modern landscape for generating extensive seminatural mature forests, especially on primary forest sites.

Revisiting the questions raised at the beginning of this chapter, we can see that detailed records of forest change from a variety of sites can provide answers to basic questions about long-term stand dynamics. Topographic position created differences in presettlement stand composition and disturbance frequencies across the landscape, but composition at sites as different as Hemlock Hollow and Chamberlain Swamp was fairly stable and presumably in balance with underlying environmental forces. With European settlement, the bulk of the landscape was converted to open pasture or agricultural use, but even the approximately 20 percent of the land that remained forested was subject to substantial structural and compositional change. Forests that appear to be stable and natural on the modern landscape have experienced natural and anthropogenic disturbance approximately every decade and have undergone major compositional changes every century. The complex disturbance regimes of these stands complicate our traditional view of successional dynamics and illustrate the flexibility that tree species have for dealing with novel environmental change.

In Chapter 8, the relative roles of local environment and site history are investigated further using detailed analyses of the correlation between modern vegetation and site characteristics. In such static studies of the modern landscape, the role of history emerges as an important contributor to the distribution of species, and it is useful to keep in mind the scale and timing of the dynamic processes observed in the historical reconstructions presented here.

CHAPTER 7

Wildlife Dynamics in the Changing New England Landscape

D. BERNARDOS, D. FOSTER, G. MOTZKIN, and J. CARDOZA

Although the forest dynamics that we have examined were profound and characterized by major shifts in land cover and forest composition, they occurred gradually within the time frame of daily life and off in the woods, where they have been easily ignored by much of the general population. The same cannot be said of the remarkable changes that have occurred recently in the wildlife of New England and much of the eastern United States (Figure 7.1). Over the past decade, nearly daily articles in newspapers and magazines have highlighted the ongoing transformation in our animal populations and the management and health issues that accompany these changes. Many larger mammals and birds that have been uncommon for decades or centuries, such as bears, beavers, fishers, moose, eagles, turkeys, herons, and vultures, are increasing and are regularly encountered in backyards, along roadsides, or in fleeting views; whereas many familiar and cherished bird species of agrarian landscapes, including bobolinks, meadowlarks, woodcock, whippoorwills, and openland sparrows, are declining or have disappeared locally (Figure 7.2). Meanwhile, species that have never inhabited the region have immigrated, in some cases across great distances, or have been introduced and are increasing in density, visibility, and ecological importance.

With these changes, wildlife managers, conservationists, and many citizens are confronted with major policy issues and ethical dilemmas. What changes are really occurring? Should we attempt to maintain populations of declining species, and if so, how? How do we control, manage, and live with the growing numbers of larger species? More fundamentally, what social, biological, or environmental factors underlie and explain these changes? A historical perspective may assist in educating and even modifying the behavior of the human population that interacts with this dynamic wildlife. In a region with immense tracts of maturing forestland, but a growing and largely suburban population, the range of issues involving wildlife extends from the desire on the part of many

Figure 7.1. A moose-crossing sign along a Massachusetts highway highlights the dynamic and increasingly wild nature of the land. In the 1850s Henry Thoreau noted that the largest wild animal in the landscape was the muskrat. Photograph by D. R. Foster.

conservationists to restore large carnivores such as wolves and cougars, to an appreciation for wildlife by the general population, balanced with concerns for human safety and personal property that appear to be posed by a handful of species.

To address these ecological questions and management issues, there is a critical need for long-term data along with a historical perspective similar to that which we have used to evaluate and discuss forest dynamics. Such an approach to these issues also offers an opportunity to answer basic ecological questions about the factors controlling wildlife populations, the nature of the feedbacks that exist between plant and animal species, and the contrasts between the dynamics of wildlife and plant assemblages.

In pursuit of a long-term perspective on wildlife dynamics, we collected historical data on animal populations for southern New England and analyzed these in order to do the following: (1) document the major trends in wildlife populations since European settlement; (2) identify the loss or arrival of major species and resulting changes in wildlife assemblages; (3) relate these dynamics to changes in the physical, biological, and cultural environments; (4) integrate this information with our historical perspectives on vegetation dynamics in order to increase our understanding of forest processes through time; and (5) provide a context and background for interpreting current changes in order to guide policy for management and conservation.

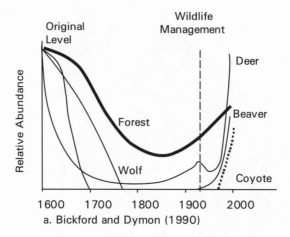

a. Bickford and Dymon (1990)

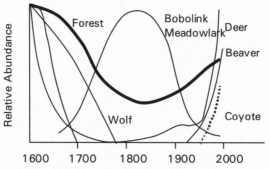

b. O'Keefe and Foster (1998; modified from
Bickford and Dymon, 1990)

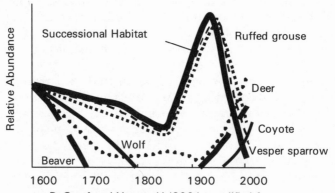

c. DeGraaf and Yamasaki (2001; modified from
Bickford and Dymon, 1990)

Approaches and Methods

The spatial scale, time frame, and species that we investigated were selected through a process that blended ecological and management interests with pragmatism. The compilation of records extending over hundreds of years requires identification of a common geographical reference area—something akin to a common denominator for data. Although many sources of wildlife data exist, most records are grouped by political unit, with state governments and agencies providing the most comprehensive recent information and town records and histories, aggregated to a state level, providing critical information for the early colonial period. In consideration of these factors, we opted for taking a statewide perspective, which offers the advantages of considerable data and ample variation in such factors as environment, vegetation, and land use. As data sources vary considerably from one state to the next, we limited our analysis to a single state—Massachusetts. From an ecological perspective this provides a good context for our other studies. Moreover, Massachusetts incorporates a wide range of the environmental and cultural gradients that are broadly representative of New England and that provide for diversity in habitat and wildlife dynamics. Finally, since much of the legislation and policy concerning wildlife originates at a state level, these data are relevant to major societal needs.

Data Sources

We utilized a variety of sources and incorporated the resulting information into our statewide GIS and database. Unfortunately, and rather surprisingly, there are no Massachusetts-wide data from archaeological studies that would provide an overview of pre-European faunal use and distribution. Consequently, though we used some archaeological sources, our earliest records focus on explorers' accounts, town and state legislation, bounty and harvest records, natural histories, scientific studies, museum collections, newspaper and other popular accounts,

---

Figure 7.2. Three related but different graphs interpreting major historical wildlife trends in Massachusetts. Panels a and b suggest that the landscape was heavily forested at the time of European settlement. Panel b (O'Keefe and Foster 1998b) is adapted from panel a (Bickford and Dymon 1990) but does not emphasize the role of management in the recovery and dynamics of species. It also includes openland bird species to underscore the fact that many species thrived and increased in the nineteenth-century agricultural landscape and subsequently declined with reforestation. Although panel c (DeGraaf and Yamasaki 2001) is also modified from panel a, it interprets the pre-European landscape as extensively covered with successional habitat due to Indian activity and natural disturbance and suggests that openland species were abundant at the time of European settlement. Modified from Foster, Motzkin et al. 2002.

and town and county histories. Although varying in taxonomic detail, data density, and accuracy, these sources provide a strong qualitative to semiqualitative picture of changing wildlife populations and help to identify some of the factors driving these changes. In particular, histories of the more than 300 towns in the state, mostly written between the mid-1800s and mid-1900s, provide a wealth of information, frequently including species lists, bounty records, anecdotal descriptions of changing wildlife populations, and dates of last and first observations. The large number of towns in the state allows us to overcome the spatial and temporal gaps in individual records and to develop fairly robust interpretations of changes through time.

In addition to town histories, many scientific studies provide overviews of the ecology, behavior, and status of individual animal species. Many also contain summaries of historical trends for particular species over time. Meanwhile, twentieth-century information on population estimates, harvests, accidental deaths (for instance, roadkills), sightings, and management activities is maintained by the Massachusetts Division of Fisheries and Wildlife in their annual reports, Game Population Trend and Harvest Surveys, and unpublished records.

Species Examined

In our melding of ecological and management interests with pragmatic constraints, we focused on species that are important to the function of forested and other natural landscapes, represent particular social and conservation concerns, and are fairly consistently represented in available records (Table 7.1). These criteria also satisfied one of our original objectives, namely, to use the history of wildlife and the New England landscape to convey a strong sense of ecological dynamics to the broad public. Thus, our data are necessarily biased toward game species, larger birds, persecuted "vermin" species, and selected animals of conservation focus. In contrast, most songbirds, amphibians, reptiles, fish, small mammals, and invertebrates are largely excluded from our analysis because they are incompletely represented in historical data and documents.

Cultural and Environmental Setting for Interpreting Wildlife Dynamics

Previous chapters provide detailed information on the climatic, vegetation, and human history of New England. Consequently, we need to expand on these only with information particularly relevant to wildlife populations. Relatively few detailed data exist on the status of wildlife for the Late Woodland period before the first European settlers. Nonetheless, it is important to underscore our earlier observations that

Table 7.1. Historical Dynamics of Birds and Mammals in Massachusetts and Adjoining Areas of New England

Regionally Extirpated or Extinct (+)

Bicknell's thrush	Eastern gray wolf	Sea mink (+)
Great auk (+)	Elk	Cougar
Heath hen (+)	Indiana bat	Wolverine
Labrador duck (+)	Lynx	Woodland bison
Loggerhead shrike	Marten	Woodland caribou
Passenger pigeon (+)	Mountain lion	

Openland and Successional Habitat (species generally declining)

Northern harrier	Brown thrasher	Upland sandpiper
Northern bobwhite	Nashville warbler	Vesper sparrow
Killdeer	Chestnut-sided warbler	Savannah sparrow
Spotted sandpiper	Prairie warbler	Grasshopper sparrow
American woodcock	Yellow-breasted chat	Bobolink
Mourning dove	Eastern towhee	Eastern meadowlark
Common nighthawk	American tree sparrow	Black-throated blue warbler
Whippoorwill	Field sparrow	Mourning warbler
Least flycatcher	Indigo bunting	
Horned lark	Red-winged blackbird	New England cottontail
Purple martin	Veery	Red fox
Bank swallow	Golden-winged warbler	Woodchuck
Barn swallow	Brown-headed cowbird	
Sedge wren	Baltimore oriole	
Ruffed grouse	Magnolia warbler	
Eastern phoebe		
Eastern bluebird		

Introduced Species (*introduction failed)

Cattle egret	Black-tailed jackrabbit	Black rat*
Mute swan	European hare*	House mouse
European starling	European rabbit	
House finch	Eastern cottontail	
House sparrow	Norway rat	
Ring-necked pheasant		
Northern bobwhite		
Rock dove		
Wild turkey (reintroduced)		

Range Expansion

Northward or eastward

Little blue heron	Northern rough-winged swallow	Golden-winged warbler
Glossy ibis	Tufted titmouse	Nashville warbler
Turkey vulture	Carolina wren	Worm-eating warbler
Black vulture	Blue-gray gnatcatcher	Northern waterthrush
Mourning dove	Northern mockingbird	Louisiana waterthrush
Barn owl	Blue-winged warbler	Northern cardinal
Red-bellied woodpecker	Cerulean warbler	
Acadian flycatcher		Virginia opossum
		Coyote

Southward

Herring gull	Bohemian waxwing	Rusty blackbird
Great black-backed gull	Magnolia warbler	Purple finch
Golden-crowned kinglet	Swamp sparrow	
Hermit thrush	White-throated sparrow	

(continued)

neous in human distribution, land cover, and wildlife habitat. In particular, today a greater proportion of the population lives in suburban areas. Differences in population density, agricultural land, and road networks across the state have increased immensely during the past century. The Berkshire Plateau and northern Worcester County remain relatively undeveloped and largely forested; eastern and southeastern Massachusetts from Boston and the South Shore toward Worcester are intensely fragmented by human activity; Cape Cod and the Islands are increasingly developed for summer homes and recreation; and the Connecticut River Valley juxtaposes farm fields, housing, and industry with wooded swamps and isolated forested mountains. Although the extent of forestland peaked in the past decade, forest age continues to increase as brushy grasslands, shrublands, and young woodlands are still being replaced by older forest. In addition, development, suburban fragmentation, and "parcelization" (progressive reduction of forest lot sizes) continues, particularly in eastern Massachusetts. Meanwhile, social attitudes toward wildlife and wildlife management and regulations have changed considerably in the past 100 years and are increasingly shifting from an emphasis on consumption to one of conservation and preservation.

General Trends in Wildlife Dynamics

Historical data for nearly 100 species highlight species-specific dynamics and long-term trends that are highly individualistic. Each of these histories can be broadly interpreted in relation to the species habitat preferences and land-cover changes, in the context of specific human pressures, especially hunting or trapping. However, despite these individualistic trends, generalization is possible. Six broad patterns are recognizable that capture the major temporal trends in wildlife dynamics (Figure 7.2b and Table 7.1).

A large group of species, including many large mammals and birds that were actively hunted and persecuted or that require extensive woodland habitat, exhibit a long-term historical pattern of decline and recovery. These species were widespread across the Commonwealth at the time of European settlement and then decreased rapidly and were either locally extirpated by the mid-nineteenth century or persisted in small, local populations. Over the past century and in distinctively individualistic patterns, they have increased, either rebounding naturally or with assistance from stocking and other forms of active management. In contrast, a group of open-land species was uncommon or absent at European settlement, increased greatly with agricultural clearing of forest to a peak in the nineteenth century, and has subsequently declined to oftentimes low abundance. This group includes many grassland, shrubland, and early-successional species that are the focus of modern conservation efforts. Some of these species are common elsewhere; others are globally rare; and for many, records are inadequate to determine whether they are native to New England. A relatively small number of animals were extirpated from the state or region, including a few that became globally extinct. Although low in number, species in this group do include animals that were particularly important ecologically or that represent major cultural features of the New England landscapes. A number of bird species and at least two mammals have naturally expanded their ranges into Massachusetts from northern, southern, or western distributions. Nonnative species have been introduced purposefully or accidentally. In some cases these new species have become naturalized and quite abundant. Finally, numerous persistent species have fluctuated through hundreds of years of settlement but have not experienced the long-term directional changes in populations noted for other groups of species.

Species Declining Historically with Recent Increases and Recovery

This rebounding group is varied but predominantly composed of birds and moderate- to large-sized mammals that depend on and use forested environments. Many of these species are notable for their rapid recent increases and their high visibility to human residents. With many mammal populations currently increasing at 3 to 10 percent annually, this group is the focus of research and management as well as frequent human conflict.

Beaver *(Castor canadensis)* and white-tailed deer *(Odocoileus virginianus)* represent two of the numerous species that decreased early in the historical period and recently have increased substantially. These two are of particular interest because of the great influence that they exert on natural ecosystems and the significant management issues and concerns that they raise in the modern landscape (Figures 7.3 and 7.4). From archaeological, ethnographic, and historical sources, it is clear that both species were widespread, common, and important in pre-European forest ecosystems and Indian economies. Both were also important prey for large carnivores, including cougar and wolf. Beaver rapidly became a focus of trapping and trading for the French, Dutch, and English in the early seventeenth century, including John Smith, who returned to London in 1616 with 1,100 skins. In Massachusetts, revenue from the beaver trade helped to finance settlement activity and encouraged exploration. The shipment of pelts paid many of the expanding colony's debts to the homeland and gradually led to local extirpation of the species. As early as the early 1630s, Governor William Bradford reported shipments of more than 12,500 pounds of beaver pelts to England. The species was eliminated from southeastern Massa-

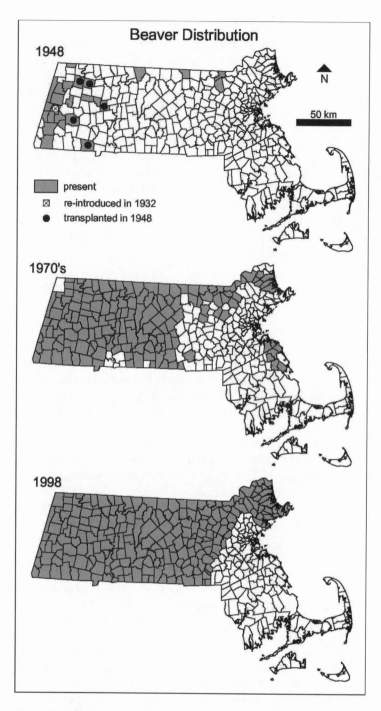

Figure 7.3. Modern expansion of the beaver population across Massachusetts following reintro-ductions in the 1930s and 1940s. The population in 2002 was estimated as exceeding 70,000 animals. Reprinted from Foster, Motzkin et al. 2002.

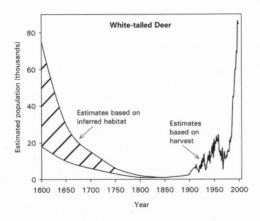

Figure 7.4. Historical changes in the population of white-tailed deer in Massachusetts. Two early scenarios are depicted that bracket the extreme high and low estimates at the time of European settlement. In the mid-nineteenth century small populations persisted in the Berkshire Hills in the west and near the base of Cape Cod in the southeast. Reprinted from Foster, Motzkin et al. 2002.

chusetts by 1635 and from most of the state, except the northern Berkshires, by 1700. As beaver were declining in southern New England, Indian and colonial trapping expanded to New York and Maine. The fur trade shifted to Canada by 1750, and beaver were completely extirpated from Massachusetts by the late 1700s.

After more than a century's absence, initial efforts were made to reintroduce beaver in the late 1920s, with the enactment of protective legislation and releases of New York beavers in western Massachusetts. Through the 1940s, numerous animals were relocated across the state, and with a population estimated at more than 300, a limited trapping season was established. The species expanded rapidly, both on its own and through continued relocation efforts on the part of state wildlife managers. By the late 1980s, the official assessment by the Division of Fisheries and Wildlife was that "beavers are deemed restored to all suitable habitats."

In the heavily wooded landscapes of central New England, with abundant streams, wetlands, and lakes, beaver are thriving, with a population recently estimated at 70,000. Increasingly, this relatively recent arrival is affecting forests by creating dams, which flood adjoining uplands, and selectively cutting trees and browsing other plant species. The ecological impact of beavers is difficult to overstate as they alter local hydrology and biogeochemistry, create wetlands, modify soils, flood and kill acres of forest, selectively alter vegetation composition, diversify landscape patterns, and create new habitats. The influence of these activities on other species is equally important. The ponds, open wetlands, dead trees, and forest openings created by beavers offer important habitats for many plants and animals in a landscape otherwise dominated by dense and continuous forests. The return and near omnipresence of beavers are major factors conveying the sense that New England is becoming wilder and more natural.

However, many effects of beaver, especially the raising of water tables and mortality of trees, have significant human consequences. Notably, the number of highway, housing, and septic system conflicts involving beavers has risen dramatically in recent years. In addition, beaver and other mammals may carry diseases such as giardiasis, a disease caused by the protozoan *Giardia lamblia,* which also causes human illness, and have the potential to spread them widely across the landscape. Despite the fact that most *Giardia* outbreaks in New England are probably of human origin (for example, from human waste fouling water supplies), beavers and other wildlife have been implicated in such outbreaks in a number of New England public water supplies, which has led to trapping, relocations, and heated public debate. In the absence of natural predators, beaver populations are continuing to expand and grow, and social conflicts are destined to increase. However, in a social environment inimical to trapping, a statewide referendum led to a ban on most leg-hold and body-gripping traps in Massachusetts in 1997. As a result, the number of animals trapped declined, from 1,136 in 1996 to 98 in 1998. With the beaver population continuing to increase and complaints to wildlife agencies continuing to rise, we face a major challenge as to how to live with or control this remarkable animal.

White-tailed deer underwent an analogous trajectory, although they were never completely eliminated from Massachusetts and did not require the same degree of active intervention to reach their current high density (Figure 7.4). Deer were important in both Indian and early colonial landscapes and economies. Deer remains are a consistent item in archaeological sites and generally constitute the most abundant vertebrate fossil. It is likely that Indians and the gray wolf were the species' major predator, as deer provided an important source of food, tool-making materials, and clothing. However, quantification of Indian impacts on deer, like population estimates for both species, is fraught with uncertainty. Heavy hunting by an expanding European population led to noticeable declines in the deer herd and the promulgation of many early, and ultimately futile, attempts at hunting regulation. In 1698, Massachusetts placed a closed season on deer between January 15 and July 15 and then enacted a three-year moratorium on deer hunting in 1718 when underenforcement of the original law and habitat loss led to further declines. Deer "reeves," one of the earliest attempts at game wardens in the United States, were appointed in each town in the Commonwealth in 1739.

Throughout the late eighteenth to late nineteenth centuries, deer became increasingly uncommon and were essentially extirpated from the central two-thirds of the state. A small population persisted in forested areas of Berkshire County in western Massachusetts and in the pine and oak woodlands in southeastern Massachusetts near the base of Cape Cod. A ten-year hunting moratorium starting in 1898, coupled with farm abandonment and a statewide increase in shrubland and woodland, produced the first rebound in the population, to an estimated 5,000 in 1905. The growing population led to crop losses, illegal hunting, and the establishment of a regulated hunting season by the Division of Fisheries and Wildlife. Although the season has been modified repeatedly in attempts to regulate the size and demography of the herd, the deer population has continued to expand, particularly since the 1940s. Currently, it is estimated at approximately 90,000 animals. On average, 10,000 deer are harvested by hunters annually, with another 7,000 killed by automobiles. However, hunting interest is declining across New England, and there are major questions concerning the potential to regulate the deer herd, especially in populated suburban areas, where hunting is often prohibited.

The expanding population of large herbivores affects forest ecosystems profoundly. Selective deer browsing alters the composition of tree seedlings, herbs, and shrubs and ultimately may have a strong and long-term impact on forest composition and structure. The sedentary behavior of deer, with adults seldom roaming more than three-quarters of a mile from their place of birth, may lead to significant local variation in effects. Landscape-level variation in deer browsing has been strikingly apparent in central Massachusetts. Active hunting throughout this broad area, including on Harvard Forest lands, has maintained a low deer population that has little effect on forest regeneration and composition. In contrast, a fifty-five-year ban on hunting in the nearby 60,000-acre Quabbin Reservation led to an extremely dense deer population and a severe understocking of seedlings, saplings, and understory trees. In many cases the forest had the open, fern-dominated appearance of a wood-pasture. Reinstitution of hunting at the Quabbin in 1991 was a controversial process that has succeeded in reducing deer densities and initiating a sustained pulse of understory recovery. Similar impacts, and management conflicts, abound across southern New England and much of the eastern United States, especially in suburban wooded areas where gardens are often the focus of impacts. The remarkable resurgence of the deer herd and the species' ability to thrive in areas heavily used by humans has been associated with unfortunate health consequences, notably the rapid spread of the tick-borne Lyme disease. As a consequence, the cultural perspective of deer is undergoing a remarkable shift in recent decades from noble and wild game animal to neighborhood pest.

Six highly conspicuous large birds are included in the group of re-bounding species (the pileated woodpecker, wild turkey, raven, osprey, eagle, and great blue heron; other species undoubtedly increasing include raptors such as the great horned and barred owls). Pileated wood-peckers, the region's largest woodpecker species (approximately 30 centimeters tall), depend on large, standing dead trees for nesting sites. Although this species declined to low numbers as forests declined and remaining stands were intensively harvested, its population has expanded greatly with recent increases in forest age and maturity. Wild turkey, a forest-dwelling species, was widespread at the time of settlement and a common food for Native Americans and early European settlers. It was extirpated across the region by overhunting but was actively reintroduced in the 1930s. Turkey have increased across much of the state, reaching populations of 20,000 in 2002 because of the excellent habitat of open oak woodlands and agricultural land. The species is hunted extensively with bow, primitive firearms, and shotguns during the fall and spring seasons. The raven, which closely resembles the smaller crow, is a northern species that commonly feeds on carrion. This formerly uncommon species has expanded naturally back into southern New England as a consequence of increased food provided by rebounding wildlife populations and roadkill.

Osprey and eagle are large raptors whose populations were decimated by indiscriminate killing and antipathy toward raptors during the seventeenth, eighteenth, and nineteenth centuries. These species were further affected by shell-thinning caused by the bioaccumulation of the insecticide DDT. As a result of the banning of DDT, active establishment of nesting platforms, general improvement in water and wetland quality, the protection of coastal habitat, and change in public attitudes toward predators, ospreys are undergoing a remarkable rebound, from a low of less than fifty in the 1970s to more than 350 today. Although heavily concentrated in coastal areas, this species should continue to expand inland up the major riverways and into the watersheds of large lakes. Eagles have been reintroduced through active hacking programs and are expanding across New England. In Massachusetts, large winter populations congregate around the Connecticut River and Quabbin Reservoir with up to fifty birds sighted in winter months and more than six breeding pairs currently established.

Great blue herons suffered along with many other showy waterfowl because of the collecting of feathers and the deterioration of wetlands and water quality. Now afforded protection and with improved water quality, heron populations have rebounded. Furthermore, the recent increase of beaver has exerted a profoundly positive effect on great blue heron, which utilizes the resulting dead trees and habitat in flooded beaver ponds.

Openland Species Increasing with
Forest Clearance and Agriculture

Forest clearance and the creation of openland habitat favored many native species that were uncommon in the forested landscape. This land-cover transformation may also have enabled grassland and shrubland species from regions including the Midwest to immigrate to New England. Species including reptiles, amphibians, diverse birds, and mammals such as the red fox, striped skunk, New England cottontail, and woodchuck peaked in abundance with maximum agriculture or during the early period of farm abandonment and forest recovery. Although exhibiting diverse population trajectories according to their habitat preferences for, for instance, grass height and density or abundance of woody vegetation, most of these species are continuing to decline as forests mature and remaining open and successional habitats become woody or are developed for human uses. As a consequence, this group includes some of the most vulnerable populations in the Northeast.

In many ways the most dramatic and interesting examples of favorable responses to historical land-use practices are witnessed in grassland bird species (Figures 7.5 and 7.6). Indeed, species such as upland sandpiper, vesper sparrow, grasshopper sparrow, meadowlark, bobolink, and savannah sparrow are a major focus of environmental concern that present a substantial management challenge and interesting ethical issues for conservationists.

The early history of these birds is uncertain. Like most passerines and nongame species, they were not recorded from Massachusetts until the publication of Peabody's *A Report on the Ornithology of Massachusetts* in 1839, which was the first attempt at a comprehensive bird list for the state. Consequently, their native status is uncertain. There is no doubt, however, that these birds proliferated in the agrarian landscape. Their response and timing of peak abundance undoubtedly varied because of subtle differences in nesting and foraging habitat. Through the height of agriculture (1830–70), the upland sandpiper, which uses large grassy areas with low vegetation, was recorded by Thomas Nutall and others as "common in Worcester County in the summer" and "abundant across the state." This is the only game species of the six and is thought to have expanded eastward from natural prairies and peaked in the mid-1800s, when it was hunted in large numbers. Vesper sparrows were so plentiful in open fields and upland pastures from Cape Cod to the Berkshires that E. H. Forbush in 1907 considered them to be the "most abundant ground sparrow in Massachusetts" after the song sparrow. The grasshopper sparrow apparently peaked slightly later in the late nineteenth century and, though never common, could be found in "spectac-

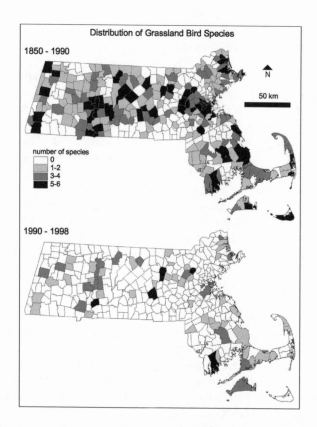

Figure 7.5. Historical and modern distributions of important grassland bird species in Massachusetts. The six species include bobolink, eastern meadowlark, savannah sparrow, vesper sparrow, grasshopper sparrow, and upland sandpiper. A broadscale reduction in distribution and abundance of all species occurred with reforestation and the loss of open and successional habitats, making these species high priorities for conservation. The status of these grassland species in Massachusetts at the time of European settlement is unknown, but most were probably quite uncommon. Reprinted from Foster, Motzkin et al. 2002.

ular abundance" on Cape Cod, Nantucket, and Martha's Vineyard and in lower numbers across the state. This remarkable upswing in now uncommon species was paralleled by some of the more familiar birds of New England's agricultural past—such as the bobolink, meadowlark, northern bobwhite, red-winged blackbird, and bluebird.

These species thrived with traditional, low-intensity agricultural practices, including extensive grazing and mowing, and they declined as open fields and pastures reverted to forest. Considered a delicacy and avidly hunted in the late nineteenth century, the upland sandpiper declined first, but by the mid-twentieth century the number of breeding sites for all grassland species was declining statewide. In the 1950s, up-

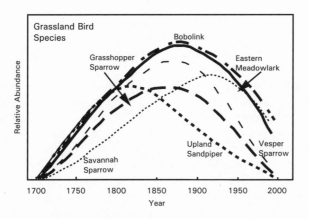

Figure 7.6. Inferred changes in the population size of important grassland bird species in Massachusetts through time. The native status of some species is uncertain. Reprinted from Foster, Motzkin et al. 2002.

land sandpipers were located in only a few localities, and by the 1970s and 1980s the three species were depending largely on artificial, highly humanized landscapes, primarily military bases, airports, and landfills. Currently the upland sandpiper is listed by the Commonwealth of Massachusetts as "endangered," whereas vesper and grasshopper sparrows are "threatened."

Ironically, ten of the eleven most important refuges for these species are industrial or military sites, where the species are being maintained only through the coordinated efforts of managers and conservation agencies to regulate mowing and burning regimes and to control disruptive human impacts. Although these species are essentially unknown by the general population, their remarkable historical dynamics, their imperiled statewide status, and their reliance on artificial habitats make them important examples of the fascinating interface between policy discussion and historical-ecological research.

Extirpated/Extinct Species

The loss of wildlife species receives considerable attention from ecologists and conservationists who seek to understand their former roles in the landscape and the effect that their disappearance exerts on current ecosystem processes. In particular, historical perspectives may provide insights into the importance of individual species for the coherence of species assemblages as well as the concept of keystone species. Remarkably, despite the massive impacts of land use, land-cover change, and human exploitation on the northeastern United States, relatively few species of plants or animals were driven to extinction. Arguably the two most important examples of colonial extinctions from the temperate forests of New England are the passenger pigeon, which was remarkably widespread and abundant, and the heath hen, which was of much lower and more localized distribution.

In contrast, a relatively large group of species remains regionally extirpated but represents potential candidates for reintroduction. Although some of these species, such as wolverine and lynx, were uncommon, were near the edge of their ranges, or were eliminated early in colonial history, a few larger predators, notably wolf and cougar, were well established and persisted into the nineteenth century. Large carnivores are underrepresented in the modern landscape but might play a key role in controlling the populations of other species and thereby influencing forest conditions. Consequently, ecological, management, and cultural ramifications of their reintroduction are increasingly discussed.

Wolves have large ranges but are habitat generalists that prefer areas with low human densities. Although their diet primarily consists of large prey such as deer, moose, and beaver they are opportunists that will feed on rabbits, other rodents, a range of small mammals, and carrion. At the time of European settlement, the gray wolf was widespread and relatively abundant across New England (Figure 7.7). However, a concerted effort to eliminate the species began almost immediately after settlement; this, combined with land-cover changes and a decline in prey species, eliminated wolves on a broad scale. The first wolf bounty was established by the Massachusetts colony in 1630, and subsequent local bounties increased this incentive and fueled active extirpation. Between 1650 and 1655, bounties were paid on 147 wolves. This number increased thereafter to a peak in the 1650s when more than 1,000 wolves were killed in four years. Although 3,043 bounties were paid between 1700 and 1737, declining populations were reported across the state, and some local bounty statutes were repealed. Intact wolf populations were apparently restricted to the Berkshires and Cape Cod at this time, prompting retention or increases in bounties in those areas. By the early nineteenth century, wolf killings became uncommon and notable in local lore. During this period, many towns noted their last sightings. A wolf that reputedly killed 3,000 sheep in the area near the base of Cape Cod over a four-year period was killed by a Sandwich man in the 1830s. This may have been the last of the species in Massachusetts.

Natural Range Extension

Throughout the historical period, but increasingly in the twentieth century, several wildlife species have naturally expanded their ranges into Massachusetts, including at least two mammals (the eastern coyote and Virginia opossum) and numerous birds that have become abundant (for example, turkey vulture, northern mockingbird, tufted tit-

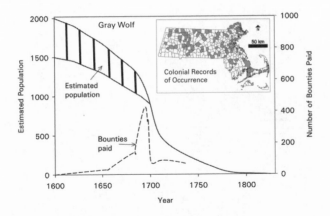

Figure 7.7. Historical distribution and decline of the gray wolf in Massachusetts. The species occurred statewide, with the exception of the coastal islands, before its eventual extirpation. Reprinted from Foster, Motzkin et al. 2002.

mouse, and cardinal). Factors underlying these expansions vary as they occur from all directions and include species with diverse habitat preferences and life histories. However, causes likely include changes in climate, the availability of new food resources (for instance, winter birdseed and roadkill), changes in competition and predation, and changes in habitat and land cover. Among these species, however, the coyote has had the most remarkable, continental-scale expansion and probably exerts the greatest influence on forest ecosystems (Figure 7.8).

Before European settlement, coyotes ranged the western prairies from central Mexico to southern Canada. With changes in land use and other wildlife, especially dramatic declines in wolves, which are predators and competitors of coyotes, coyotes began expanding significantly in the nineteenth century and moved slowly eastward. Coyotes reached Massachusetts by the late 1950s, when the first animals were shot in the western part of the state. The new population rapidly expanded in size and geographical area, reaching 500 by 1979 and an estimated 3,000 to 4,000 in 1996. Although a coyote hunting and trapping season was initiated in 1981, the number of animals killed each year is relatively low, and the species now occupies the entire state except for a few of the coastal islands. Currently, the population is continuing to increase.

Although the coyote represents the first large canine predator to roam the Massachusetts landscape widely in nearly three centuries, it does not replace the wolf ecologically. Eastern coyotes are larger and do have a greater tendency toward packlike behavior than their western counterpart. Nonetheless, they are smaller than wolves; are more adaptable to a range of habitats, including human environments; and typi-

Coyote Distribution

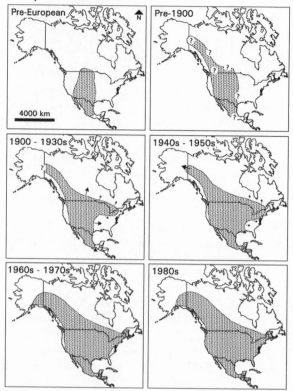

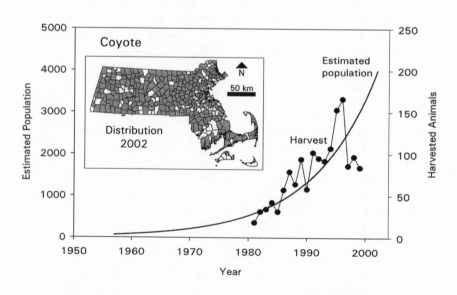

cally forage on smaller prey, such as rodents, birds, amphibians, and small mammals. Eastern coyotes do take deer, though the magnitude of this activity is uncertain; despite this, coyotes appear to lack the capability of controlling the population growth of large animals such as beaver, deer, and moose that were historically preyed on by wolves, cougar, and Native Americans.

Introduced Species

As a consequence of intentional or accidental introductions, a wide range of organisms, from microbes to mammals, have been added to the state's biota. Although most introductions fail or have only local effects, a number have led to widespread populations that have exerted major impacts on natural ecosystems. The earliest introductions were accidental, including the house mouse and black rat, two species that have become common. A second group was game animals, primarily fish and a few birds, introduced to replace disappearing native species in intriguing attempts to fill what were perceived as vacant niches, or in an effort to provide new hunting and fishing opportunities. Rainbow trout, European brown trout, and carp are examples of nonnative species that were stocked from hatcheries and have established naturalized, reproducing populations. Similarly, pheasant, eastern cottontail, black-tailed jackrabbit, and European rabbit have maintained low to widespread populations as a consequence of intentional, and in some cases (for example, pheasant) persistent, human efforts. Ironically, the same public agencies that today try to restrict the introduction of exotic plant and animal pests were responsible for the active introduction and maintenance of nonnative game species. The ecological consequences of these introductions are poorly understood as research effort has been focused on successful establishment rather than attendant consequences to habitat quality or native species.

Finally, driven by a range of motivations, numerous nongame bird species have been introduced and naturalized. One notable example is the European starling, which was released into New York's Central Park in 1890–91 as part of an attempt "to introduce all of the birds mentioned in Shakespeare's plays." Reaching Massachusetts in the early 1900s, the population peaked at more than 500,000 by the 1930s but has subsequently declined somewhat, apparently due to winter mortality. Another common species, the house sparrow, was released in 1858 and

Figure 7.8. Range extension of the coyote and its local increase in population in Massachusetts. The coyote is expanding and is currently the largest top predator in the region. Reprinted from Foster, Motzkin et al. 2002; North American maps modified from Moore and Parker 1992, with permission from G. Moore.

then introduced in 1868–69 in an attempt to reduce gypsy moths and other insect pests. Although the species peaked between 1890 and 1915, it is still common and widespread and is considered disruptive to native species, including the eastern bluebird.

Persistent Species

In contrast to the long-term directional dynamics discussed above, numerous species have remained relatively common over the past 300 years despite periodic fluctuations. For example, raccoons have fluctuated with trapping, severe weather, and disease, such as the rabies epizootic in the early 1990s; porcupines presumably varied with changes in forest cover and predators, including the fisher; and bobcat populations have changed with prey availability, including rabbits. Similarly, crows and gray squirrels have been reported as common from the days of Thomas Morton and William Wood to the present, despite bounties from the mid-1600s to the 1800s. Interestingly, the gray squirrel population dropped sharply between 1910 and 1920, evidently in response to the widespread mortality of chestnut trees, which succumbed to the blight that was spreading across the land. Squirrels disappeared completely in some localities, for example becoming rare in Petersham for more than two decades. However, they recovered strongly in the 1930s, and the species is apparently as abundant and widespread today as before the blight. Overall, these and many other species exhibit no major long-term trends.

Ecological Implications and Social Consequences of Wildlife Dynamics

It is apparent that wildlife populations have been highly dynamic in response to historical changes in landscape conditions and habitat availability, active human persecution, and a range of indirect activities of people. An understanding of these dynamics provides a useful background for policy decisions and affords interesting insights into the functioning of the forest landscape of New England.

A basic question emerging at the community level for both plants and animals concerns the coherence of species assemblages through time. At the most fundamental level, ecologists are interested in whether assemblages of plant and animal species exhibit continuity through time or whether species operate fairly independently of one another. Although natural history texts are replete with examples of tight relationships among specific sets of species (for example, specialized plants and their animal pollinators; predator-prey cycles), the question remains whether such examples are representative of broad interrelationships or whether the majority of species actually form loose and highly malleable associations. Paleoecological and archaeological studies of plants and animals indicate that through the postglacial period, individual plant and animal taxa responded quite independently to the many different climatic and environmental changes and settings that arose. Necessarily, such individualistic behavior resulted in a sequence of quite different assemblages of organisms through time. Thus, the very long-term, though incomplete, postglacial record indicates that the suites of species that we see in the landscape today are novel and have no great historical continuity.

The historical data on changes in wildlife distribution and abundance certainly confirm this pattern. Clearly there are strong interactions among many species, and historical studies yield examples of patterns of change over time that may be generalized for groups of species. However, it is also clear that each species is unique in its dynamics, that animal distributions and assemblages have changed continuously through time, and that modern conditions and modern assemblages of organisms are distinct. Few species exhibit closely linked dynamics, as each responded independently to the unusual combinations of habitat and human activity that the landscape experienced in the past 200 years. Although this individualistic behavior may be accentuated by the selective focus of humans on specific animals—either promoting or persecuting them—it is apparent that the linkages among the organisms that we have examined are relatively loose.

Drawing this observation and the individualistic notion together, we can conclude that the very natural appearance of the modern forest landscape, including its populations of coyotes, fishers, bears, moose, deer, and turkeys, is culturally conditioned and is not analogous to pre-settlement conditions. Thus, the process of forest growth and the return of many forest animal species is very different from a simple restoration of past conditions. Though perhaps obvious on serious reflection, recognition of the strong element of direct and indirect cultural control over our modern landscape is critical for successful conservation and ecological understanding. Conveying the scale of recent dynamics and their linkage to human and landscape history is also a critical element in public education and ongoing policy development.

The changes that have occurred recently in wildlife populations, including the reappearance of moose, fisher, and bear in much of New England, remind us of the sizable and important lags that are inherent in ecological response; they also alert us to anticipate future changes, even if additional human activity were to cease (Figure 7.9). Forests, once established, take decades or centuries to mature; similarly, animals, even when highly mobile, require time to migrate and expand their populations when the landscape, environment, or cultural setting changes. Examples like turkey, beaver, coyote, and white-tailed deer illustrate this process and underscore the potential changes that await the moose, os-

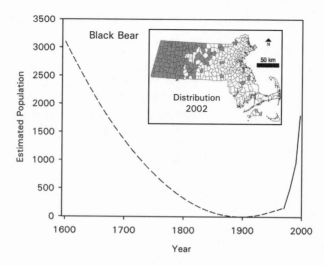

Figure 7.9. Historical dynamics and major recent increases in bear populations in Massachusetts. The species continues to expand its range eastward from local populations that persisted in the nineteenth century in the western part of the state and is increasing at approximately 8 to 10 percent annually. Reprinted from Foster, Motzkin et al. 2002.

prey, bear, eagle, and great blue heron populations. However, population growth is also strongly affected by mortality. Hunting is especially effective at controlling population growth where population levels are low, as was true for white-tailed deer through much of the twentieth century.

The New England landscape, which has already changed quite dramatically in the past century, is undergoing additional alterations as the plants, animals, and ecosystem processes respond slowly to changes in historical disturbance and habitats. Indeed, the historical perspective underscores the fact that wildlife assemblages at any given time are composed of species undergoing strikingly different trajectories. Many animals that can thrive in our newly reforested and maturing landscape are well-established; some are just arriving and becoming established and are poised to flourish; others are yet to arrive but may eventually get here naturally or through human intervention and may yield unforeseen impacts. In contrast, species that were common in our agricultural past are in the process of a long decline that may be inevitable as a consequence of ongoing changes in the condition of the landscape. Therefore, at any given time the assemblage of animals on the landscape includes many species, each of which is on a different ecological trajectory in response to past and ongoing changes: some are increasing, some declining, others perhaps are exhibiting few changes. In order for ecologists to evaluate species' roles or for conservationists to develop effective management strategies, it is critical to be able to identify the specific trajectory associated with each species.

Ecologically, many questions and challenges still remain in our understanding of the consequences of the wildlife dynamics that we have highlighted. Currently, at least twenty large or important forest species that were present at the time of European settlement are absent from New England. It is challenging enough to determine the role and influence of new species that have arrived such as coyote, but how do we evaluate the consequences of the absence of historically important species on the functioning of modern ecosystems? What role did passenger pigeons play in the dispersal of trees and the dynamics of New England forests, and how would our landscape differ in the presence of million-bird flocks and their dense and extensive roosts? What impact would the reintroduction of wolves or cougar have on other animal populations, and, in turn, how would these effects ripple out into the structure, composition, and function of the forests? What effect will an expanding moose population have on forest regeneration, understory composition, and nutrient cycles? As we draw on paleoecological and historical data for our understanding of long-term forest dynamics, how do we incorporate our emerging knowledge of the faunal changes that have occurred? The loss and the addition of new species provide an unusual opportunity, and an important research mandate, to investigate the role that individual species play in the structuring and functioning of ecosystems.

On the cultural and policy side, awareness of the magnitude and rate of wildlife changes over time provides a useful perspective for conservation and management. On the basis of past changes and trajectories, we can anticipate future declines in some species, major increases in others, and some of the consequences of newly arriving species. Clearly, at least two major and interrelated issues face wildlife managers. Foremost is the observation that the trend toward a more forested and wild landscape with large forest animals along with an expanding suburban human population will lead to increasing conflicts between human safety and appreciation for wild nature. At the very least, this means we need to educate people about wildlife, nature, and its history and modify some of our behaviors, such as removing bird feeders in the early spring when the bear population emerges from winter dens. In the case of many of the larger mammals (for example, bear, moose, beaver, and coyote) the social carrying capacity of the landscape (that is, the density and distribution of a species that humans can tolerate or accommodate) is ironically declining as the natural carrying capacity of the land is increasing. Modification of human behavior would enable greater populations to be tolerated more safely. Second, because the modern fauna is dominated by relatively few, large species and lacks major predators, we need to provide more ability to control wildlife populations, either through direct management or through well-conceived introductions of additional species. This is a formidable task for a human population that is generally poorly informed about nature and wildlife dynamics and is largely opposed to the most ready means of wildlife regulation: hunting and trapping.

Wildlife brings immeasurable ecological and social benefits but may also disrupt and damage human property and occasionally even pose direct or indirect threats to the health of humans, as well as domestic animals. Beavers cut trees and flood cellars, roads, and sewer systems; deer and moose can alter forest composition, damage human property, and present a major hazard on highways; coyotes and bears may become too accustomed to people for the welfare of either species; and a range of diseases from giardiasis and Lyme disease to rabies and West Nile virus can be transmitted or promoted by animal vectors. Current U.S. expenditures to deal with Lyme disease alone are estimated to exceed $500 million per year, and southern New England states such as Connecticut experience as many as 75,000 cases of the disease annually.

Evaluating the benefits and costs of wildlife and developing socially acceptable measures of control will be a major challenge for New England's future. The evaluation of historical trends, although it may supply few direct solutions, can assist in defining the issues, anticipating conflicts, and developing strategies for long-term changes. A historic perspective can also provide intriguing insights that aid in informing managers, the public, and scientists of some of the major changes that are occurring around us.

The Modern Forest Landscape
Legacies of Historical Change

CHAPTER 8

Forest Landscape Patterns, Structure, and Composition

G. MOTZKIN, D. FOSTER, A. ALLEN, K. DONOHUE,
and **P. WILSON**

The forests of New England are incredibly varied, from 400-year-old hemlock stands in dark ravines to young hardwood forests that stretch across the rolling uplands, and from spruce swamps overlying deep peats to stunted pitch pines on dry ridge tops. Earlier we related much of this regional variation in forest composition to broad gradients in environment and disturbance history and illustrated that vegetation composition has been highly dynamic over time in response to changing climate and a wide range of natural and anthropogenic disturbances. However, as we shift our attention from this regional and long-term perspective to the variation in vegetation that we observe across the modern landscape, we need to consider the factors that control local differences in forest composition and structure. Why is it that we find mature hemlock stands immediately adjacent to younger hardwoods or gradual transitions between forests in one locality and very sharp breaks separating stands in other areas? Do the factors that control the distribution of tree species in a particular forest or landscape also control understory plants, or are different life-forms of plants sensitive to different environmental and historical factors? How does a region's disturbance history play out locally, and does it exert a persistent influence on vegetation patterns, even long after the disturbances have ended?

To answer these questions, we have investigated modern forest patterns on a range of sites and have studied the mechanisms that control these patterns. A major goal has been to test the commonly held notion that variation in vegetation primarily responds to differences in current site conditions as opposed to historical factors. In particular, we wished to determine the extent to which modern vegetation patterns reflect wind, fire, and especially land-use disturbances over past centuries, rather than current environmental conditions. Given the tendency for scientific research to become increasingly reductionist, understanding the relative importance of historical legacies versus current conditions provides important insights to our science.

Background

Studies in North America and Europe have found that modern vegetation composition and species diversity are strongly influenced by historical factors as well as modern site conditions. In particular, several investigations, including many originating from the Harvard Forest, have suggested that land-use activities may alter species composition or richness for many centuries, long after the disturbances have ended. Disturbance history may influence subsequent species composition in several ways as disturbance may

- alter physical site conditions directly and thereby influence subsequent plant performance and competitive interactions, or change the suite of species that can potentially occupy a site;
- allow for the establishment of species that are then able to persist for long periods of time, even on sites where they would, in the absence of disturbance, be unable to establish or compete effectively; or
- remove species that are extremely slow to recolonize, even in the absence of altered resource conditions.

For instance, plants with slow rates of dispersal or establishment may be absent from a site for decades or centuries, not because the site is unsuitable, but simply because the species was removed by a prior disturbance and has not had sufficient time or opportunity to recolonize. Because species' distribution patterns may be controlled by current environmental conditions and/or historical factors, a major challenge for ecological study is to evaluate the relative influence of these factors on modern vegetation, and to determine how the relative importance of contemporary conditions versus history changes with time since disturbance.

The possibility that legacies of past disturbance may endure to control forest structure and composition long after the direct evidence of the disturbance has disappeared has important ramifications for our understanding of forest ecosystems and for our attempts to manage them for resources and conservation values. However, evaluating the influence of historical disturbances is difficult because specific land-use activities and other disturbances are typically restricted to particular types of sites. For instance, even if we are able to determine that certain species occur primarily in areas that were used historically for crop cultivation, it is often difficult to know whether these species occur on these sites because they were formerly cultivated or because the species prefer the types of soils and locations that farmers selected historically for intensive agriculture.

We adopted a number of approaches to address this problem of confounding factors. On sites where soils and other physical conditions are relatively homogeneous but where land-use history varies, we were able to test land-use effects that are independent of variation in environment.

Figure 8.1. The Montague Plain is a flat and sandy outwash plain dominated by forests of pitch pine, oak, and scrub oak. The homogeneous soil conditions enable us to examine the effect of land use and other disturbances on vegetation patterns. Photograph by D. R. Foster.

Basically, this approach is akin to an experiment in which all factors but the one of interest (in this case, history) are controlled and held constant. On more complex sites, which is the typical situation in New England, we developed very detailed information about history and site conditions in order to help us interpret the relative contribution of each.

Our intensive study of the Montague Plain (Figure 8.1), a level outwash plain in the Connecticut River Valley that has limited environmental variability, allowed us to evaluate the effects of historical disturbances without the confounding effects of differing initial site conditions. Results from that investigation document the overwhelming influence of historical land-use activities on modern vegetation patterns. They also identify several species that have not recolonized former agricultural lands in 50 to more than 100 years since these farmlands were abandoned. Using these results, we initiated detailed studies of nutrient cycling and population demography at Montague to determine the mecha-

Figure 8.2. The locally varying environment on the rolling uplands including the Prospect Hill tract makes it more difficult to separate the influences of environmental factors and disturbance history on vegetation patterns. This northward view from the canopy tower near Hemlock Hollow toward the fire tower on Prospect Hill captures some of the topographic variation and the subtle changes from hardwood (lighter tones) to white pine and hemlock (darker). Photograph by D. R. Foster.

nisms by which historical factors may exert persistent influences on vegetation patterns and ecosystem properties.

We conducted similar investigations on the Prospect Hill tract of the Harvard Forest (Figure 8.2), a more complex upland site with highly variable physiography and a complicated history of both natural and human disturbances. This site is particularly appropriate for such research. It is representative of a much broader region in terms of vegetation, site conditions, and disturbance history, and ninety years of research enabled us to compile unusually detailed historical records. Thus, on Prospect Hill we can evaluate the influence of history and environment on modern vegetation in a way that is not possible for most sites.

Here we present some of the major findings of these community and population studies. We begin with an overview of disturbance history on Prospect Hill and present our understanding of the factors that control modern vegetation patterns on this site. We then present a similar analysis for the Montague Plain and compare disturbance histories and vegetation responses among these two sites. The results of population studies that seek to identify the mechanisms by which disturbance may exert persistent influence on vegetation is presented through a case study of wintergreen *(Gaultheria procumbens)*, a species whose modern distribution is, on sites such as Montague Plain, strikingly controlled by historical land-use practices. Finally, we conclude with some thoughts on the implications of our findings for understanding modern community patterns in regions that have long and varied histories of natural and human disturbances.

Prospect Hill
Disturbance History

Before European settlement, the Prospect Hill tract was largely forested, and there is no evidence to suggest land clearing by native people (see Chapters 4 and 6). The wealth of historical records, including ownership and land-use maps, census data, and extensive stand records and forest inventories, enabled us to develop a detailed picture of the changing landscape (Figure 8.3). We sought to corroborate these historical sources in our extensive fieldwork by recording observations on land-use artifacts such as plow mounds, stone piles, dams, and cellar holes. In addition, we examined soils to determine the degree to which they had previously been disturbed by agricultural practices. On sites that were formerly cultivated, a distinct plowed surface soil horizon (Ap horizon) remains visible for many decades, enabling us to identify and map former cultivated fields. By combining field observations and historical sources, we determined that approximately 16 percent of the area remained wooded in the mid-nineteenth century. Approximately 7 percent has deep soil disturbance, indicating crop cultivation, and an additional approximately 15 percent has shallow soil disturbance, suggesting that it was lightly plowed or harrowed historically and most likely used as improved pasture. Fields (plowed areas and improved pastures) were established primarily on well-drained soils, whereas rough pastures, which lack any visual evidence of soil disturbance but were known from historical records to have been cleared, occurred on a wider range of soils, both drier and wetter. The permanent woodlands were largely restricted to swamps, poorly drained soils, or rocky sites. Beginning in the second half of the nineteenth century, widespread reforestation occurred, and by the early 1900s, most of the fields had been abandoned and allowed to revegetate naturally or had been planted to conifer plan-

ratio of carbon to nitrogen (C:N ratio) in the soil. The current vegetation is strongly related to land-use history, with, for example, hemlock stands largely restricted to primary (never-cleared) sites, red maple stands mainly on old unimproved pastures or woodlots, oak-maple stands on former pastures and some tilled fields, and pine-oak stands and plantations on old pastures and plowed fields but generally not on continuously wooded sites. For some species, both spatial distribution and abundance are strongly influenced by nineteenth-century land use (Figure 8.5). For instance, white pine, staghorn clubmoss *(Lycopodium clavatum),* and Canada mayflower *(Maianthemum canadense)* occur more frequently or abundantly on former agricultural fields than in continuously forested areas. In contrast, several species are more frequent or abundant on continuously forested sites, including hemlock *(Tsuga canadensis),* witch hazel *(Hamamelis virginiana),* and several bryophytes (for instance, *Calypogeia fissa* and *Brotherella recurvans).*

These findings support studies from eastern North America and Europe that have identified species that are largely restricted to continuously forested sites and that are slow to recolonize areas once they have been removed. Several mechanisms may contribute to this pattern. For some species, dispersal, establishment, or other biological limitations may restrict colonization (see the case study of wintergreen below). In addition, the abundance of hemlock, which casts deep shade and forms deep, acidic litter, in continuously forested sites significantly influences the understory environment and composition. In some instances this relationship is quite strong, as in the case of the liverwort *Lepidozia reptans,* which is found on Prospect Hill almost exclusively in association with a hemlock overstory.

SPECIES RICHNESS

Studies that have evaluated the influence of historical land use on species richness (the number of species per unit area) have differed in their results. Whereas several studies, including those of George Peterken in Britain, have found that primary woodlands or old secondary woodlands adjacent to primary woodlands are species-rich relative to more recent and isolated secondary stands, others have found no difference across differing land-use histories, or that more recently or intensely disturbed sites may actually support a greater number of species. However, it should be noted that in some instances where total species richness does not vary by land use, the number of "true woodland species" (that is, species that are restricted to woodland habitats) may still differ. In our study, the number of bryophyte (moss and liverwort) species does not differ according to historical land use, whereas the number of trees, shrubs, and herbaceous species does differ, with the fewest species in continuously forested stands. Although some of this effect may result from the persistence of a few weedy species on former

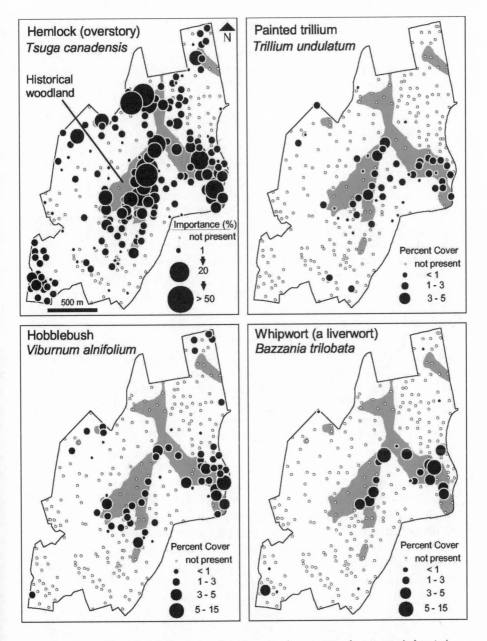

Figure 8.5. Distribution and abundance of plant species characteristic of continuously forested sites on the Prospect Hill tract of the Harvard Forest. These species are less abundant and less common today on sites that were cleared for eighteenth- and nineteenth-century agriculture. Based on Motzkin, Wilson et al. 1999.

agricultural sites, it is likely that the inability of many species to tolerate low light in dense hemlock stands is a major factor limiting richness in primary woodlands. Thus, historical land use can affect species richness and composition both directly and indirectly. Interestingly, continuously wooded sites with a history of least intensive land use have the smallest number of species.

Many species also vary in distribution according to differences in soil drainage and C:N ratios. Soil drainage strongly influences the mosses and liverworts, whereas many tree species tolerate a wide range of moisture conditions, from relatively dry to fairly wet. Some species vary according to C:N ratios, which are highest in hemlock stands on permanently wooded sites and lower in vegetation on former agricultural fields. In general, gradients in calcium and magnesium concentrations and pH do not strongly influence species distributions in this landscape, perhaps because soils are fairly uniform. These results differ somewhat from those of other studies (for example, the work by Peet and Christensen 1980, Balter and Loeb 1983, and Cowell 1993) that examined a broader range of site conditions than those that occur on Prospect Hill.

In addition to those factors that strongly influence modern vegetation, it is instructive to consider variables that, rather surprisingly, are not closely associated with forest patterns. For instance, the 1938 hurricane clearly had a major effect on forest structure and overstory composition, as it damaged most stands and generated dramatic structural effects, including downed trees, windthrow mounds, and damaged crowns that are apparent sixty years later. However, despite dramatic wind impacts on both overstory composition and structure, our study indicates that the degree of 1938 hurricane damage helps to predict the modern distributions of only a few species such as thread moss *(Atrichum angustatum)*, which is frequently found on old tip-up mounds. What explains this seemingly contradictory result? Several factors offer partial explanations: (1) understory species may be relatively independent of overstory structure and composition, particularly as herbs, shrubs, and small trees may be undamaged by wind or capable of rapid resprouting; (2) sufficient time may have passed since the hurricane to obscure many of the initial and successional effects; (3) hurricane damage may have been sufficiently widespread and varied that our stand-level data are inadequate to detect effects on the understory flora; and (4) logging and related salvage activity, which was widespread after the hurricane, may have obscured wind effects.

In addition, we suspect that part of our inability to detect significant hurricane effects may be a common problem in ecological sampling: we are able to measure current environmental conditions and vegetation patterns, but we are unable to evaluate directly the difference between what the landscape is and what it would have been if the disturbance had not occurred. Thus, effects of widespread disturbances such as wind, logging, or pathogens, though important, may go undetected.

Land Use, Soils, and Vegetation on Montague Plain

The difficulty of separating historical and environmental effects on vegetation in the complex landscape of Prospect Hill prompted us to investigate more homogeneous sites. We hoped that by limiting variation in topography, soils, and drainage, we could study the mechanisms by which historical disturbances influence vegetation in the long term. Several of these studies were conducted on the Montague Plain, a large sandy plain in the Connecticut River Valley (Figure 8.1). This site is ideal for such studies because it is homogeneous with respect to soils and yet supports a range of vegetation and prior land use. Using a combination of historical sources and field studies similar to those on Prospect Hill, we pieced together the site history, which is somewhat different from that of most upland areas. Although the land on Montague Plain was first divided and assigned to owners in the mid-eighteenth century, historical records indicate that widespread forest clearing did not occur until after 1830. During the mid-nineteenth and early twentieth centuries, much of the plain was used for agriculture. In fact, because plow horizons are distinct in these soils, we were able to determine that approximately 80 percent of the site was plowed historically (Figure 8.6). By the late 1930s, most of the site had been agriculturally abandoned and was reforesting naturally. Because this is an extremely sandy site, pitch pine, which is tolerant of dry conditions, was the dominant overstory species to become established on these old fields.

Interestingly, even though the fields were generally abandoned 75 to more than 100 years ago, the modern forests on these sites still differ substantially from adjacent areas that were never plowed (Figure 8.7). Some weedy species are more common today in the forests that established on fields than in nearby unplowed sites, whereas some characteristic "pine barrens" species are abundant on historically unplowed sites but have not successfully recolonized areas that were cleared for agriculture (Figure 8.8). We suspect that pine barren species such as wintergreen and huckleberry were formerly widespread across the entire plain, that they were eradicated from most of the area by plowing, and that they are currently restricted because of severe limitations in their ability to recolonize. In fact, on this and similar sites in the Connecticut River Valley, the presence of wintergreen and huckleberry is a strong indication that the site was never cleared and plowed. Consequently, boundaries between unplowed and formerly plowed sites are often visible as distinct breaks in understory vegetation, with the restricted species having spread only a few meters onto former fields.

Among the species whose modern distributions are strongly deter-

Figure 8.6. Soil profiles from two sites on the Montague Plain with contrasting land-use histories. Top: Pitch pine forest on a site that was cleared and plowed for agriculture during the nineteenth century. A deep, homogeneous Ap "plow horizon" extends to 20 centimeters and persists despite 100 years of forest growth. Bottom: Scrub oak stand on a continuously forested site. The undisturbed and shallow upper (A) soil horizon grades gradually downward in appearance and color. Photographs by A. Allen, from Motzkin et al. 1996, reprinted with permission from the Ecological Society of America.

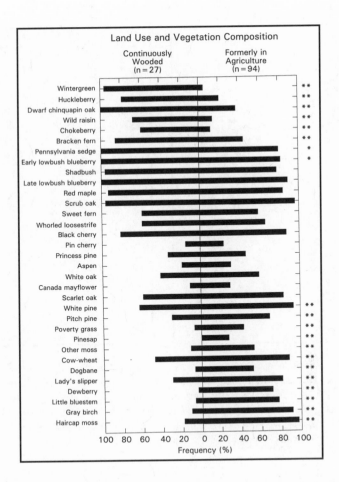

Figure 8.7. The influence of land-use history on current vegetation composition on the Montague Plain. The frequency of occurrence of common species (occurring in more than 20 percent of plots) illustrates that some species are highly restricted to areas continuously in woodland, others are preferentially found on sites previously in agriculture, and others are widespread and relatively indifferent to prior land use. Asterisks indicate significant differences; *$p < .05$, **$p < .01$. Modified from Motzkin et al. 1996, with permission from the Ecological Society of America.

mined by past land use, wintergreen *(Gaultheria procumbens)* is the most striking (Figure 8.9). We found this creeping species in 96 percent of plots that had never been plowed but in fewer than 5 percent of plots on former fields. Because our soil and environmental analyses indicated that most conditions were identical across these land-use boundaries, we looked in detail at the population demography of wintergreen to determine why it recolonizes these sandy old fields so slowly.

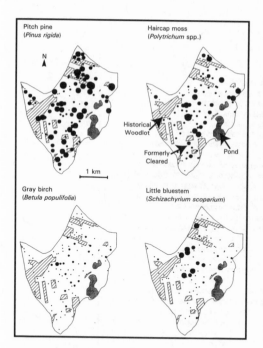

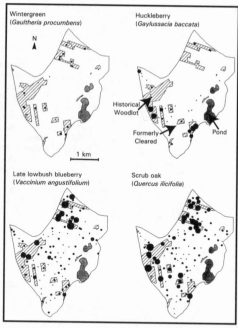

Figure 8.8. The influence of land-use history on patterns of species distributions on the Montague Plain. Distribution and abundance of species characteristic of formerly plowed areas (left) and sites that were never cleared for agriculture (right). White areas were cleared and plowed for agriculture but were abandoned and currently support forest that is 50 to more than 100 years old. Diagonal shading indicates continuously wooded areas. Modified from Motzkin et al. 1996, reprinted with permission from the Ecological Society of America.

We hypothesized that if plowing and agriculture had altered resources such that formerly plowed sites are now less suitable for wintergreen, then it should exhibit decreased size, vigor, or growth rates on these sites. In order to test this hypothesis, we compared plant characteristics that are associated with population establishment and growth (such as flower and fruit production, seed germination, growth, and survivorship) on wintergreen growing on plowed versus unplowed sites. We found no evidence of successful wintergreen colonization of former fields through sexual reproduction. In fact, we determined that on formerly plowed sites, wintergreen is restricted to areas adjacent to unplowed sites, from which it has spread vegetatively an average maximum distance of only 10 to 14 meters in the 50 to more than 100 years since these fields were abandoned. However, our findings do not support the notion that former plowed fields are unsuitable for wintergreen. In fact, contrary to this prediction, we found that for many demographic

Figure 8.9. Wintergreen *(Gaultheria procumbens)*, an ericaceous herb that is slow to expand from primary forests into previously plowed areas in the sand plain environment (see Figures 8.7 and 8.8) but is quite widespread across forests of different land-use history on the mesic uplands like the Prospect Hill tract. Photograph by J. Gipe.

parameters, wintergreen performed as well on formerly plowed sites as on unplowed sites, or in some instances actually performed better on the old fields. For instance, germination, seedling longevity, and rhizome growth rate were greater on plowed versus unplowed sites.

Thus, it seems that on the sand plain, wintergreen's absence from many former fields is not a result of agricultural modification of environmental conditions, but rather comes from inherent biological limitations on the species' ability to colonize new sites. The current distribution of wintergreen on this and similar sites thus apparently results from historical factors rather than from current site conditions. Interestingly, wintergreen does not display the same degree of restriction to continuously wooded sites on the Prospect Hill tract or other moist upland areas. On these sites it more readily spreads from old woodlands onto recently abandoned land and is now widespread in forests across many former land uses. This indicates that even for individual species, responses to historical factors are complex and may vary according to local site conditions.

Comparing Prospect Hill with Montague Plain

Prospect Hill and Montague Plain may be thought of as representing points on the continuum of land-use impacts. Although approximately 80 percent of Montague Plain was plowed historically, our analyses suggest that this disturbance resulted in relatively minor long-term alteration of physical and chemical soil properties, presumably because of the short duration and low intensity of historical agricultural use of this marginal site. However, even without prolonged, intense disturbance, the long-term effects of land use on modern vegetation are striking as a result of species differences in colonization ability. It is reasonable to conclude that on sites where the duration or intensity of agricultural use was greater, resulting in altered site conditions, subsequent vegetation patterns would reflect the confounding effects of recolonization limitations and altered site conditions. Such an interaction is thought to be significant in many European ecosystems and undoubtedly applies to broad areas of North America that have a long history of intensive agriculture. Our results suggest that this may also be the case for much of Prospect Hill and the broad uplands of New England. Although the total percentage of land cleared is comparable on Prospect Hill and Montague Plain, crop cultivation on the former was less widespread but more intensive and of longer duration, resulting in more pronounced, long-term soil modification. This contributes to the persistent influence of historical land use on modern species distributions, despite the considerable length of time (75 to 160 years) that has passed since agricultural abandonment, a time during which species could have shifted their distributions in response to more recent disturbances or to more closely reflect variation in contemporary site conditions.

Conclusions

Why is it that on some sites the influence of history on modern vegetation is so striking, whereas on others it is more difficult to detect? Several factors are apparently important. The history of land use across the northeastern United States is varied and complex. Whereas some areas were cultivated intensively for many years, others were never entirely cleared for agriculture but were managed as woodlots and were cut repeatedly. Even on individual sites, the sequence of land uses is complex and is generally impossible to reconstruct fully. For instance, many sites that were cultivated for crop production were subsequently used as pastures and, after farm abandonment and succession to forest, were cut repeatedly or burned. Thus, despite our most rigorous efforts to determine historical land-use practices, our knowledge of the history of any particular site is always incomplete and, because of its complexity, is not easily quantified. In addition, environmental variability differs considerably among sites. Although sites like Montague Plain are extremely homogeneous, most of the northeastern landscape is much more like Prospect Hill, with its varied terrain, drainage patterns, and environmental conditions. The fact that land-use history is frequently correlated with environmental variability tends to restrict our ability to identify clear land-use impacts, even on sites where they occur. Finally, our flora are quite varied and each species responds to this complex history as well as to environmental gradients in different ways and at different rates, making it difficult to generalize to any great extent. Similar to wildlife assemblages, when we sample New England's vegetation at any point in time, we are observing species at different positions in their response to and recovery from historical disturbances. For instance, although some species may have largely recolonized sites that were occupied before the agricultural period, others have moved only a few meters onto old farmland. For species with severe limitations on dispersal or establishment, decades or centuries may be needed for them to spread onto old fields.

Thus, while there is no simple answer to the question of the relative importance of environment versus history in controlling modern vegetation patterns, we can clearly find evidence for the strong influence of both factors. When we look closely with a view that is strongly informed by historical information, we see that landscape patterns of species assemblages and individual species distributions in New England continue to be influenced by historical land-use activity. This is true despite the passing of a century or more since the abandonment of most former agricultural sites, and the occurrence of many more recent disturbances, such as the 1938 hurricane, fire, numerous pathogens, and twentieth-century logging, that could have obscured the effects of earlier land-use practices. We conclude that modern assemblages and species distribution patterns are only partly the result of contemporary environmental conditions and that for some species, modern distributions may be controlled more by historical factors than by variation in current resource conditions.

These results have important implications for the use of standard community ecological data and analyses to interpret community composition and dynamics. Studies at the Harvard Forest and throughout the northeastern United States have documented the dramatic landscape transformations that occurred during the historical period as a result of changing land-use practices and natural disturbances. In fact, it is likely that the distribution and abundance of every species on the modern landscape have been significantly altered by historical disturbances. Our limited ability in complex landscapes to identify such influences results in part from the factors described above, and also from the limitations of our approaches in quantitative community ecology that typically enable us to measure current environmental conditions and vegetation patterns well but do not allow us to adequately evaluate what the landscape would have been if historical disturbances had not occurred. Thus, the effects of widespread and complex disturbances such as historical land use, wind, or pathogens, though important, may go undetected. We do not suggest that these factors be ignored as untractable; on the contrary, they must be rigorously analyzed using as complete historical and environmental investigations as possible. However, we must also recognize the potential importance of factors that, though not easily quantified, may exert a strong and persistent influence on the landscape.

CHAPTER 9

Land-Use Legacies in Soil Properties and Nutrients

J. COMPTON and R. BOONE

The preceding chapters have shown that common historical land-use activities in New England such as plowing, pasturing, and logging have persistent impacts on the composition and structure of our forests. A basic question naturally follows from these results: To what extent and for how long does the function of these forest ecosystems depend on past land use? Because the nature and intensity of land management affects vegetation trajectories and the amounts and availability of soil nutrients, historical land use may have persistent effects on such fundamental ecosystem characteristics as the storage of soil carbon and nitrogen, quality of organic matter, activity of microbes, and mineralization of nutrients from organic matter. Here we describe our studies that examine the influence of land-use history on soil processes across the Prospect Hill tract and Montague Plain. Our research has shown that the type and duration of land use, as well as inherent site characteristics, can affect the persistence and nature of these impacts and that nineteenth-century history has an enduring influence on modern ecosystem processes.

Influence of Nineteenth-Century Agriculture on Soil Processes

While initial clearing of forests by logging and burning can cause important changes in soil organic matter and nutrient availability, long-term cultivation of soils on these cleared sites dramatically alters the carbon content, microbial populations, and nutrient cycling. For example, modern studies have shown that after several decades of plowing, soil carbon on formerly forested sites generally decreases by an average of 30 percent, and nitrogen and phosphorus contents may be reduced by more than 20 percent. Organic matter losses result from numerous processes: accelerated decomposition in plowed soils, removal of plant biomass through harvest or mowing, and erosion of the organic-rich surface horizons. With plowing, relative losses of 50 to 75 percent have been commonly reported for microbial biomass and the light (or easily decomposed) carbon fraction of soils. Such changes in organic matter and microbes are generally also reflected in the nitrogen cycle. Mineralizable nitrogen is often reduced by plowing, but the net production of nitrate is typically high on most tilled sites probably because of increased soil pH from repeated liming, higher ammonium availability from fertilization, and/or limitations on microbial nitrogen immobilization imposed by the relatively low availability of carbon.

While the alteration of soils following the conversion of natural vegetation to various types of agriculture has been documented extensively, soil changes during reforestation of agricultural land have received much less attention. This is surprising given that historical reforestation has been a common process in the temperate and tropical zones and is continuing in many parts of the globe. For example, because of newly enacted European Union policies in response to surplus agricultural production, extensive reforestation of arable land and pastureland will occur in the near future in many regions of Europe. The conservation and forestry opportunities for countries such as Denmark, where forest area is projected to double in the next two decades, are great, as is the need to anticipate how the resulting forest ecosystems will reflect their land-use history. Depletion and recovery of carbon, nitrogen, and phosphorus levels may have long-term implications for the productivity of forests, composition of plant species, sequestration of carbon, and capacity of these ecosystems to retain and respond to atmospheric pollutants. For example, forests on previously plowed soils, perhaps because of reduced amounts of organic matter and lower soil C:N ratios, may have a lower capacity to retain nitrogen than those never tilled. The factors that determine the capacity of forests to retain nitrogen are still incompletely understood, but assessing the role of site history may be an important step toward improving this understanding.

The broadscale changes in the New England landscape over the past 300 years clearly influenced soil properties and microbial processes. Widespread agricultural practices, including deforestation, pasturing, and plowing of hillsides, led to local erosion and declines in soil fertility. Evidence for soil movement across even gently rolling topography is observed throughout central Massachusetts, including the Harvard Forest. This is particularly noticeable where stone walls run across slopes through the forest. Upslope of the stone walls, the upper "A" soil horizons are deep, homogeneous, and stone-free as they are formed from fine material that was washed downslope and accumulated against the wall over the decades when the site was in open agriculture. In contrast, on the downslope side of the stone walls, the soils are rocky and heterogeneous. By the mid-1800s, various soil amendments including potash, manure, and gypsum were widely used to fertilize fields, thereby maintaining productivity. Although soil organic matter may have been depleted through early land clearance and crop removal, manure applications presumably helped to reverse this trend somewhat and to restore soil nitrogen and phosphorus levels, resulting in a relative increase in soil fertility.

Clearly, the range of potential impacts of land-use history on modern forest ecosystems is wide but will depend on many interacting factors. In order to examine these land-use legacies on ecosystem processes, we examined forests on agricultural lands abandoned 50 to 120 years ago in central Massachusetts and focused on several major questions:

- Does land-use history have a long-term effect on the amount of carbon, nitrogen, and phosphorus in the soil?
- How does land-use history influence the cycling and storage of nitrogen after reforestation has proceeded for nearly a century? Is this influence important for our understanding of current forest function and response to disturbance?
- Are there long-term effects of land-use history on the quality of organic matter that are driven by the composition of the plant community or availability of nutrients?
- How long do these effects persist? By what pathway does recovery proceed?

We conducted studies on Montague Plain and Prospect Hill, in conjunction with the vegetation studies described previously. At Montague we examined nitrogen cycling in the sandy outwash soils where the duration of agricultural use was short, only several decades, and the homogeneous soils allowed us to ascribe differences in nutrient cycling to land-use history or vegetation rather than site factors. We also examined carbon, nitrogen, and phosphorus dynamics on Prospect Hill, where agriculture persisted for at least 150 years on relatively hilly, moist, and more variable soils.

The Contrasting Study Sites

As detailed information on the Montague Plain and Prospect Hill tracts is provided in previous chapters, we concentrate here on characteristics important to soil and ecosystem processes (Table 9.1).

Montague Plain

On the outwash delta of sands and gravel that forms the 2,000-acre Montague Plain, the soil is highly permeable and prone to drought, and the water table is approximately 20 meters below the surface. The soils developed in siliceous sand and gravel; the texture of the upper 15 centimeters of soil is loamy sand to loamy fine sand and is fairly homogeneous.

Logging was the primary land use until the mid-1800s, when large

Table 9.1. Descriptions of the Two Study Areas

	Montague Plain	Prospect Hill
Topography	Flat, homogeneous sand plain	Rolling, varied upland
Soil series	Hinckley, Windsor	Canton, Scituate
	Typic Udorthents	Typic Dystrochrepts
Soil drainage	Excessively drained	Well drained
Historical land use	82 percent cultivated; no pasturing; 18 percent primary woodland	15 percent cultivated; 65 percent pastureland; 20% primary woodland or swamp
Woodlot vegetation	Pitch pine, scrub oak	Mixed hardwoods

sections of the plain were plowed for corn and hay until the early 1900s. There is no evidence of permanent residences and little indication of intense pasturing on the plain itself, presumably because of the dearth of surface water. Consequently, we suspect that most plowed lands were not heavily fertilized with manure or other fertilizers.

The history of use, as either cultivated field or woodlot, strongly controls the structure and composition of the modern vegetation. Forests range up to 120 years old on primarily woodland sites that were last cut or burned in the late nineteenth century and on the oldest secondary forests. More than 97 percent of pitch pine stands occur on formerly plowed sites, and 89 percent of scrub oak stands are located on primary areas that were never cleared. Other vegetation types and associated prior land use include grasses and aspen (*Populus* spp.) on plowed sites that were abandoned as recently as 40 to 55 years ago, pitch pine *(Pinus rigida)* and white pine *(Pinus strobus)*–scarlet oak *(Quercus coccinea)* stands on plowed fields abandoned 55 to more than 100 years ago, and scarlet oak and scrub oak *(Quercus ilicifolia, Quercus prinoides)* stands on sites never plowed. Many aspen stands contain pitch pine, white pine, and oak regeneration and will probably develop into pine-oak communities in the future.

Prospect Hill

The rolling terrain of the Prospect Hill tract supports more loamy and mesic soils and experienced the full range of land-use histories typical for upland areas. We selected twelve plots for study on the Prospect Hill tract from a subset of those used for the vegetation study described in the previous chapter. We examined three major land uses in currently forested areas: woodlot (primary forest), formerly pastured, and formerly plowed. Land-use history was determined using the field and historical evidence described in the vegetation study and by incorporating the extensive historical research of R. Fisher, W. Lyford, H. Raup, S. Spurr, and others. Field indicators of agricultural use other than grazing include the presence of an Ap (plow) horizon, absence of large surface stones, smooth microtopography, absence of ancient mounds and pits, and nearby stone walls composed of small rocks indicative of field improvement. Sites that had been plowed have 16- to 20-centimeter-thick homogeneous surface soil horizons (Ap horizons) with abrupt boundaries to the deeper soil below. We used historical records to differentiate rough pastures from woodlots as there is often little apparent difference in soils.

Caveat

Reconstructing the detailed history of a 20-by-20-meter plot subject to multiple human and natural disturbances over 200 years is nearly impossible, even through intensive site analysis backed by the excellent historical records in the Harvard Forest Archives. Our land-use categories consequently represent general types and broad differences in intensities of prior use, with an emphasis on the most intensive use of a site over time. We use this simple framework of woodlot, pastured, and cultivated land use to determine whether there are persistent effects from these management practices.

It is important to recognize that inherent soil differences among sites with different histories can confound the interpretation of site factors versus history in controlling modern soil differences. For example, soil properties and prior land use are somewhat related at Prospect Hill, in that farmers tended to avoid the poorly drained areas and converted the level, better drained sites to cultivation. In addition, prior land use can alter vegetation trajectories by affecting dispersal, establishment, and growth rates. We tried to minimize such inherent site differences by keeping slope, soil series, and soil drainage constant across all sites and by stratifying by vegetation type at Prospect Hill. These interactions were much less of a concern for Montague Plain because there is so little variation in soil properties that land-use boundaries appear to be a function of historical ownership rather than of inherent soil properties.

Soil Legacies on Montague Plain

On Montague Plain, more than 80 percent of the area had been cultivated and subsequently abandoned 40 to more than 100 years ago. Of the plots studied in detail, the two youngest secondary forests were on fields that had been abandoned 40 to 60 years ago. These had the lowest carbon content in the upper mineral soils (0 to 15 centimeters), approximately 30 percent less than the average for the unplowed soils on primary forest sites (Figure 9.1). Soil carbon concentration was also significantly higher in unplowed soils than in all formerly plowed soils.

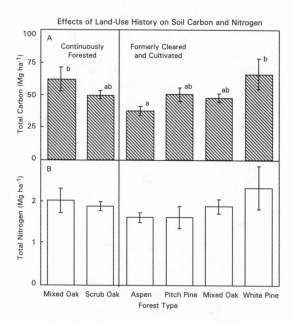

Figure 9.1. The influence of land-use history and present vegetation on the carbon (A) and nitrogen (B) contents of forest soils at Montague Plain. All sites currently support forest that is older than seventy-five years; however, some sites were continuously forested and others were previously plowed for agriculture. Within each land-use category, columns that share a letter are not significantly different ($p > .05$) from one another. There are no significant differences in nitrogen across land-use and vegetation types. Bars represent standard errors between plots within each vegetation type ($n = 2$ or 4). Modified from Compton et al. 1998, 539: fig. 1, with permission from Springer-Verlag (copyright 1998).

However, the higher bulk density in formerly plowed soils offset these differences, and there were no significant differences in total carbon content between plowed and unplowed soils. While the actual amount of soil nitrogen was not affected by land-use history or vegetation type, net nitrogen mineralization showed much greater variation. Net nitrogen mineralization measured over the month of August was more strongly related to present vegetation than to land-use history or soil nitrogen content (Figure 9.2) and varied nearly fortyfold among stand types; it was lowest in pitch pine and white pine stands (−0.13 and 0.10 kilograms of nitrogen per hectare, respectively), intermediate in scrub oak stands (0.48), and highest in aspen and mixed oak stands (1.34 to 3.11, respectively).

Appreciable net nitrification was observed only in the most recently abandoned aspen plot (0.82 kilograms of nitrogen per hectare), which was the youngest stand and the only area where we believe that agricultural lime added to the soils by farmers in the twentieth century in order to enhance productivity may continue to influence nutrient dynamics. The C:N ratios increased and pH declined with stand age or time since abandonment. Higher bulk density and C:N ratios and slightly lower carbon concentrations in the surface mineral soils persist today as long-term legacies of historical agriculture on Montague Plain. However, the relatively brief duration of agriculture and the low initial carbon and nitrogen concentrations in these nutrient-poor, sandy soils appear to have limited the long-term effects of agriculture on soil carbon and nitrogen content and nitrogen cycling.

Although long-term effects of land-use history on vegetation patterns at Montague Plain were striking (see the previous chapter), agricultural effects persisted in only a subset of the soil properties examined. The increase in soil bulk density associated with plowing is certainly a dramatic long-term change in these soils that could influence soil water availability, root distribution, and biological activity well into the future. The unplowed soils had higher C:N ratios and slightly more total soil carbon per gram of soil. Only the most recently abandoned sites, which had received modern liming and fertilizing treatment, show residual impacts of cultivation on soil carbon content and nitrification. After reforestation proceeds for more than fifty years, carbon and nitrogen levels are quite similar to those of unplowed areas, and net nitrogen mineralization is controlled largely by tree species composition, presumably through differences in litter quality, instead of direct land-use effects. Although plowing undoubtedly had important short-term ef-

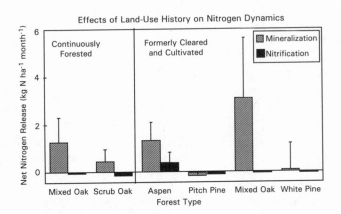

Figure 9.2. The effects of forest type and land-use history on nitrogen cycling (net nitrogen mineralization and nitrification) on Montague Plain. All sites currently support mature forest or scrub oak vegetation. Bars are standard errors between plots within each vegetation type ($n = 2$ or 4). Modified from Compton et al. 1998, 540: fig. 3b, with permission from Springer-Verlag (copyright 1998).

fects on soil carbon storage, long-term effects were not observed, indicating that the marginal agriculture at Montague Plain did not dramatically influence soil properties other than bulk density. The impact of soil disturbance on the vegetation, notably the establishment of white pine and pitch pine stands in old fields, indirectly resulted in important long-term differences in net nitrogen mineralization.

Soil Legacies on Prospect Hill

Even 70 to 140 years after agricultural abandonment, the impacts of different land-use histories on the forests on Prospect Hill are apparent in soil carbon, nitrogen, and phosphorus contents. In forests on old plowed fields, the amount of soil carbon was approximately 15 percent lower than in permanent woodlots, while nitrogen and phosphorus contents were slightly higher (Figure 9.3), especially in hardwood forests. The C:N ratios were lower on the plowed soils, as was the amount of what we call light fraction or easily decomposed organic matter. These results confirm that the levels and quality of organic matter are altered for a very long time after cultivation and reforestation on these loamy soils.

On this rolling upland site both vegetation composition and land-use history influence the amount and ratios of important soil elements. The C:N ratios in mineral soils were lower in formerly cultivated and hardwood soils than in permanent woodland or conifer forests (Figure 9.4). Such long-term alteration of soil nutrient ratios can have profound consequences on a range of forest ecosystem processes. In particular, they could increase the rate of nitrogen cycling through plants and microbes and increase decomposition rates. Such changes would tend to maintain lower C:N ratios and thus could be self-perpetuating.

The cycling of nitrogen is also influenced by both land-use history and present vegetation (Table 9.2). Net nitrogen mineralization is primarily controlled by present vegetation and was higher in hardwood sites. This result agrees well with the observation of the lower C:N ratios of hardwood soils. In contrast, nitrification was more strongly influenced by land-use history and was greater in cultivated sites. Nitrification potential and the biomass of nitrifying microbes were higher in both pastured and cultivated soils, yet only cultivated soils had higher net nitrification, suggesting that higher immobilization in pastured soils may reduce net nitrification.

Higher soil nitrogen and phosphorus and higher nitrification rates in the plowed areas 70 to 140 years after abandonment suggest that the addition of manure to these crop fields in the 1800s had a remarkably persistent influence on these sites. At least 65 percent of the land area across Prospect Hill and most of the Massachusetts uplands were used for pasture in the eighteenth and nineteenth centuries, leading to a con-

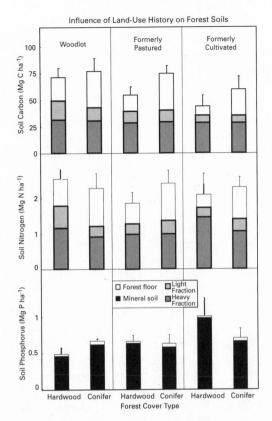

Figure 9.3. Total carbon, nitrogen, and phosphorus in the forest floor and top 15 centimeters of mineral soil by prior land use and modern vegetation on the Prospect Hill tract. The light fraction is organic matter that is relatively recent in origin; the heavy or mineral-associated fraction is more recalcitrant and older. Standard error bars ($n = 2$) are shown for the forest floor plus mineral soil. Based on Compton and Boone 2000.

tinual removal of nutrients and carbon from these areas. Farmers then transferred small portions of these nutrients to the plowed cropland through the process of manure applications. In order for a given plot of land to maintain productivity over the more than 100-year period during which it was typically farmed in New England, organic matter amendments were often applied. This movement of nutrients may have produced a wider range of variation in soil nutrients across the landscape than might be expected in native forests.

Soil-vegetation feedbacks on litter quality have occurred since the reestablishment of the forest, as indicated by our results that forest floor C:N, carbon-to-phosphorus, and nitrogen-to-phosphorus ratios are still low in former agricultural sites. Interactions between land-use history and vegetation were important in determining soil carbon, nitrogen, and

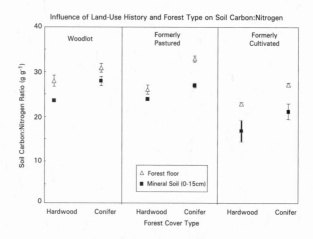

Figure 9.4. The ratios of total carbon to nitrogen in the forest floor and top 15 centimeters of mineral soil on Prospect Hill by prior land use and present vegetation. Standard error bars are shown ($n = 2$). Based on Compton and Boone 2000.

phosphorus contents as well as nitrification, and the hardwood sites generally exhibited greater residual impacts of prior land use. Our findings lead to the conclusion that the effects of agriculture on soil carbon, nitrogen, and phosphorus contents; microbial populations; and nitrogen cycling can persist for more than a century after abandonment. These effects may result from residual impacts of agricultural practices such as manuring or from the influence of land use on the pattern of vegetation recovery during reforestation.

Table 9.2. Effects of Land-Use History on Nitrogen Cycling on the Prospect Hill Tract

	Nitrogen Mineralization (kg ha^{-1} yr^{-1})	Nitrification (kg ha^{-1} yr^{-1})	Nitrification (% of N mineralization)
Conifer			
Cultivated	7.9 (3.5)	0.8 (0.5)	9.3 (2.4)
Pastured	8.2 (0.6)	0.5 (0.2)	6.0 (2.1)
Woodlot	7.7 (0.1)	0.1 (0.0)	1.6 (0.1)
Hardwood			
Cultivated	18.1 (0.7)	4.3 (0.3)	23.7 (2.5)
Pastured	11.8 (1.6)	0.6 (0.2)	4.6 (1.3)
Woodlot	14.7 (0.0)	0.6 (0.1)	4.1 (0.5)

Source: Reprinted from Compton and Boone 2000, with permission from the Ecological Society of America.
Note: Growing season net nitrogen mineralization (net ammonium plus nitrate accumulation in buried bags), nitrification, and nitrification as a percentage of nitrogen mineralization are shown. The time period is May 11 through October 31, 1995. Standard error of the mean of two plots is shown in parentheses.

Consistent Patterns and Differences between Sites

Collectively, these results indicate that three important factors play major, though variable, roles in controlling modern soil characteristics and important aspects of forest biogeochemistry: inherent soil properties, historical land-use practices, and vegetation composition. These factors interact through time to regulate the long-term recovery of soil characteristics and processes after agricultural abandonment, logging, or other activity. As noted by New England settlers and scientists alike, trees and then forests quickly reclaim pastures, hay fields, or croplands once the grazing, mowing, or plowing ceases. However, the recovery of critical soil characteristics such as soil organic matter and microbial populations may lag decades behind the development of the new forests. Thus, the appearance of forests can be deceptive and may hide major differences in ecosystem function resulting from historically distant activities. From our comparison of contrasting sites, we can conclude that the recovery of soil carbon lost due to deforestation and cultivation requires at least 70 years of reforestation and that the rate of carbon sequestration may be slowed by other consequences of agriculture, including a self-perpetuating increase in the quality and decomposition of soil organic matter on some sites. Forests on former agricultural sites have persistently lower soil C:N ratios. These results fit a general pattern that is emerging from our work as well as the few studies available from other ecosystems. For instance, recovery of soil carbon appears to require at least 50 years after abandonment of grasslands in Colorado and forests of the southern Appalachians, while more than 200 years may be necessary to regain soil carbon lost in the more slowly growing forests of New Hampshire.

Our work also illustrates that the type and intensity of land use are quite important in influencing the subsequent recovery pattern. On the poor-fertility soils of Montague Plain, agricultural use was short in duration and probably did not include intensive amendment of cultivated areas with animal manures or other fertilizers. After more than fifty years of reforestation, soil carbon levels are approaching 90 percent of levels observed in soils that were never plowed, and nitrogen cycling is very similar between stands with different land-use histories. However, the legacy of prior land use on the forest vegetation is more pronounced than that on soil nitrogen cycling at Montague Plain, and in turn, present vegetation appears to have a much stronger effect on soil properties and processes.

The Prospect Hill farms have been abandoned for a longer time than the Montague farms, yet because of the more intensive and more prolonged agricultural use at Prospect Hill, the long-term effect of past agriculture on soil C:N ratios and nitrate production is enhanced (Figure 9.5). In fact, our results suggest that intensive agriculture actually in-

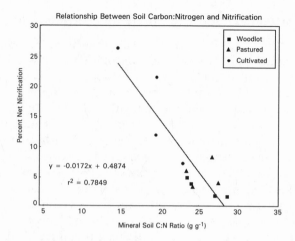

Figure 9.5. Relationship between nitrification and C:N ratios for forest soils with different land-use histories on the Prospect Hill tract. Based on Compton and Boone 2000.

creased nitrogen cycling rates, probably as a result of incorporation of animal manures into the soil in pastured and cultivated sites. These additions are still evident as slightly higher soil nitrogen and phosphorus contents, lower C:N and carbon-to-phosphorus ratios, and higher nitrification rates in cultivated soils. Our results clearly do not support the commonly held notion that early New England farming depleted soil fertility; rather, because nineteenth-century farming practices moved organic materials from one area to another within the farm, some areas may have been enriched in carbon, nitrogen, and phosphorus, while others were depleted.

These findings indicate that prior land use has persistent, important effects on the dynamics of soil organic matter and on nutrient storage and availability. Cultivation and manure amendments appear to have long-lasting effects on soil properties and microbial nutrient processing, resulting in a pattern of recovery that is decidedly different from that observed for other disturbances such as logging, fire, and windthrow. Given the possibility, however, that not all cultivated lands were treated with the same type and amounts of organic amendments, the simple land-use categories described here may not completely reflect the complex history of a given plot of land. In addition, the patterns of nitrogen availability following reforestation are not just the direct effects of land use but are also driven by the resulting indirect changes in vegetation and feedbacks to forest floor decomposition.

How Will Land-Use History Influence the Forest's Response to Future Changes?

In order to understand important ecosystem processes such as forest growth and carbon storage or the ability of forests to retain added nitrogen, future researchers should carefully examine land-use history. In our comparisons of forests with different land-use histories at Prospect Hill, intensive cultivation increased soil nitrate production by decreasing the C:N ratio (and perhaps through increases in nitrogen through manure additions). This effect was maintained for a very long time. Because of this long-term legacy of cultivation and organic matter amendments, it is possible that forests regrowing on cultivated sites could be more susceptible to the effects of nitrogen saturation than pastured or logged sites. Recognizing the differences in land-use history of the chronic nitrogen experimental plots (see Chapter 12) has proved to be important in understanding the responses to nitrogen additions. The formerly cultivated sites could have an accelerated response to nitrogen deposition, even overcoming the presumed roles of species differences (pine versus hardwood) in controlling soil and ecosystem processes.

Forest regrowth on former agricultural lands is a major sink for atmospheric CO_2 in the eastern United States. High nitrogen availability on former farms may allow rapid regrowth of forests, allowing these ecosystems to serve as a major carbon sink for decades after agricultural abandonment. The long-term influence of specific land-use practices on forest regrowth (cultivation versus pasture versus logging) is not yet clear, but the legacies could also drive the variations in this important mechanism of carbon storage across the landscape and region.

CHAPTER 10

Exchanges between the Forest and the Atmosphere

J. W. MUNGER, C. BARFORD, and S. WOFSY

In this chapter we move beyond the forest and terrestrial ecosystems to consider the interactions between the landscape and the atmosphere. The atmosphere provides major constituents of the physical environment for forest vegetation, including climatic variables (wind, temperature, humidity), nutrients, and toxins. In turn, the atmosphere is continuously altered and modified in many ways by the vegetation and physical features of the landscape. For example, the dark foliage of a forest absorbs considerably more solar radiation than the light-colored surface of a dry grain field or snowy hillside, thereby providing more energy to heat the atmosphere, evaporate water, or support thermal convection. Evaporation of water from the surface of vegetation and through transpiration cools and humidifies the atmosphere, which leads to an observable increase in cloudiness. The dense and rough canopy of a forest slows the wind and increases surface turbulence, enhancing the removal of pollutants and aerosols, whereas soils and vegetation are themselves significant sources of many of the trace gases occurring in the atmosphere.

Human activity has fundamentally altered essential characteristics of both the land surface and the atmosphere in New England over a period of several centuries. The change from forested to intensely agricultural (eighteenth century) to reforested (nineteenth century) to urbanized (twentieth century) implies significant shifts in the structure of the land surface presented to the atmosphere and therefore important changes in the efficiency of chemical deposition and energy transfer between the two. Changes in the biomass, species composition, and soil chemistry through time in regrowing forests have resulted in major storage of carbon in forest ecosystems over the past century and have significantly altered balances of important trace gases. At the same time, unprecedented emissions of CO_2, nitrogen oxides, and a wide variety of pollutants from industrial sources across the eastern United States and beyond affect regional atmospheric chemistry and the biological function of forests. Understanding the interaction between the forest and the atmosphere and the roles played by human activities on local to broad scales to alter these processes provides critical insights into both the functions of forest ecosystems and a host of global environmental issues linked through the dynamics of the atmosphere.

Global CO_2 and the Carbon Balance of Forests in Central New England

Atmospheric CO_2 is a globally important gas closely linked to forest dynamics. Carbon dioxide is a major greenhouse gas and a primary product of the combustion of fossil fuels. In the 1990s, about 25 percent of the CO_2 from fossil fuel combustion was absorbed by the ocean and roughly 40 percent stayed in the atmosphere. The remainder was taken up by terrestrial vegetation. Analyses of $^{13}C/^{12}C$ isotopic ratios in CO_2 and concentrations of oxygen in the atmosphere, along with patterns of CO_2 distribution over the globe and data from forest inventories, point to significant storage of carbon in forest vegetation and soils. According to the analysis of air retrieved from polar ice and snow cores, the process of major storage of carbon by the global terrestrial biosphere has been important only over the past few decades. Before that, the biosphere was neutral, or possibly a source of CO_2 to the atmosphere.

Many attempts have been made to explain why the biosphere is currently a net carbon sink, including

- Fertilization by increasing concentrations of CO_2 in the atmosphere and nitrogen deposition to forests, both resulting from human activities
- Longer growing seasons and northward forest expansion due to climate warming
- Reforestation of former agricultural lands in the eastern United States, Canada, and Europe; most of these regrowing forests are still increasing in wood volume

Deciding among these three major causes for increased carbon storage in temperate forests has important geopolitical implications. Fertilization by CO_2 might be expected to continue indefinitely, but nitrogen deposition can eventually lead to negative as well as positive effects on tree growth (see Chapter 12). Whether global warming results in increased uptake or release of CO_2 depends on whether photosynthesis or respiration is most sensitive to changes in temperature and the length of the growing season, as well as how each response is affected by changes in the availability of water. On the other hand, reforestation may represent only a temporary potential for storage, since forests reach a balance between growth and decay as they mature; but management options can strongly influence how, or even if, this balance is attained. Harvesting of timber could truncate increases in carbon storage in forests or could enhance forest growth and storage. The fate of products derived from forest harvesting (for example, rapidly decomposed paper versus long-term usage in construction) and the methods and rotation pattern of harvesting are important components of this calculation of net balance.

Motivation for Long-Term Atmospheric Measurements at the Harvard Forest

Understanding forest carbon dynamics requires a combination of models and observations of net forest exchange at timescales appropriate to the instantaneous response of plants to temperature and light as well as the longer-term consequences of climatic variations and successional change. Similarly, atmospheric chemistry responds to short-term factors as well as long-term changes driven by emissions control regulations, urban development, technological changes, and climate variability. At the Harvard Forest we set out to answer the following questions with a comprehensive suite of observations that cover all of these timescales:

- *Carbon balance:* How much carbon is being sequestered in the aggrading, actively growing forests of the northeastern United States? What factors regulate the magnitude of this uptake? How is this uptake (amount and timing) related to the effects of land-use history; forest composition; climate variations on seasonal, annual, and decadal timescales; nutrient deposition; and pollution?
- *Atmospheric chemistry:* How is the chemistry of the atmosphere affected by emissions from, and deposition to, the forest? What influence does the forest have on the long-range transport of pollutants from the industrial Northeast and Midwest? How does this transport and deposition of pollutants affect the growth and health of the forest?

These questions have been addressed at the Harvard Forest by a comprehensive suite of observations carried out continuously since 1989 at the Harvard Forest Environmental Measurement Site (EMS, Figure 10.1). The core measurements of atmosphere-biosphere exchange have been made using the relatively new eddy covariance method and have been augmented by a set of field observations, process studies, and modeling. The first five years of the Harvard Forest project were largely devoted to developing the eddy covariance method and establishing its reliability so that it could be applied elsewhere. We have applied the technique in parallel studies in Manitoba, Canada, and Brazil. An even broader context is provided by the AmeriFlux, Euroflux, and other networks that support more than 100 flux-tower sites worldwide.

The Harvard Forest Environmental Measurement Site

The Harvard Forest EMS provides a long-term record at a rural continental site of trace-gas concentrations and surface-exchange fluxes, along with supporting measurements of physical environment

Figure 10.1. Installation of the gas sampling lines on the Environmental Measurement Site (EMS) tower. The facility was established by hand with minimal site disturbance. Branches and other vegetation near the tower were left intact. Photograph by J. W. Munger.

Table 10.1. Measurements Made Routinely at the Environmental Measurement Site

Sensor	Inlet/Instrument Height	Determined Quantity	Symbol
Sonic anemometer	30 m	Horizontal and vertical wind Flux of momentum and heat	u, v, w, T F_{MOM}, F_{HEAT}
NO_y (catalyst→NO)	30 m	Flux and concentration of NO_y	F_{NO_y}
High-speed CO_2–H_2O IR absorbance	30 m	CO_2 flux H_2O flux and concentration	F_{CO_2}–F_{H_2O}
High-speed O_3 C_2H_4 chemiluminescence	30 m	O_3 flux and concentration	F_{O_3}
Slow CO_2 IR absorbance	30, 24, 18, 12, 6, 3, 1, 0.05 m	CO_2 vertical profile	
Slow O_3 UV absorbance	30, 24, 18, 12, 6, 3, 1, 0.05 m	O_3 vertical profile	
Slow NO_x (photolysis, O_3 chemiluminescence)	30, 24, 18, 12, 6, 3, 1, 0.05 m	NO, NO_2 vertical profile	
Thermistor, thin-film capacitor	30, 22, 12, 6, 3 m	Temperature and relative humidity profiles	
Thermistors	Surface (6 reps), 20 cm, 50 cm	Soil temperatures	
Quantum sensor	30,12 m	Photosynthetically active photon-flux density	PPFD
Net radiometer	30 m	Net radiative heat flux	R_{net}
CO gas–filter correlation IR absorbance	30 m	CO concentrations	
Gas chromatograph–Flame ionization detector (GC-FID)	30 m	CH_4 concentrations*	
Two-channel GC-FID	29, 24 m	C_2–C_6 hydrocarbon concentrations and gradients	
Four-channel GC–electron-capture detector	29 m	Halocarbons, N_2O, CO, CH_4, SF_6 concentrations†	

Note: Numerous other measurements are made with specialized equipment that is installed for shorter durations. The tower and analytical equipment are inspected and serviced routinely every two to three days.
*Measurements by P. Crill, University of New Hampshire.
†Installed September 1995; official station in the NOAA halocarbon-monitoring network.

and biological processes (see Table 10.1 for a complete list of measurements). The EMS is located near the eastern boundary of the Prospect Hill tract, surrounded by Harvard Forest and private lands covered by typical upland forest dominated by mixed hardwoods, especially red oak and red maple, with scattered hemlock and pine (see Figure 2.8).

Atmospheric composition and trace-gas exchanges on a local scale in a forest are modified by the regional surroundings in prevailing upwind directions, primarily to the northwest and southwest in New England. Within 100 kilometers of Petersham, the surrounding area is largely rural, with a mixture of mostly small (population up to 10,000) and a few medium-sized (population less than 100,000) towns surrounded by forested lands. However, extensive urban areas to the southwest have relatively high pollutant emissions densities within 100 to 500 kilometers (Figure 10.2). Forested regions with interspersed local agriculture and low population density extend for hundreds of kilometers to the northwest. As we will see in the following sections, the variation from densely urbanized to rural landscapes at different directions from the EMS is reflected clearly in variations in atmospheric measurements through time. In general, the Harvard Forest receives extremely clean continental air when winds blow from the northwest and quite polluted air when the winds are southwesterly. This variation makes the site ideally positioned for determination of the influence of pollution on the forest and of the forest on pollution.

The central facility of the EMS is a 30-meter tower mounted with sensors and sampling inlets located above, within, and below the approximately 24-meter canopy. A small companion building shelters instruments and data-acquisition equipment. Electrical power and communication lines to the site are buried beneath the 1.5 kilometers of woods road extending back to Shaler Hall. The EMS was placed at this remote forested site in order to be distant from paved roads or other cul-

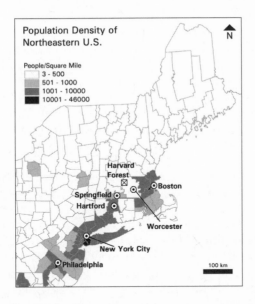

Figure 10.2. Population density in the northeastern United States illustrating the sharp gradient between the densely populated urban-industrial zone along the coast and the mostly rural interior. As a consequence of this pattern, northwesterly winds bring relatively clean air from extensively forested areas to the Harvard Forest, whereas southwesterly winds bring air that is affected by recent industrial and automotive emissions. Sampling and discriminating these two prevailing wind directions provide insights into human influences on atmospheric composition. Population data from the U.S. Census.

tural activity. To minimize disturbance, all site installation activities, including the digging of tower foundations, were done by hand, and only small limbs that directly obstructed the tower and its guy wires were cut during installation.

Eddy Covariance and Net Flux Measurements at the EMS Tower

We measure the fluxes or exchanges of trace gases between the atmosphere and the forest using the eddy covariance method, in which the differences in concentration and vertical velocity between updrafts and downdrafts above the forest are measured and are then used to calculate the overall direction and magnitude of fluxes (Figure 10.3). Updrafts bring to the sensor air that has just been in contact with the forest understory and canopy and hence is depleted relative to downdrafts in substances that are taken up by or deposited in the forest (for example, CO_2 during the day). They are enriched in substances emitted by or produced in the forest (for example, CO_2 at night).

Turbulence in air movement is an important regulator of the forest environment and flux rates and is often described in terms of the frequency in oscillation between updrafts and downdrafts. For instance, shifts between updrafts and downdrafts at 1-second intervals would have a frequency of 1 cycle per second (also denoted in hertz). The length of turbulent eddies is determined by frequency and wind speed; a 1-hertz eddy in a 1-meter-per-second wind would be 1 meter in length. In actual practice, turbulence at the top of the forest canopy exhibits a wide range of variation, from the rapid fluctuations (0.1 to 10 cycles per second) imparted by fast winds moving across the rough surface of the canopy, to very large eddies with slow frequencies (0.01 cycles per second) associated with convective weather cells and clouds in the planetary boundary layer. For typical wind speeds, the horizontal extent of turbulent eddies that affect the transport of materials and energy to forest canopies ranges from a few meters to several hundred meters.

In order to distinguish updrafts and downdrafts at the finest scales and to determine accurate fluxes, many tower instruments take readings of chemical concentrations and wind speed and direction at least once per second. A long-term or running average is subtracted from the instantaneous values of concentration and vertical wind speed to determine the fluctuations. The product of the fluctuations is averaged over an interval long enough (typically approximately 30 minutes for forests) to provide a valid sample of updrafts and downdrafts. An important assumption of the eddy covariance method is that averages over time at the sensor location are equivalent to spatial averages across the surrounding landscape in the upwind direction; that is, the concentrations and winds measured at the tower are typical of a larger area upwind. De-

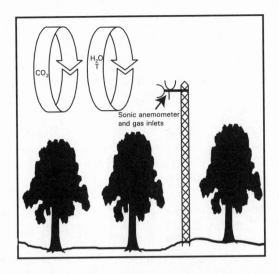

Figure 10.3. Drawing of the eddy covariance measurement of atmosphere-biosphere exchange fluxes at the Harvard Forest. The sonic anemometer, which measures wind velocity in three dimensions, and inlets for gas measurements sit atop a 30-meter tower above the 22- to 24-meter canopy and make high-frequency (four to eight times per second) measurements of the wind, air density, and concentrations of CO_2, ozone, total oxidized nitrogen (NO_y), and other gases. These measurements allow the direct determination of fluxes between the forest and the atmosphere. Eddy covariance was originally developed to measure CO_2, water, and energy fluxes over crops; in the early 1980s it was extended to tall vegetation where it was used for periods of a few days or weeks and was very labor intensive. In collaboration with colleagues at SUNY-Albany, we developed fully automated instrumentation that could define net ecosystem growth over a year by summing hourly data without prohibitive accumulation of error.

termining the validity of this assumption is an important criterion for identifying good data periods before making further analyses.

In addition to measuring the fluxes of trace gases such as CO_2, water (H_2O), and ozone (O_3), eddy covariance provides the fluxes of sensible heat (the product of vertical wind speed and temperature variations) and momentum (the product of vertical and horizontal wind speed fluctuations). Sensible heat flux is the exchange of warmer (or cooler) air between the canopy and air. The sum of sensible heat flux and latent heat flux, which is the heat exchanged by evaporation of water, should balance the energy gained (or lost) from the system by radiation. Momentum flux measures the influence of friction at the canopy surface to slow the wind speed. It depends on both the vertical gradient in wind speed and the strength of vertical mixing. Micrometeorologists frequently use the term *friction velocity* (denoted by the symbol u*, the square root of [−1 × momentum flux]). When friction velocity is large, turbulent exchange of air between the canopy and the atmosphere is efficient and accounts for most of the canopy-atmosphere exchanges. When friction velocity is low, turbulent mixing is weak and canopy-atmosphere exchange may occur by other processes that are not easy to measure. We will use friction velocity throughout this chapter to distinguish periods when the air at the canopy interface is thoroughly mixed from periods when it is poorly mixed.

The flux measured just above the canopy at 30 meters may not account for all of the CO_2 taken up or released by the forest; some additional CO_2 may stay in the air spaces of the canopy and contribute to changes in concentrations there that we also need to measure. Consequently, the true net ecosystem exchange of carbon (NEE) is the measured flux at the top of the tower plus the change in CO_2 mass contained in the 30-meter column of air below the flux sensor (that is, the change in mean concentration of CO_2 multiplied by the height).

Net ecosystem exchange is the sum of gross ecosystem exchange (GEE, equivalent to gross photosynthesis) by autotrophs (plants) minus the total respiration efflux (R) by both autotrophs and heterotrophs (for example, microbes, soil invertebrates, and other animals) and is generally expressed in units of mass per unit time and area (for example, micromoles of CO_2 per square meter per second, or kilograms carbon per hectare per year). Atmospheric measurements are referenced to a vertical scale with zero at the ground and increasing positively with height. By this convention, GEE is a negative flux (that is, CO_2 is moving "downward" from the atmosphere into the forest), and R is positive.

At the Harvard Forest we use the nighttime measurement of NEE to determine R because at night photosynthesis or production by plants is zero in the absence of sunlight whereas respiration activity by both plants and heterotrophs continues. Daytime R is estimated from the ambient temperature on the basis of the relationship between nighttime NEE (which is the same as R) and temperature. GEE may then be calculated from NEE (measured by eddy covariance) and R calculated by the simple relationship with temperature (Figure 10.4).

The NEE measurements have inherent uncertainties, attributable to the nature of the instruments and their occasional failure due to lightning strikes, power outages, and other events and to a variety of measurement artifacts caused by the nature of air movement under different environmental conditions over complex terrain such as occurs in central Massachusetts. We have critically evaluated the sources of uncertainty and their effect on annual carbon balances and have divided them into three types: (1) uniform systematic error, associated with equipment, calibration gas mixtures, and data processing; (2) selective systematic error, due to the inability of the tower/sensor system to sample the forest adequately under certain conditions or in certain wind directions; and (3) sampling uncertainty, associated with periods of missing

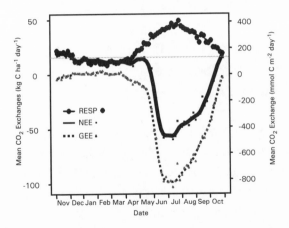

Figure 10.4. The mean course of measured respiration (RESP), net ecosystem exchange (NEE), and gross ecosystem exchange (GEE) for 1992 through 1999 at the Harvard Forest. Respiration is derived from the relationship between nighttime temperatures and NEE. Gross ecosystem exchange is calculated by subtracting RESP from NEE. Negative values represent downward fluxes of carbon into the forest ecosystem, and positive values represent releases from the forest into the atmosphere. Measurable GEE is observed in early April because of the scattered conifers and understory vegetation that greens up early at the Harvard Forest, but respiration continues to dominate until the full forest canopy develops in late May.

data. Discussion of these uncertainties and the manner in which we address them provides important, though frequently underappreciated, insights into the scientific process (Box 10.1).

CO_2 Exchange and Net Ecosystem Production

SEASONAL AND DAILY CYCLES

During the growing season, NEE is positive during nighttime hours when respiration is the dominant process, becomes negative at dawn as photosynthesis increases, and then is positive again at dusk (Figure 10.5). Because the vertical mixing of air depends on thermal convection induced by solar heating, the friction velocity (u^*) also rises abruptly at dawn and drops sharply at dusk. Net storage of CO_2 within the canopy air spaces occurs during still periods at night, and this stored carbon is lost in the early morning as wind speed and frictional velocity increase (Figure 10.5). Because of this, the flux of CO_2 across the top of the canopy (F_{CO2} in Figure 10.5) varies slightly from the actual rate of carbon production or consumption by the forest (NEE, Figure 10.5).

Interestingly, the quantum yield or amount of carbon fixed per amount of incident radiation is slightly lower in the late afternoon than during the morning for a given light level because of mild water stress

Box 10.1.
Analysis of Error in Eddy Covariance Measurements

Uniform systematic errors in the EMS measurements are caused by underestimation of the total CO_2 mass flux by the eddy covariance measurement method. We uncovered and quantified this bias by making comparisons between the CO_2 flux and latent heat flux, which is measured by the same instrument using eddy covariance. The frequency characteristics of our measurements of latent heat flux suggested that the high-frequency signals were being damped, thus masking part of the heat flux. Analysis of the total energy budget, of which latent heat is a part, supported the notion that latent heat flux was being underestimated. By analogy, the same high-frequency damping and resultant underestimation occurs in the measurement of CO_2 flux. To avoid this systematic error, we have developed and applied a correction factor based on sensible heat flux data.

Selective systematic errors occur particularly on still nights, when the EMS underestimates respiration fluxes. This bias develops especially under calm conditions, with friction velocities less than 0.2 meters per second. Without large energetic eddies driven by surface heating, other processes—including small high-frequency eddies induced by the rough canopy, cold-air drainage flows, and intermittent gusts, which are usually relatively unimportant—become the dominant mixing processes. All of these secondary processes are extremely difficult to measure accurately. Because of the relatively calm conditions and lack of vertical air movement, CO_2 storage in and beneath the canopy increases and fluxes above the canopy at 30 meters decrease. However, comparison of NEE and calculated R for calm nights indicates that CO_2 storage fails to compensate for the reduced upward flux, leaving "missing" CO_2. The morning efflux of CO_2 from the forest to the atmosphere due to resumed atmospheric mixing is also smaller than expected. This conundrum is addressed by replacing NEE data from calm nights with estimates based on regression relationships between soil temperature and respiration. A friction velocity of 0.17 meters per second has been established as the threshold between windy and calm at this site, because nighttime CO_2 efflux becomes independent of air turbulence above this value. Correction for this systematic error is very important as it reduces our estimates for annual carbon sequestration at the Harvard Forest by 0.5 to 1.0 tons of carbon per hectare (see, for example, the differences between Wofsy et al. 1993 and Goulden et al. 1996a).

Sampling uncertainty arises when incomplete data sets must be summed to find the total annual carbon balance. Gaps in eddy covariance data result from interruptions for routine calibration, maintenance, and data transfer as well as equipment malfunction. Our summation technique accommodates gaps by first dividing the year into short segments (each generally about four days long) in which environmental conditions are relatively well correlated. Within each segment, missing CO_2 flux data are replaced using empirical relationships between CO_2 exchange and climate variables. Calculated fluxes are then averaged for each hour of the day within the segment, and the hourly fluxes are summed. Finally, the short segments are summed to yield annual carbon balance. The sampling uncertainty embedded in this approach was evaluated for the carbon balance in 1994 using a Monte Carlo simulation. The 90 percent confidence interval for sampling error was +0.3 tons per hectare compared with the overall balance of −2.1 tons per hectare. The analyses indicated that replacing missing data introduces less uncertainty than the alternative approach of assuming that the days with valid data are representative. Not surprisingly, simulations also showed that a single long data gap creates disproportionately more uncertainty than do a number of short data gaps of the same total duration spread through the year.

The sampling uncertainty of CO_2 flux is smaller than for oxides of nitrogen (NO_y) because of the skewed distribution of daily deposition rates; more than 50 percent of the summer NO_y input is deposited during extreme events that occur on only 20 percent of the days. Because of this excessive skewness, simulations with random subsamples containing less than 50 percent of possible data yielded highly variable NO_y deposition estimates; however, the variability between random subsamples of the data dropped below 15 percent when more than 50 percent of the available data were included. These results corroborate the uncertainty analysis of the carbon measurements. They also confirm the value of continuous, unattended monitoring combined with prompt correction of all system malfunctions.

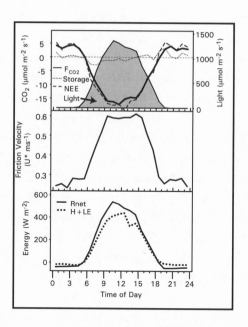

Figure 10.5. Mean daily course of carbon exchange and light (photosynthetic photon flux density; shading) is shown in the top panel. Separate lines show the measured CO_2 flux (F_{CO2}), the storage term, and their sum, net ecosystem exchange (NEE). The monthly average light levels are shown by the stippled line. Friction velocity, a measure of the vertical mixing intensity of air, is shown in the middle panel. Energy exchange by radiation (Rnet) and the sum of sensible (H) and latent (LE) heat flux is shown in the bottom panel. The data illustrate average values from July 1998 at the Harvard Forest.

Table 10.2. Annual Net Ecosystem Exchange (NEE), Gross Ecosystem Exchange (GEE), and Ecosystem Respiration (R) at the Harvard Forest as Measured by Eddy Covariance Methods at the Environmental Measurement Site

Growing Year	NEE	GEE	R
1991–1992	−2.0	−11.4	9.4
1992–1993	−1.9	−13.3	11.4
1993–1994	−2.0	−12.3	10.3
1994–1995	−2.5	−12.3	9.9
1995–1996	−2.0	−13.2	11.3
1996–1997	−2.1	−13.9	11.8
1997–1998	−1.2	−12.1	10.9
1998–1999	−2.3	−13.9	11.6
1999–2000	−2.1	−14.3	12.2

Note: Values are for metric tons of carbon per hectare per year on a growing year basis that starts on October 28 and runs to the following October 27. By examining the carbon budgets over growing years, we avoid splitting up the winter dormant season. Positive values denote emissions from the forest.

and decreased ambient CO_2 concentration, which decrease photosynthesis, and increased soil and air temperatures, which increase respiration. A smaller drop in quantum yield occurs between early and late summer, presumably on account of the changing chemistry and declining efficiency of the aging leaves. Net uptake of carbon by the forest during a typical summer day (in 1992) was −14 to −19 moles per square meter per second, whereas flux out of the forest due to nighttime respiration was 3 to 5 moles per square meter per second. During the leaves-off period between October and April, carbon effluxes ranged from 0 to 5 moles per square meter per second during both night and day.

Summed for the entire year (Table 10.2), GEE has ranged between −11 and −14 tons carbon per hectare and occurred largely between late May and late September. The magnitude and timing of the carbon and ecosystem exchange cycle each year have depended closely on the timing of leaf emergence in the spring and leaf senescence and fall in the autumn (see below). GEE exceeds R throughout the growing season, when the forest gains 30 to 60 kilograms of carbon per hectare each day. During the largely dormant months when deciduous species are leafless, the forest loses 10 to 20 kilograms carbon per hectare each day, even though low soil and air temperatures reduce respiration. During warm sunny periods in winter, we can detect some CO_2 uptake by evergreen conifers, which make up approximately 25 percent of the tree basal area around the tower.

The net result of these seasonal differences in production and respiration processes has been annual carbon sequestration (net ecosystem production, or NEP) of −1 to −3 tons of carbon per hectare, which is 10 to 30 percent of annual photosynthesis (GEE, Table 10.2). This uptake can be attributed largely to the developmental age of the forest; namely, it is maturing and growing actively after a history of intensive cutting and agriculture. More than half of the organic matter fixed during the past few years (about 60 percent) was stored in the trunks and larger branches and roots of live trees (1.5 of 2.2 tons of carbon per hectare per year). Red oak, which constitutes half of the aboveground woody biomass, made the largest contribution to this carbon storage. Examination of growth by oaks in 1998 to 2000 suggests consistent carbon sequestration between large- and small-NEP years. In contrast, carbon sequestration by the other dominant species in the forest, red maple, was much smaller and more variable. Further study of tree growth will help define the relationship between NEP and ongoing changes in species composition at the Harvard Forest and enable better predictions of future sequestration.

In addition to providing net annual carbon fluxes, the high-frequency information obtained by eddy covariance methods provides a totally new kind of data on whole-canopy physiology that is valuable for either parameterizing or validating existing models of canopy processes. A number of models have been used in this way, including Jeff Amthor's "big-leaf" physiologically based model of deciduous canopies that predicts hourly CO_2 and O_3 uptake; Richard Waring's "quantum efficiency" model for mixed-conifer and deciduous hardwood canopies that predicts monthly gross ecosystem productivity; and John Aber's PnET II, a whole-forest model that predicts monthly GEE, NEE, and carbon allocation. After validating a model with data from a particular site, such as the Harvard Forest, the model can be extended and further evaluated across larger regions, as for example the application of PnET-II to our entire New York/New England study region (see Chapter 17).

CARBON BALANCE MODULATED BY ENVIRONMENTAL CONDITIONS
AND CLIMATE VARIATION
Some of the most interesting features of the long-term measurements for GEE, NEE, and R at the EMS are the shorter-term responses to variations in climate. Overall, annual uptake of carbon varied by up to a factor of two over the nine-year period of observations (Table 10.2). Analysis of this variation provides important information on the factors that regulate the rate of carbon sequestration in this maturing and actively growing forest. Because CO_2 is the currency of both carbon fixation and oxidation, control of carbon exchange between the forest and the atmosphere involves all the environmental regulators of photosynthesis and respiration, notably light, soil and air temperatures, wind, humidity, and soil moisture. The number of these physical and chemical factors, their potential to compensate or interact with one another, and the importance of their timing with respect to daily and seasonal cycles all contribute to the natural variability and complexity of forest ecosystems. Continuous

monitoring over nearly a decade has yielded the data required to begin untangling the factors regulating carbon exchange on a forest level and to observe interannual variation in NEE.

In nine complete years of measurement (1992 to 2000), climate conditions included cold and mild winters, snowy and snowless winters, hot and cool summers, and wet and dry summers. In response to these changing conditions, NEE varied up to 40 percent between successive years, with a total range from 1.2 to 2.5 tons of carbon per hectare per year (Table 10.2). However, these shifts in annual NEE resulted from the major effects of climate anomalies during specific intervals when the forest was particularly sensitive, rather than differences in annual mean conditions. For example, GEE was 10 percent less on average for 1992, 1994, and 1995 compared with 1993, 1997, and 1999–2000. These decreases in GEE corresponded with shorter growing seasons, resulting from delays of six to ten days in spring leaf emergence. In contrast, in 1992 and 1993, delayed leaf senescence and leaf fall due to relatively warm autumn nights boosted gross production by about 500 kilograms of carbon per hectare per year (12 percent) compared with other years. Given the importance of intercepted light for carbon uptake, it is not surprising that summer cloudiness also affects GEE. This effect is seen in mid-July 1992, mid-August 1992, and August 1994, when cloudy periods reduced gross production by around 400 kilograms of carbon per year. These results reflect the disproportional influence on annual NEE of weather during the growing season, as opposed to the dormant months.

Water relations provide an example of compensating effects on NEE that were unexpected at the initiation of the study. Annual GEE declined during a severe drought in late summer 1995, but only by a modest 10 percent. However, the concurrent decline in total forest respiration was much greater (1,000 kilograms of carbon per hectare), and the amount of sunshine was above average. Consequently, net carbon sequestration for 1995 was relatively large (Table 10.2) despite increased water stress on the trees. This result may appear counterintuitive as "sequestration" normally connotes carbon storage as wood in growing trees. However, the soil carbon compartment plays a large role in net carbon storage. This result suggests that whereas trees were able to tap into deep soil water via roots, the microbially dominated respiration flux was sensitive to dry surface conditions and declined proportionally more than photosynthesis. This contrast in susceptibility to drought between production and respiration is likely a characteristic of mesic forest ecosystems, in contrast to grasslands, savannahs, or similar ecosystems.

Closer examination of the late summer drought of 1995 reveals additional features of the relationships among forest respiration, soil temperature, and rainfall. The strong dependence of soil respiration, and by extension forest respiration, on soil temperature is often expressed as what is termed a "Q_{10}" value. Q_{10} specifies the change in soil respiration

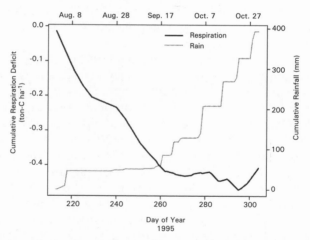

Figure 10.6 Relationship between ecosystem respiration and rainfall for the dry (days 213 to 258) and wet (days 259 to 304) study periods in 1995. Cumulative respiration deficit is the difference between measured nocturnal CO_2 efflux and the flux predicted by a Q_{10} relationship between soil temperature and ecosystem respiration fitted for all of 1995 nocturnal periods with adequate momentum flux for reliable measurements ($u^* > 0.2$ m second^{-1}).

for a 10°C change in temperature. Previous studies have shown a Q_{10} value of approximately 2.2 for Harvard Forest soils. This value was used to fit an exponential relationship between soil temperature and CO_2 efflux for all sufficiently windy nights in 1995. The resulting equation was used to predict forest respiration during the latter part of the dry period, between August 1 and September 15, and during a wet period between September 16 and October 31 (Figure 10.6). Respiration during the dry period was 400 kilograms of carbon per hectare less than expected, while total respiration during the wet period was roughly equal to expectation despite marked changes in the rate during and after rain events. The deficit in the respiration of carbon resulting from low rates during the dry period accounts for a significant fraction of the net uptake of carbon in 1995. Evidently, because of the sensitivity of forest respiration to daily weather patterns, accurate description of whole-system respiration requires continuous measurements such as those provided by the EMS, with careful attention to respiration and decomposition.

Changing weather can also increase CO_2 efflux, as in the cases of high winds and warmer soil temperatures. Strong winds, however, may produce increases that are not associated with an actual increase in respiration. During the winter of 1992–93, high winds coincided with intermittent CO_2 efflux totaling 1.6 to 2.0 tons of carbon per hectare, as compared with total annual respiration of 11.4 tons. Soils did not freeze during that winter, and some of the efflux was likely due to the flushing of CO_2 from soil pores by the wind. In other years, smaller increases in winter, spring, and fall respiration have been correlated with anomalously warm soil temperatures such as fall 1993, when a 2°C increase in soil temperature corresponded to an increase in respiration of 200 kilograms carbon per hectare. In winter 1994, higher-than-normal respiration was observed despite colder-than-normal air temperatures as deep snow insulated the soils and kept them much warmer than the air. The winter of 1997–98 illustrates the complexity of weather effects on soil processes. Although respiration rates exceeded 30 kilograms carbon per hectare per day during an unusually warm late fall, respiration rates were lower than average during mid-winter months because of the lack of snow cover and frozen soils (see Figure 10.8).

Interannual variations in NEE and climate indicate that carbon sequestration at the Harvard Forest increases significantly in response to warmer springs, warmer autumn nights, diminished snow pack, and decreased cloud cover. Each of these trends has been observed over northern continents in recent decades, except for cloud cover, which has increased.

Comparison of the Harvard Forest with an Old-Growth Boreal Forest

Further insight into the processes that control carbon storage is gained by comparisons with other quite different forests for which comparable data exist. Since 1994, Harvard Forest researchers from the Department of Earth and Planetary Sciences at Harvard have operated a second eddy-flux tower site near Thompson, Manitoba, as part of the NASA-supported Boreal Ecosystem Atmosphere Study (BOREAS). The Thompson site differs strikingly from the site in central Massachusetts in forest conditions, environment, and carbon dynamics (Figure 10.7). This area of Canada is dominated by old black spruce forest on organic soils 1 to 2 meters deep, overlying poorly drained lake clays that were deposited in glacial Lake Agassiz. The ground cover is a deep carpet of sphagnum and feathermosses.

Overall, carbon fluxes (both uptake and effluxes) in the spruce forest are less than at the Harvard Forest (Figure 10.8). This result is consistent with the colder temperatures, shorter growing season, and nutrient-poor, waterlogged soils at the northern site. Seasonal patterns of ecosystem function at the two sites are also quite different. Surprisingly, despite the cold and long winters, the shift from CO_2 efflux to uptake by the forest in the spring occurs in Manitoba *before* Harvard Forest because the evergreen spruce forest is able to begin photosynthesis as soon as the surface soils have thawed and air temperatures are consistently above freezing. Furthermore, the wet peat soils are slow to warm at depth because of their insulating moss layer and evergreen forest

Figure 10.7. Installation of instruments on the tower of the NASA-supported Boreal Ecosystem Atmosphere Study (BOREAS) old black spruce site, which was established in 1992 at a site 50 kilometers west of Thompson, Manitoba. The vegetation is dominated by black spruce with a ground cover of feathermoss and sphagnum. The site is underlain by peat deposits at least 1 meter deep, with a clay layer below. Air samples are drawn from the top of the 30-meter tower. Photograph by J. W. Munger.

canopy. Consequently, soil respiration rates remain low in the boreal forest until well into the summer.

In contrast, the surface soil at the mostly deciduous Harvard Forest begins to warm under direct sunlight in the spring as soon as the snow cover is gone. On warm spring days before the leaves emerge, the litter and surface soils are actually warmer than they are in the middle of the summer, and respiration rates increase sharply in response. Some herbs and scattered hemlocks and pines do begin to photosynthesize at the Harvard Forest during this early spring period, but the CO_2 uptake by this small amount of vegetation doesn't compensate for the large increase in respiration. Consequently, net CO_2 uptake does not usually begin in deciduous-dominated forests of central Massachusetts until leaves in the canopy emerge in late May or early June.

By early to mid-summer the deciduous vegetation at the Harvard Forest has reached maximum net uptake rates that are three to four times those in Manitoba (Figure 10.8). Smaller net CO_2 uptake by the black

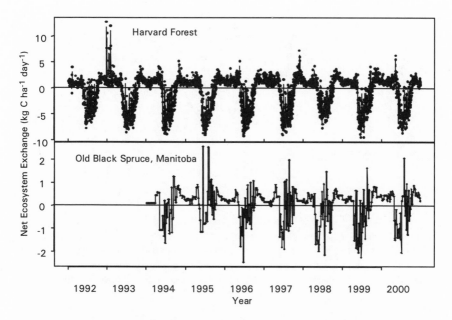

Figure 10.8. Mean daily net ecosystem exchange (NEE) of carbon for the Harvard Forest and the BOREAS old black spruce site in Manitoba. Note different vertical scales. Although the northern site is colder, it begins net carbon uptake earlier in the spring because the conifers are able to photosynthesize as soon as the surface soils thaw. Declining NEE during mid-summer at the BOREAS site is a consequence of the slow onset of increased soil respiration as the deeper peat layers warm. At the Harvard Forest the deciduous vegetation is able to take advantage of the warm summer temperatures for photosynthesis, and there is no reservoir of readily decomposable organic matter to be respired.

spruce forest results from the lower photosynthetic efficiency of the conifer foliage and very high soil respiration rates that develop when the deep peat layer has warmed up in mid-summer. Large oscillations between uptake and efflux in Manitoba result from the approximate balance between relatively large rates for GEE and R. When GEE is reduced slightly by a period of cool cloudy or hot dry weather, the balance shifts to net CO_2 efflux. In fact, in some years the boreal forest ceases to take up carbon at about the same time the deciduous forest at the Harvard Forest is achieving its maximum uptake rates.

The cumulative carbon uptake at the Harvard Forest exhibits a steady sawtooth pattern of large wintertime efflux and generally larger summertime uptake (Figure 10.9). Perturbations in the rates of efflux and uptake within and between years are largely due to climatic factors. Overall, the forest took up nearly 18 metric tons of carbon between 1992 and 2000, which is consistent with our expectations for a healthy middle-aged forest stand. Measurements of tree growth and accumulated

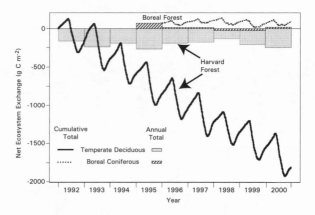

Figure 10.9. Annual trajectory and sum of carbon dynamics at the Harvard Forest and boreal forest site. The annual sums for net ecosystem exchange are shown as bars along the top of the graph, whereas lines show the trajectory of carbon storage or release. The second-growth temperate forests at the Harvard Forest are rapidly accumulating carbon, whereas the old boreal forest is nearly at equilibrium.

woody detritus corroborate this assessment of the average net carbon balance at the Harvard Forest. In contrast, carbon fluxes at the boreal site have been nearly in balance over the period of observation. Small shifts in the carbon exchange rates and annual net balance depend on the interplay between temperature and moisture at this site. Historical and ecological differences between the two stands are also key determinants of their carbon dynamics. The Harvard Forest stand, like much of New England, is relatively young and still recovering and growing after agricultural clearing, logging, and the 1938 hurricane. The old black spruce forest stand has not experienced disturbance by wildfire in approximately 150 years; consequently, a thick organic soil-and-moss layer retains many nutrients and insulates the soil, reducing productivity. Slow tree growth and accumulation of carbon in surface peat are balanced by decomposition of carbon in deep peat layers.

Chapter 19 will discuss the implications of these observations in a broader regional and historical context. We will see that many environmental and historical factors regulate the carbon cycle at the Harvard Forest. These factors can be divided into categories, from those that act on the short term (days, hours; for example, temperature and sunlight) to those that control the forest on timescales of decades or longer (for example, succession and soil organic matter). To understand the Harvard Forest in a global context, we have to understand *quantitatively* how these factors interact, a difficult scientific, intellectual, and practical challenge.

Exchanges of Nitrogen Oxides, Ozone, and
Reactive Hydrocarbons

FORMATION OF NO_x, NO_y, AND OZONE IN THE ATMOSPHERE

Fossil fuel combustion emits nitric oxide (NO), which rapidly converts in the atmosphere to nitrogen dioxide (NO_2) and nitrate radical (NO_3). This group of nitrogen compounds, which collectively are denoted NO_x, reacts in the atmosphere on a timescale of hours to days to form nitric acid (HNO_3) and nitrate (NO_3^-) aerosol, which fall on terrestrial ecosystems as precipitation or dry deposition. Inputs of human-produced nitrogen compounds augment the natural deposition of nitrogen. Some NO_x reacts with hydrocarbons to form peroxyacetyl nitrate (PAN) and other organic nitrates. Since nitrogen is a limiting nutrient and these anthropogenic inputs from the atmosphere can be large relative to natural levels, the deposition of these nitrogen compounds is an important ecological process. At the EMS we measure directly the input of all of these nitrogen compounds using a sensor that reduces them collectively to NO on a gold catalyst and then quantifies this total, which is denoted as NO_y.

During the oxidation of reactive hydrocarbons in the atmosphere, NO_x radicals catalyze the formation of ozone (O_3), the irritant in smog. Ozone reacts with many materials, especially cell membranes. It damages the photosynthetic apparatus of plants and is harmful to human breathing ability. Reactive hydrocarbons are naturally emitted from vegetation during the growing season (see the following discussion) and are usually abundant in the rural atmosphere throughout the summer. Consequently, NO_x is often the limiting factor for O_3 production. The modern increase in surface O_3 levels and heavily polluted and smoggy air is therefore a direct result of the expansion of fossil fuel combustion by industry and especially automobiles. Together, NO_x, NO_y, and their byproduct O_3 represent the most abundant air pollutants that affect vegetation.

DAILY CYCLES OF CONCENTRATION AND DEPOSITION

Typical daily patterns observed for NO_x, NO_y, and O_3 concentrations in the atmosphere at the Harvard Forest are shown in Figure 10.10. Throughout the year, concentrations of NO_x and NO_y are at least twice as high when winds are from the southwest as opposed to the northwest and north, because of the large urban source areas of pollution along the East Coast. The pattern is different for O_3, which is a secondary pollutant produced in the atmosphere from other reactants. In winter months, O_3 concentrations are lower in polluted southwestern air because pollutants initially consume O_3, and, under cold temperatures and reduced light, O_3 production proceeds slowly. There is not enough time for abundant O_3 to be generated before the air masses reach central Massachusetts. In summer, however, reactions speed up, and polluted air from the southwest is markedly enriched in O_3 compared with the air associated with northwesterly winds, which has been depleted by deposition and through dilution with cleaner air.

Concentrations of NO_x and NO_y increase at night as pollutants from regional sources are trapped in the stable air near the ground. After sunrise, solar heating drives the convective mixing of air throughout the lower atmosphere above the forest, and NO_x and NO_y concentrations decrease as the surface layer of air is diluted with cleaner air from above. Ozone concentrations at the surface vary in the opposite sense; they are highest in daytime, since sunlight drives production, and low at night when O_3 is destroyed by reaction with the vegetation and soil. Stronger daily cycles of reactive trace gas concentrations are observed in the summer than in winter. Contributing factors are the presence of a leaf canopy to react with the pollutants and the deeper planetary boundary layer for the mixing of air in the atmosphere during summer.

Fluxes of pollutants to the forest depend on atmospheric concentrations, the rate of vertical mixing of air above and into the forest canopy, the reactivity of the gas, and characteristics of the canopy and plant surface affected. The typical daily cycles for reactive nitrogen and O_3 deposition to the forest canopy show generally higher fluxes in the daytime (Figure 10.10, second panel). Enhanced deposition rates and higher ambient concentrations of NO_y are associated with winds from the southwest rather than the northwest in both summer and winter. Fluxes of NO_y during the winter are independent of time of day. If NO_y deposition were limited by vertical exchange, the maximum flux would correspond to the midday peak in friction velocity. Similarly, if photochemical production of HNO_3 limited NO_y deposition, the fluxes would track solar radiation. Instead, the eddy fluxes of NO_y increase sharply after sunrise to a maximum before noon and then decline during the afternoon to low nighttime values. Peak NO_y fluxes precede both the maximum in friction velocity and photochemical activity. This pattern represents evidence for the conversion of NO_x to HNO_3 at night on the surfaces of small particles in the atmosphere. This process produces about 20 percent of atmospheric HNO_3 in summer and is by far the dominant oxidation process occurring in the lower atmosphere during the winter months.

Eddy fluxes of O_3 during winter at the Harvard Forest have a distinct midday maximum, which is independent of wind direction and coincident with maximum vertical transport into the canopy. In summer, O_3 fluxes are identical for both clean and polluted wind sectors, despite the 50 percent difference in O_3 concentration (Figure 10.10). Peak fluxes generally occur near or just before noon, coinciding with peak solar irradiation (PPFD), and there is a strong relationship between O_3 flux and both radiation and canopy conductance (uptake of gases by vegetation). These results are consistent with findings from other ecosystems that O_3

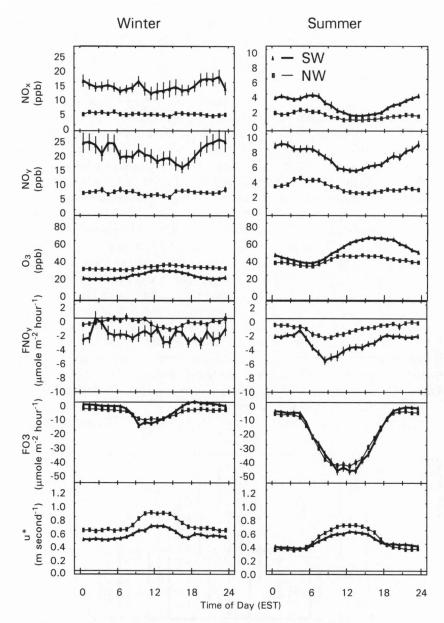

Figure 10.10. Hourly variations of concentrations for nitrogen oxide radicals (NO_x = NO, NO_2, and NO_3) (top panel), total oxidized nitrogen (NO_y = NO_x, peroxyacetyl nitrate, nitric acid, and other nonradicals) (second panel), and ozone (O_3, third panel) are shown for winter (left) and summer (right) as a function of prevailing wind direction. Thick lines indicate southwest (SW) winds coming over urbanized and industrial areas, whereas clean northwestern (NW) winds are shown as thin lines. Note that the y-axis scales are larger in winter for NO_x and NO_y concentrations. Hourly variations of fluxes of total oxidized nitrogen (FNO_y, fourth panel), ozone (FO_3, fifth panel), and momentum (sixth panel; u* equals the square root of negative of horizontal and wind covariance) are also shown. Modified from Munger et al. 1996, 12649, 12651, with permission of the American Geophysical Union (copyright 1996, American Geophysical Union).

is actively taken up by vegetation, along with the CO_2 necessary for photosynthesis, through stomates, pores in leaf surfaces controlled by plants.

Annual Cycles and Fluxes

The seasonal cycle of NO_y concentration paralleled that of NO_x: highest in winter and lowest in summer (Figure 10.10). In contrast, concentrations of the secondary pollutant O_3 peaked in the summer (Figure 10.10), when the twenty-four-hour mean concentrations were twice those in winter. More important for vegetation health, the extrema (for example, 90th percentile O_3 concentrations between 10 A.M. and 6 P.M.) increased even more sharply, exceeding 70 parts per billion (the threshold for damage to plants) ten to twenty times during the growing season in most years (see also models of O_3 effects in Chapter 17). Ozone levels are higher still in the adjoining Connecticut River Valley 30 kilometers to the west, where pollution plumes from the south are entrained and trapped.

Ozone damage to plants is related to uptake into leaves rather than to ambient concentration. Uptake is a major component of total deposition rates. For example, average daily deposition rates (Figure 10.11) are fairly constant from November through March at doses less than 200 micromoles per square meter per day and increase as the understory foliage emerges and coniferous vegetation begins to photosynthesize in April. Rates continue to rise through the spring in parallel with increasing foliar activity and hours of daylight to a maximum during June of 500 micromoles per square meter per day.

However, despite measuring high concentrations of O_3, we cannot de-

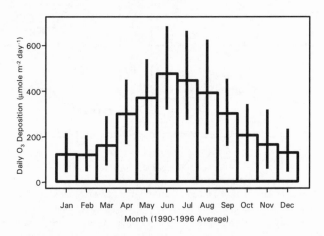

Figure 10.11. Median daily deposition flux of O_3 at the Harvard Forest, showing that the maximum pollutant inputs occur during canopy development during the spring and the growing season. A vertical segment at each point indicates the spread between the 25th and 75th percentile of the data.

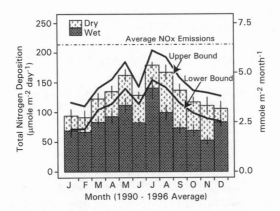

Figure 10.12. Monthly mean values of reactive nitrogen at the Harvard Forest from dry deposition (light shading) and precipitation (dark shading). Sampling uncertainty for each monthly estimate is indicated by thin vertical segments. The combined uncertainties of the total deposition are shown by solid lines. The average NO_x emissions within 250 kilometers of the Harvard Forest (from emissions inventories of the Environmental Protection Agency) are shown by the horizontal dashed line. Reproduced from Munger et al. 1998, 8359, with permission of the American Geophysical Union (copyright 1998, American Geophysical Union).

tect a negative effect of O_3 concentrations on photosynthesis, indicating that any such influence must be modest. Since O_3 concentrations and fluxes are correlated with wind direction and sunlight, it is difficult to discern various contributions to the observed variance of CO_2 fixation that occurs with different weather patterns. Consequently, we cannot document for certain that O_3 pollution is affecting photosynthesis or canopy development (see, again, Chapter 17). Furthermore, ozone damage to plants tends to be cumulative, and reduction in CO_2 uptake might lag behind the O_3 concentrations or depend on the cumulative sum of O_3.

The observations at the Harvard Forest illustrate the patterns and processes that control nitrogen deposition across New England. Total inputs of reactive nitrogen amount to 6.7 kilograms of nitrogen per hectare per year (Figure 10.12), which is small relative to the total pool of fixed nitrogen in the forest or to the net mineralization rate (80 to 100 kilograms nitrogen per hectare per year). The summertime input of reactive nitrogen by wet and dry deposition is about twice the wintertime rate and is comparable to the regional mean NO_x emission rate. Oxidation of NO_x to HNO_3 or other depositing (organic nitrogen) compounds is the principal factor determining the rate for removal of reactive nitrogen from the atmosphere. Wet or dry weather affect the relative contribution between wet and dry deposition, but not the total deposition.

Oxidation of NO_x to HNO_3 by O_3 occurs through a complex set of reactions in which the final step can be catalyzed by particles in the atmosphere. This happens mainly in winter and on summer nights because the intermediates are rapidly photolyzed and destroyed during daylight hours. Both the hydroxyl radical (OH) and O_3 were important oxidants of NO_x throughout the year. The enhanced rate of NO_y deposition in the morning hours observed at the Harvard Forest during the summer provided strong evidence for the importance of this oxidation pathway. Also, we found that significant reactions of NO_x with the products of oxidation of biogenic hydrocarbons (isoprene and terpenes), forming depositing species, were needed to account for observed nitrogen deposition.

From the observations at the Harvard Forest we can derive the mean lifetimes for oxidation of NO_x and deposition of the products over the year. We estimate that 45 percent of anthropogenic NO_x in the boundary layer of the northeastern United States is removed in 1 day during summer but that the rate drops to 27 percent in winter. Removal of 95 percent took 3.5 and 5 days in summer and winter, respectively. Hence, in order to be transported to remote regions, NO_x must either be pumped from the boundary layer or be converted to more stable species (for example, PAN) and then be pumped to the upper troposphere, where stability is increased by the cold ambient temperatures. These results indicate that regions more than 1 day downwind from a major emissions region could receive more nitrogen deposition in winter than summer, possibly affecting acidic runoff from snowmelt.

Forest Emissions of Reactive Hydrocarbons

Forests not only receive reactive compounds from the atmosphere, but also act as important sources for hydrocarbons like isoprene and terpene. This is an important process and consideration for the Harvard Forest and much of southern New England because the present vegetation assemblage is dominated by oak (*Quercus* spp.), which is a strong emitter of isoprene. Measurement of isoprene fluxes for an entire summer revealed that temperature, light, and growth stage of the canopy control emissions. Young leaves did not emit isoprene for the first two weeks after emergence and did not reach their maximum emission rate until the fourth week. After normalizing for the influence of temperature and light, we see that the basal isoprene emission rate remained nearly constant for the height of the growing season in July and August and then decreased steadily through September and October when emissions ceased (Figure 10.13). The seasonal changes in isoprene emission rates significantly affect rates of regional O_3 production because isoprene is a natural catalyst for O_3. Isoprene emissions also influence the fate of NO_x because of the reactions with intermediates in the isoprene oxidation path. Although isoprene is the dominant light hydrocarbon species emitted by this forest, we also have documented small but significant production of ethylene, propene, and 1-butene. Al-

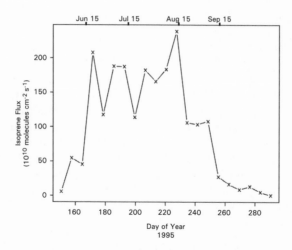

Figure 10.13. Seasonal pattern of measured isoprene flux based on weekly means of fluxes measured between the hours of 10 A.M. and 3 P.M. EST. Isoprene is released by oak leaves and is an important precursor of ground-level ozone. Modified from Goldstein et al. 1998, 31054, with permission of the American Geophysical Union (copyright 1998, American Geophysical Union).

though we have no data for terpene concentrations or emissions, we expect them to be emitted by conifers as well. The large variations in compounds emitted and the emission strengths among tree species suggest an important mechanism by which vegetation change may affect the atmosphere.

Effects of Exchanges on Forests and the Atmosphere

Long-term observations show that net fluxes of oxides of nitrogen and O_3 to stands at the Harvard Forest are generally moderate. Inputs of nitrogen compounds represent about 8 to 10 percent of the overall mineralization rate of nitrogen for this ecosystem. Ozone fluxes and concentrations at the Harvard Forest occasionally reach levels that are known to be associated with acute vegetation damage, but to date, any effect on forest photosynthesis is undetectable.

In contrast, the influence of the forest on the atmosphere may be more significant. The forest removes considerably more O_3 than would be taken up by an ecosystem dominated by low-stature vegetation or by the artificial surfaces that predominate in urban areas. This uptake depletes O_3 concentrations in the planetary boundary layer, thereby reducing pollutant exposures downwind. Deposition is probably less important as a factor in regulating ambient concentrations of NO_x and NO_y, but emissions of reactive hydrocarbons by the forest appear to be quite important as a source for reactivity and as a sink for NO_x through formation of organic nitrates.

The observational perspective and unique data sets provided by the Harvard Forest EMS materially aid the analysis of regional air pollution and improve understanding of the transport of pollutants to the global environment. The long-term nature of the record has been particularly important in this regard. We have learned that NO_x is rather short-lived in the lower atmosphere in both summer and winter. To determine the quantities of reactive species exported to the global environment, we evidently must focus attention on the critical importance of the formation of long-lived, nondepositing species. We have advanced our ability to quantify the effects of forest vegetation on the atmosphere and the influence of pollutants on the forest.

IV

Understanding Forest Ecosystem Dynamics through Long-Term Experiments

It is clear from our review of landscape history over past millennia that temperate forests are dynamic ecosystems that have been shaped by climate change, natural disturbance, and human activity. This retrospective view has also underscored the remarkable resiliency of these forests to a wide range of physical disturbances, including wind damage, fire, pathogens, and intensive land use. However, the forests of New England are now experiencing a suite of novel changes that are the indirect rather than direct consequences of human activity. This raises a new set of questions. In particular, we need to know whether stresses resulting from changes in the Earth's atmosphere, including ozone events, increased nitrogen and sulfur deposition, and increases in global temperature, will generate fundamental shifts in forest ecosystem processes or forest structure and composition. Collectively, ecologists, land managers, policy makers, and the public are confronted with the fundamental question of whether our forests are as resilient in the face of (or resistant to) these chemical and climatic stresses that are arising at a rapid rate as they are to the types of physical disturbances that they have experienced in the past.

To help answer these questions, we have developed a set of large-scale, long-term experiments that begin to compare ecosystem responses to historical disturbances with those initiated by novel stresses. The suite of experiments devised complements our historical and modern process studies well. A knowledge of historical disturbance patterns is used to identify experiments that will mimic those events. Measurement of similar processes across experiments allows the direct comparison of ecosystem response to historic and novel stresses.

The disturbances and stresses that we selected for investigation include the following:

- Hurricane blowdown, an infrequent, catastrophic disturbance to New England forests
- Canopy gap blowdowns, the most prevalent form of disturbance in this forest type
- Chronic additions of nitrogen, a novel stress resulting from fossil fuel combustion
- Soil warming, an important component of projected global climate change
- Long-term changes in litter inputs, which could result from any of the above

Assessing Ecosystem Response

To compare forest response to these factors, we developed an ambitious set of experimental manipulations with a core set of ecosystem measurements made on each manipulated area and on adjacent control areas. To develop meaningful comparisons, we designed experimental treatments that simulated closely the characteristics of hurricane damage, nitrogen deposition, soil warming, and gap formation in typical second-growth forests with similar vegetation and site conditions. Each experiment was necessarily designed to be long term in order to allow for the often gradual or lagged response of soils and vegetation to impacts and for the cumulative effects of chronic stress.

The measurements we used assess important ecosystem processes and characteristics, including

- Fluxes of nutrients, other chemicals, moisture, and energy
- Ecophysiological performance
- Population dynamics
- Vegetation structure and composition
- Ecosystem productivity

We particularly emphasized nitrogen cycling and the exchange of gases such as methane, CO_2, and nitrous oxide between forest soils and the atmosphere. These fluxes play an important role in regulating the Earth's climate and also integrate across several important ecosystem processes. The value of long-term ecological research is borne out by the approach taken, as initial, short-term responses are often different in size, or even in direction, from long-term effects.

CHAPTER 11

Simulating a Catastrophic Hurricane

D. FOSTER, S. COOPER-ELLIS, A. BARKER PLOTKIN, G. CARLTON, R. BOWDEN, A. MAGILL, and J. ABER

Background and Major Questions

In 1635, fifteen years after the establishment of Plymouth Plantation on the eastern shore of Massachusetts, Governor William Bradford awoke one August morning to witness the arrival of one of the great hurricanes in New England history: "It began in the morning a little before day, and grew not by degrees but came with violence in the beginning, to the great amazement of many. It blew down sundry houses and uncovered others. . . . It caused the sea to swell to the south of this place above 20 foot right up and down, and made many of the Indians to climb into trees for their safety. It blew down many hundred thousands of trees, turning up the stronger by the roots and breaking the higher pine trees off in the middle. And the tall young oaks and the walnut trees of good bigness were wound like a withe, very strange and fearful to behold."

Since the day that this awe-inspiring storm struck, New Englanders have been occasionally reminded of the impacts that tropical storms can exert on their landscape. Through time this awareness has led to many questions concerning the importance of hurricanes in controlling the structure, composition, and function of forest ecosystems across the region. As our understanding of the historical frequency and distribution of these storms has improved through our reconstructions and modeling, we have also sought to understand the patterns of forest damage from wind disturbance, the rate of recovery of the vegetation and ecosystem function, and the differences between hurricane damage and the effects of other types of natural and human disturbances. This interest in the ecological role of hurricanes has been paralleled by a desire to apply this information to the conservation and management of forests and other ecosystems across New England.

Interpretations of the effects of intense wind disturbance on the northeastern United States have been strongly shaped by observations and experiences related to the Great Hurricane of 1938, the most destructive storm to hit the region in the past 175 years. This storm struck Long Island, New York, on September 21 and then passed northward across central Connecticut and Massachusetts to the northwest corner of Vermont, bringing wind speeds exceeding 130 miles per hour as well as 6 to 14 inches of rain. Forest damage, which included the destruction of more than 4 billion board feet of timber along a 100-mile-wide path, was most intense east of the storm track where the forward momentum and rotational velocity of the storm coincided (see Figure 1.4). In the aftermath of the hurricane, the federal government organized a regional timber program that salvaged more than 1.5 billion board feet of lumber. This salvage operation sought to reduce the immediate threat of wildfire and recoup lost investment, but it also had the indirect consequence of promoting extensive road construction, soil scarification, and burning of logging slash.

At the Harvard Forest, research on wind disturbance evolved around the effect of this storm. Even before 1938, R. T. Fisher and his students had recognized that natural disturbance by wind, as well as fire, ice, snow, and pathogens, was important to the maintenance of a shifting mosaic of vegetation and the retention of early-successional species in the landscape. However, the unanticipated severity of the 1938 storm and its catastrophic effect on forests, research experiments, and regional economic prospects promoted a rethinking of forest-management approaches and ecological interpretations across the northeastern United States (Figures 11.1 and 11.2).

Surveys conducted immediately after the hurricane concentrated on such practical issues as efficient salvage methods and fire prevention, but research attention gradually turned toward seeking silvicultural and ecological insights from the widespread pattern of uprooted and broken trees. Three questions loomed above all others: Was there a predictable return interval for these storms, and if so, how did this vary across New England? Did strong relationships exist between forest composition and age and susceptibility to wind damage? What were the patterns of forest recovery and the long-term effects on forest conditions? Answers to all three questions were deemed critical if ecologists were to understand the spatial and temporal variations in vegetation across New England and if foresters were to incorporate an understanding of tropical storms into long-term forest planning. While a graduate student (and future leader of temperate forest silviculture) named David (D. M.) Smith addressed the first question in a classic master's thesis at Yale University (1946), his counterpart at Harvard, Willett Rowlands, clambered through the tangle of windthrown trees on the Harvard Forest to address the second (1941), and a succession of graduate students and researchers at Harvard tackled the last.

Although Smith's work was never published beyond the original thesis, it is a beautiful piece of regional reconstructive research. In fact, his

Figure 11.1. The old-growth forest on the Harvard Forest Pisgah tract in southwestern New Hampshire before (top) and after (bottom) the 1938 hurricane. Photographs from the Harvard Forest Archives.

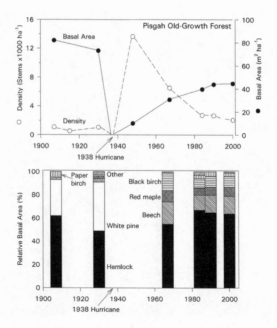

Figure 11.2. Changes in forest structure (top) and composition (bottom) at the Pisgah Forest from the early 1900s to 2000, showing the forest response to the 1938 hurricane. The original old-growth forest was composed of widely spaced but massive white pine and hemlock, which were nearly all blown down by the storm. A young and dense stand of hemlock, beech, red maple, and birch has been undergoing a process of thinning in density and gradual increase in basal area over the past sixty-five years. White pine was essentially eliminated from the stand by the windstorm. Updated from Foster 1988b and Foster, Orwig, and McLachlan 1996 by A. Barker Plotkin (Harvard Forest Archives).

study presages our hurricane modeling work by using a knowledge of historical sources and meteorology to interpret all storms since 1635. Smith's study clearly defined the south-to-north regional gradient in hurricane return interval and average intensity. Meanwhile, Rowlands looked at sites across Petersham that were topographically exposed to the strong winds from the hurricane and documented the type and extent of damage to trees in forests of different age and composition. He discovered that the even-aged white pine that were abundant on old-field sites were much less windfirm than the hardwoods. For all species, he found that wind damage increased linearly with tree height and that open-grown trees survived better than those growing in densely stocked stands. Rowlands also concluded that even-aged stands were more wind-resistant than mixed-aged stands as a consequence of lower turbulence in the canopy. Furthermore, he suggested that soil moisture was correlated with a tree's proneness to uprooting. As moisture increased, rooting strength declined, and the trees toppled over more readily. Rowlands's results had immediate implications for forest managers concerned with producing windfirm stands, and they provided new insights into the natural dynamics of forests in response to wind.

The extensive and preferential damage to white pine forests gave impetus to additional studies and interpretations of forest dynamics. The emphasis of most early New England forest management, including much of the research at the Harvard Forest, was aimed at the sustained production of white pine. By 1938 it was evident that most of the mature white pine forests that uprooted were the product of natural succession on abandoned agricultural fields. What would replace the blown-down pine forests? This important ecological and silvicultural question was addressed by Ralph Brake and Howard Post, graduate students at the Harvard Forest, who assessed patterns of natural regeneration on sites with different soils, contrasting amounts of prehurricane pine and hardwoods and varying intensities of posthurricane logging. They found that despite heavy damage to hardwood seedlings and saplings by the hurricane and subsequent salvage logging, the windthrown areas were often densely filled with rapidly growing hardwoods. The density, species composition, and type of tree regeneration varied by soil type. For example, red maple, red oak, and various birches predominated on mesic sites, whereas gray birch and white pine were more common on sandier soils. Nonetheless, despite this excellent work many questions remained concerning explicit connections between local site conditions and revegetation, the long-term changes that would occur in the forests and their environment in the years after a storm, and the influence of windstorms on the natural, hardwood forests.

One critical element was also largely missing from these early studies of the 1938 hurricane: investigations of disturbance impacts on fundamental ecosystem processes such as carbon accumulation, hydrology, and nutrient dynamics. Some related issues were partially addressed by Jim Patric's ingenious investigation of long-term response of river flow to the 1938 hurricane (Figure 11.3). Patric, a soil scientist and Bullard Fellow at the Harvard Forest in the 1960s, reasoned that the broadscale reduction in forest canopy caused by the storm might produce a regional change in river discharge. By comparing long-term flow records from the Connecticut River, whose watershed was intensively damaged by the storm, with those from the Androscoggin River, where little damage occurred, he was able to demonstrate that a substantial increase in river flow was sustained for five years after the 1938 hurricane. Patric concluded that interception of precipitation by the vegetation, evaporation from leaf surfaces, and transpiration of moisture from the forest canopy were all reduced when the storm abruptly altered the forest cover. With less moisture returning to the atmosphere, more was delivered to groundwater and small tributaries and eventually into the larger river. The cumulative effect on regional hydrology of changes in forest struc-

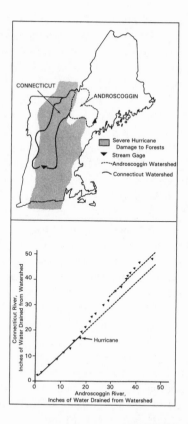

Figure 11.3. Analysis by Jim Patric (1974) of changes in major river flow as a long-term consequence of the 1938 hurricane. Top: Distribution of severe damage in relation to New England forest watersheds. Bottom: Cumulative summer flow (July to September) for the Androscoggin and Connecticut Rivers. After the storm there was a substantial increase in flow from the damaged Connecticut River watershed that persisted for nearly five years. The Androscoggin watershed was undamaged. Reprinted from Foster and Boose 1995, with the permission of Cambridge University Press.

ture across this enormous watershed was a major consequence of the hurricane (Table 11.1). Patric likened the phenomenon to the increased runoff measured at watersheds like Coweeta and Hubbard Brook after experimental logging, and he described the hurricane as a "forest treatment of regional proportions." Patric's findings on stream flow, observations of increased nutrient export associated with experimental cuttings and herbicide treatments at Hubbard Brook, and documentation of increased fluxes of trace gases after Hurricane Hugo in Puerto Rico have raised many new questions about the effects of hurricanes on hydrology, soil chemistry, and ecosystem dynamics. All of these results and emerging questions motivated the Harvard Forest hurricane experiment.

Table 11.1. Increased River Flow and Decreased Evaporative Loss from Hurricane-Damaged Watersheds in Comparison to the Undamaged Watershed of the Androscoggin River (inches per hydrologic year)

Year	Flow Increase		Evaporative Decrease	
	Connecticut	Merrimack	Connecticut	Merrimack
1939	+4.8*	+5.1	−5.8	−5.0
1940	+2.4	+2.9	−1.6	+1.6
1941	+2.5	+3.8	−3.4	−4.2

Source: Patric 1974, copyright 1974 by the Society of American Foresters, reproduced with permission of the Society of American Foresters in the format Other Book via the Copyright Clearance Center, Inc.
*Significant at the .05 level of probability.

Rationale for the Hurricane Experiment

Few disturbances have the potential to rearrange forest biomass on a broad scale as abruptly or as dramatically as severe windstorms. Unfortunately, our understanding of the ecological effects of these events has been based almost exclusively on post-hoc assessments of storm impacts on sites where little baseline data have been available. With the exception of Hurricanes Hugo and Mitch in Puerto Rico, where the existence of the Luquillo LTER project enabled a range of continuous measurements, the lack of baseline information has precluded good integrated studies of forest response to hurricanes. Consequently, to investigate vegetation and ecosystem response to severe wind disturbance in a controlled setting, we designed an experimental hurricane as a core part of our comparative study of ecosystem dynamics. The hurricane experiment, like our other manipulations and long-term measurements, was situated in the most prevalent forest type in our landscape, a mature red oak–red maple stand. To maximize the realism of the manipulation, we based the damage design on empirical studies of the 1938 hurricane, especially those by Willett Rowlands, Howard Post, and Ralph Brake, and we initiated the experiment during the height of hurricane season.

Experimental Design

The hurricane experiment was designed to examine both forest responses to and recovery from major wind damage in a comprehensive way. We sought to (1) identify changes in important plant resources such as light, moisture, and nutrients resulting from changes in forest structure; (2) document patterns of species damage, repair, regeneration, and growth; and (3) explore the relationships among vegetation dynamics, environmental conditions, and ecosystem processes. To quantify changes accurately, our design allowed for extensive measurement before the manipulation in both the disturbed area and in a nearby control area.

Measurements were made to further our understanding of the relationship between wind disturbance and ecosystem function, with an emphasis on such critical environmental processes as the movement of materials among the biota, the soil, and the atmosphere. Included were measurements of soil environment conditions such as moisture and temperature that can affect decomposition, respiration, and other belowground processes; changes in soil nutrient availability and cycling, with an emphasis on nitrogen; and concentrations and fluxes of the trace gases nitrous oxide, CO_2, and methane, which are important greenhouse gases. The experiment also provided an opportunity for us to examine vegetation responses, including the type and distribution of damage and patterns of mortality among tree species; changes in the structure of the canopy and forest floor; the importance of different modes of tree regeneration (for example, seedlings versus saplings or sprouts) in controlling species composition and abundance; and changes in understory composition as a result of damage to the overstory and forest floor structure. Finally, in addition to detecting changes in ecosystem functioning across the area, we were also interested in quantifying environmental resource conditions such as light and soil moisture at the scale of the microsites created by the disturbance. This we sought in order to contrast the patterns of resource availability in the experimental study area with those in the undisturbed forest understory and to relate these to plant performance.

Mimicking a Hurricane

The seventy-five-year-old red oak–red maple forest selected for the experiment is typical of the most common forest type in central New England. In addition to oak and maple, the relatively homogeneous stand includes white pine; white ash; black, yellow, and paper birch; hickory; and other species. The site occupies a gentle northwest slope, and soils are moderately well-drained with a discontinuous hardpan. We surveyed the rectangular experimental area, which is roughly the size of a soccer field (0.8 hectares), and the nearby control area and mapped and tagged all trees. We established plots on transects across the experimental and control areas to monitor regeneration and understory vegetation and made baseline measurements of the forest environment, soils, and ecosystem processes.

In designing the experiment, we sought to mimic damage from the 1938 storm, when approximately 80 percent of canopy hardwood trees and all mature white pine on exposed sites were damaged. We used data collected by Rowlands to determine the proportion of trees to pull down, based on the size and type (conifer or hardwood) of tree. Using observations from 1938 and some preliminary attempts by our woods crew to simulate the damage from a hurricane, we determined that

Figure 11.4. The edge of the experimental "blowdown" in the second year after the manipulation. Most of the trees uprooted, producing large mounds, and many of these produced thick crowns of leaves the following summer. Photograph by M. Fluet.

pulled trees account for fewer than 40 percent of total trees damaged and that the majority of broken and uprooted trees are actually damaged indirectly as pulled trees fall against other, generally smaller, trees in their path. In early October 1990, we pulled more than 235 canopy trees toward the northwest using a cable attached to a large winch on a logging skidder located outside the study area. This method simulated the lateral force of wind against the crown of a tree and allowed the structure of the tree to determine whether the actual damage would be by uprooting or breakage. By keeping heavy equipment off the site and by generating a range of direct and indirect damage to plants and soil, this treatment effectively reproduced many of the natural consequences of a hurricane (Figure 11.4). We made no effort to pull trees beyond their initial point of repose if they were hung up on other trees, and we left all trees, including those indirectly damaged by falling trees, where they fell.

*Assessing the Consequences and Limitations
of the Manipulation*

As expected, the manipulation produced dramatic changes in forest structure, including a redistribution of biomass from a canopy nearly 30 meters tall to a thick tangle of subparallel and prostrate boles and branches resting a few meters above the forest floor (Figure 11.5). Nearly 70 percent of all trees were damaged, and the basal area and density of standing trees declined by more than 50 percent. Interestingly, the proportions of uprooted and snapped trees (approximately 70 to 30) closely approximated the damage produced by the 1938 hurricane and documented by Rowlands. A range of new tree and ground-surface structures was produced by the manipulation, including bent and broken understory trees and a heterogeneous array of forest floor microsites. Although more than two-thirds of all trees were uprooted, the majority of these came to rest in the crowns of other trees or on mounds. Consequently, the damaged trees lay at a wide range of angles with respect to the ground surface. Uprooted boles covered about 13 percent of the ground surface, and disturbed soil from the resulting complexes of pits and mounds covered about 8 percent.

The experimental area appeared to produce initial consequences that were strikingly similar to those from the 1938 hurricane. However, there are differences between our manipulation and a natural storm that may be important to consider as we follow the experiment through time. Hurricanes are often preceded or accompanied by heavy rains that, depending on timing and amount, may affect the response of the soil and vegetation to intense winds. It was not feasible to simulate this effect, although we were able to conduct the manipulation during the first week

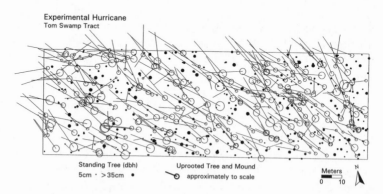

Figure 11.5. The experimental hurricane study area showing standing trees (solid circles; dbh, diameter at breast height), uprooted trees (open circles), and the length and orientation of downed trees. Mound sizes and downed tree lengths approximate their true size. Modified from Cooper-Ellis et al. 1999, with permission from the Ecological Society of America.

of a rainy October hurricane season. In addition, we did not replicate the effect that high winds have in stripping foliage and small branches from remaining trees and producing a wide range of damage to tree crowns. The manipulation, however, took place within two weeks of seasonal leaf-fall, and most of the surviving trees did sustain some damage to their branches as surrounding trees fell. Otherwise, the overall pattern of damage, including the percentage of damaged trees, the distribution of damage among size classes, and the orientation and distribution of boles and woody debris, matched the effects of the historical hurricane. Finally, the size of the resulting gap greatly exceeded the single-tree gaps of many simulations, although it was smaller than many damage patches created by hurricanes. Indeed, the sharp transition between the edge of the experimental area and the surrounding forest is more like the pattern produced by a tornado. Despite these differences, the large extent of the experimental area allowed us to measure most of the important attributes of a posthurricane environment.

Vegetation Response

We were interested in documenting the effect of the manipulation on trees and understory plants. The extent and type of damage were of immediate interest, as these clearly set the trajectory for long-term changes in forest structure (Figure 11.6). Understanding the variation in damage by species and size would also help predict the initial impacts of hurricanes in other areas with similar species composition. In addition, the initial and long-term rates of mortality among injured trees could have important consequences for ecosystem processes. Finally, the capacity of various plant species to respond opportunistically to altered conditions created by the disturbance helps to determine both the rate of stabilization of environmental conditions and the degree of change in composition of the postdisturbance community.

The majority of the largest trees in the predisturbance forests were red oaks up to 60 centimeters in diameter, whereas red maples were typically 5 to 10 centimeters in diameter. In general, the proportion of trees damaged increased with size up to 15 centimeters across and then decreased above 15 centimeters. Smaller trees tended to bend, larger trees uprooted, and trees that were previously standing but dead or decaying mostly snapped. Nearly 95 percent of the damaged red oak uprooted, and damage to red maple was more evenly distributed between uprooting, snapping, bending, and leaning (Figure 11.6). This variation in damage type by species suggests that differential survival of the different sizes and species of trees will be important in determining long-range compositional changes for the stand.

Despite the dramatic reorganization of forest structure caused by the manipulation, the vegetation exhibited a remarkable capacity for sur-

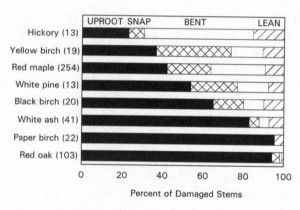

Figure 11.6. Distribution of damage types among major tree species in the experimental hurricane for all damaged stems greater than 5 centimeters in diameter at breast height. Numbers in parentheses indicate tree sample sizes. Modified from Cooper-Ellis et al. 1999, with permission from the Ecological Society of America.

vival and growth under the altered conditions. In fact, the capacity of injured hardwoods to survive for one to many years despite intense damage and then to generate sprouts from the roots, root crown, or trunk was a major factor in maintaining vegetation and canopy continuity through time. Although the manipulation displaced more than 80 percent of canopy trees from their vertical positions, nearly 90 percent of these survived the first growing season, and many leafed out again even in a prostrate position (Figure 11.7). Mortality occurred gradually but progressively; despite high survival rates the first season, after six years fewer than 20 percent of damaged trees survived. Persistence varied strongly by species, with approximately 30 percent of oaks and 70 percent of red maples releafing or sprouting after six years. Because oaks were generally larger, however, these differences have not resulted in major changes in the relative importance of species in the stand. In contrast to the New England forests of 1938, our study area contained only a handful of white pines, most of which were understory trees. These largely uprooted, and overall survival for pines was low.

Environmental conditions at the ground surface did not change much in areas other than the pits and mounds. This is because a dense, though low, canopy was maintained by the strong sprouting and releafing response of damaged trees, the growth of saplings already present in the understory, and the establishment of new seedlings. The density of sprouts and saplings increased from fewer than 6,000 per hectare before the manipulation to nearly 25,000 three years after the pulldown, and then decreased to 17,500 per hectare in 1999. These understory trees now compose slightly more than a third of the total woody basal area of the stand (Figure 11.8). Although nearly half of the saplings present af-

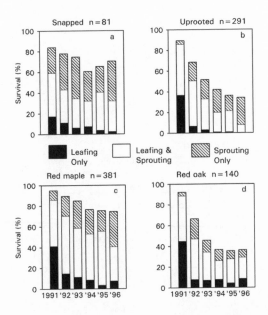

Figure 11.7. Survival (releafing and sprouting) of trees by damage type (a and b) and species (c and d) in the experimental hurricane area. Considerable variation in these responses underscores the potential for great variation in vegetation response to disturbance depending on the type and intensity of damage and the original forest composition. Modified from Cooper-Ellis et al. 1999, with permission from the Ecological Society of America.

ter six years became established after the manipulation, they represent a minor part of the stand basal area since they grow much more slowly than either the older saplings or new sprouts. If we assume that natural mortality will continue to thin these crowded understory stems, density will continue to drop rapidly. However, because many birch seedlings became established on tip-up mounds where they will have a height advantage and reduced competition, birch may increase in importance in the overstory.

With the majority of standing trees felled, the new forest canopy was initially characterized by multilevel, heterogeneous cover, with broad areas of thick foliage generating dense shade that alternated with occasional open areas where light penetrated to the forest floor. The understory vegetation responded strongly to these changes, especially increases in light. Saplings have increased from less than 10 percent to nearly 40 percent cover. Shrubs have also increased in cover and average height. Ferns present before the experiment spread to form dense patches in some areas. Meanwhile, the vegetation at the forest floor has remained relatively stable. Understory species richness increased for two years but decreased by year four. Prior land-use activity, primarily pasturing, had generated a seed bank of disturbance-related species that has persisted

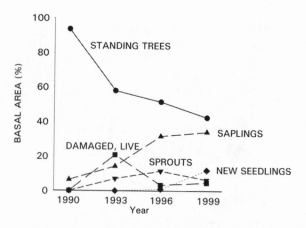

Figure 11.8. Through time, the relative contribution of survival versus different types of regrowth (damaged trees, saplings, sprouts, and seedlings) has varied in the experimental hurricane. Initial high levels of tree survival and rapid regrowth provided great control over environmental factors and ecosystem processes in the developing forest. Updated from Cooper-Ellis et al. 1999.

for more than a century. Thus, scattered colonists such as blackberry, raspberry, staghorn sumac, pin cherry, European honeysuckle, and fireweeds appeared in the pulldown, mostly in areas with soil disturbance. However, only the blackberries, raspberries, and some tussock-forming sedges have persisted. Consequently, compositional changes in the understory, like the overstory, have been minor.

To summarize, the hurricane manipulation initiated profound changes (Figure 11.9). The extent and types of damage paralleled the dramatic effects of the 1938 hurricane, but environmental conditions were stabilized by the persistence of damaged hardwoods and the rapid growth of sprouts and saplings. Forest structure, which was dramatically changed by the dislocation of standing trees, has become more complex and multilayered and continues to reorganize over time. However, the composition of the forest community has remained remarkably stable. Therefore, the imprint of the disturbance—as evidenced by the presence of uproot mounds and pits, bent and misshapen trees, and scattered weedy or early-successional species—will likely be evident well into the future, despite the fact that the dominant species will remain largely unchanged.

Heterogeneity in Environmental Resources: A Major Consequence of Wind Damage

The disturbance created a range of microsites that vary in important characteristics that affect ecosystem processes and the ability of

Figure 11.9. The edge of the experimental blowdown in 2002, nearly twelve years after the manipulation. The uproot mounds have eroded considerably, and the regrowing vegetation is tall and thick. Photograph by D. R. Foster.

plants to establish and grow (Figure 11.10). However, overall differences between the manipulation and the undisturbed forest understory were relatively minor as a consequence of the small extent of ground area directly affected by tree uprooting (about 8 percent), the releafing of damaged vegetation, and vigorous growth of new plants. Nonetheless, the different microsites—uproot mounds, pits, and areas of dense fern—constitute distinct environments from the intact forest understory in terms of light levels, substrate type, stability, and composition and thus contribute to the small-scale heterogeneity of the forest environment. They may provide contrasting opportunities for different species in the postdisturbance forest community.

Although light levels within 50 centimeters of the forest floor increased immediately in some places and remained elevated for at least three years, light was extremely variable among the different microsites. Soil moisture, organic matter content, and nitrification rates were rela-

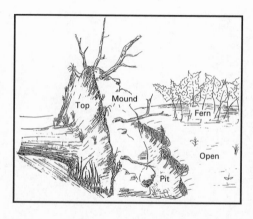

Figure 11.10. The range of microsites that develops after the uprooting of a large canopy tree. Resources such as light, nutrients, moisture, and competition among plants vary tremendously among these small-scale habitats. Modified from Carlton and Bazzaz 1998a, with permission from the Ecological Society of America.

tively similar between the disturbed and undisturbed forest, but also varied substantially among microsites. Similarly, although the soil in pits had lower temperatures and respiration rates than that on mounds or intact sites, the limited area of pits and mounds resulted in virtually no impact on ecosystem-wide rates of CO_2 flux from soils to the atmosphere.

Mounds and tops of tip-ups had contrasting patterns of resource availability, despite being very close to one another spatially (Figure 11.11). Mounds had high levels of light availability but much lower levels of soil resources, for instance, water and nitrogen. In contrast, tops received very low amounts of light, mainly because they were north-facing, and had high rates of nitrogen mineralization and nitrification. Pits generally had average resource levels (compared with open sites) but had much thicker litter layers than other sites because of the ongoing erosion of material. Contrasting patterns of resource availability indicated that resource congruency was lower in blowdown plots than in undisturbed forest, especially for light versus soil resources.

Because light, CO_2, and soil resources vary at different temporal scales after disturbance, plants are exposed to various combinations of resources over time. Differences in the ability among species to tolerate or to capitalize on these resources will influence their survival, growth rates, and future success. Fast-growing, colonizing species like paper birch thrive on the mineral soil seed bed, high light conditions, and reduced competition characteristic of the mounds; they thereby gain an early height advantage over seedlings established on the intact forest floor. Yellow birch, which is more shade tolerant and was present as a seedling in the understory before the experiment, occurred frequently

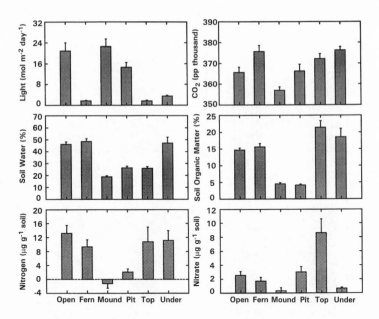

Figure 11.11. Resource availability in hurricane blowdown microsites. Means for each microsite were pooled across experimental plots. The availability of plant resources varies tremendously across the different microsites produced by the windthrow of trees in the experimental hurricane. Modified from Carlton and Bazzaz 1998b, with permission from the Ecological Society of America.

on the mound "top" (that is, the side of the root plate closest to the tree). These seedlings become elevated as the tree falls and creates a mound, and respond to the increased light and leaning orientation with a vertical (perpendicular) growth spurt. The relative hospitality of the different microsites varies considerably depending on overall soil and climatic conditions. Thus, in a particularly dry climate mounds may be less favorable seed beds than we observed. Conversely, in areas where accumulation of litter and water and slumping of mound materials into pits are not as extensive as in our study, pits may provide favorable resource conditions and establishment opportunities.

Soils and Biogeochemistry

The massive rearrangement of forest structure, the large number of broken and damaged tree boles, and the striking appearance of the pit-and-mound topography strongly suggested that the blowdown would greatly alter physical soil conditions and biogeochemical cycling. On the basis of the extent of forest damage and the results of studies from the Hubbard Brook watershed and the Puerto Rican landscape after Hur-

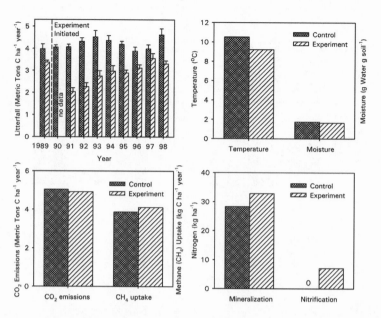

Figure 11.12. Summary of the major ecosystem and biogeochemical changes after the experimental hurricane blowdown. As a consequence of releafing of the windthrown canopy and rapid recovery of leaf area (litterfall), there were only slight changes in microenvironment (temperature and moisture) and nitrogen and carbon dynamics. Modified from Foster, Aber et al. 1997; unpublished litterfall data from A. Magill.

ricane Hugo, we anticipated that the experimental manipulation might be disruptive to the soil environment and reduce biotic control of biogeochemical processes. Thus, we looked closely at the soil environment, including soil temperatures and moisture content, changes in the rate of nitrogen cycling and available nitrogen content, and altered fluxes of gases between soils and the atmosphere.

Remarkably, few ecosystem-level changes in soil conditions or biogeochemical processes were detected (Figure 11.12). Average annual soil temperature in the upper 2.5 centimeters of soil was unchanged, probably because of shading of the soil surface by the leafing and resprouting of downed trees during the summer after the blowdown. Survival of damaged trees and the rapid recovery of leaf area across the experimental blowdown maintained evapotranspiration, which kept soil moisture at predisturbance levels. The stabilization of physical conditions at the soil surface by living vegetation mitigated effects of the disturbance on soil nutrient processes. Soil respiration, a function of both root respiration and organic matter decomposition, was unchanged. Active root systems in damaged trees ensured the continued contribution

of root respiration to total soil respiration rates. In addition, the manipulation did not affect organic matter decomposition, which is particularly sensitive to temperature and moisture availability in temperate forests.

We did record changes in nitrogen processing, but these were local in scale and minor in terms of the overall nitrogen economy of the site. Net nitrogen mineralization (conversion of organic nitrogen to ammonium and nitrate) was unchanged in the summer immediately following the manipulation. We detected a large relative increase in the net conversion of soil ammonium to nitrate (nitrification); however, given the low absolute levels of nitrification in forests in central Massachusetts, such changes were extremely modest. Soil nitrate generally remained low, whereas available soil ammonium increased to levels nearly double those in the control site. This change was probably due to decreased production of leaf litter, which declined by about 60 percent in the first two years after the experiment and recovered to nearly 75 percent by the fourth year. Net release of nitrous oxide by soils to the atmosphere was lower in the experimental area than in the control, but again, overall flux rates were low compared with the range of levels measured in other temperate and tropical forests and after other disturbances.

Had nitrogen cycling been altered to a large degree, soil fluxes of nitrous oxide to the atmosphere would have increased dramatically. Another strong indicator that the physical and biogeochemical environment was not severely altered by the experiment was the overall stability in methane levels. Methane uptake by soil is quite responsive to the availability of ammonium and to changes in soil moisture, which controls methane diffusion through soil pore spaces. There was no detected reduction in methane consumption.

Consequently, despite the physical damage to the forest and the soil by the experimental hurricane, including the massive vertical redistribution of biomass, a number of forest responses mitigated changes in the soil environment. In turn, as the survival and releafing of damaged trees and rapid growth of sprouts and understory plants maintained the soil microenvironment, they also led to a rapid recovery of leaf area and root biomass and presumably active root uptake. These biotic responses provided strong biological control over biogeochemical processes, resulting in relatively minor changes in ecosystem processes. In the face of apparent forest destruction, many fundamental processes were maintained.

Interpretation and Insights from the Experimental Hurricane

In important ways, the results of the blowdown experiment have led us to modify our interpretations of the response of temperate forest ecosystems to intense wind events and to reevaluate the role of tropical storms in structuring our forest landscape. The surprising vege-

tation and ecosystem responses also required us to reexamine some earlier interpretations of the impact of historical events such as the 1938 hurricane. Collectively, these results enable us to offer some insight into the consequences of management efforts after natural disturbances.

Three aspects of the vegetation response to the experiment stand out: the high rate of initial survival in damaged (including broken and uprooted) trees; the importance of vegetative reproduction during the initial period of forest recovery; and the minor compositional changes that have occurred. Survival by damaged trees, sprouting along the trunk, stump, and roots, and growth of advanced regeneration produced a large leaf area and maintained fairly constant soil and microenvironments. In turn, this continuity in microenvironmental conditions and the relatively small amount of soil disturbance minimized the importance of fast-growing early-successional trees in the revegetation process. The resulting forest, therefore, includes a wide range of growth forms and age cohorts: relatively undamaged survivors from the mid-canopy and understory of the original stand; a wide array of bent, misshapen, and reiterating stems; and sprouts and seedlings concentrated on the ground, tip-up mounds, and pit margins (see Figure 11.8).

This response stands in striking contrast to the conditions that developed after the 1938 hurricane, in which even-aged stands of paper birch and pin cherry were common. The differences may be largely attributed to two historical conditions arising from human land use: the predominance of young stands of old-field white pine in New England in 1938 and the regionwide salvage logging that followed the hurricane (Figure 11.13). Given the history of farm abandonment and ongoing forest cutting into the twentieth century, forests in 1938 were generally young and even-aged. Across this landscape, the forests that were most susceptible to wind damage were the extensive stands of white pine that had established on old fields. More than half of the volume of timber damaged by the 1938 storm was pine. White pine forests differ in important ways from the mixed hardwood stands that predominate naturally in this region and that were studied in our experimental manipulation. Unlike the hardwoods, white pines die rapidly after uprooting or breakage, and they lack the ability to sprout or releaf. In addition, in fairly dense stands of young pines, advanced regeneration is limited and generally doesn't include the shade-intolerant pine. Consequently, following the intensive damage experienced by pine stands in 1938, continuity of forest cover was interrupted, as the pine overstory died abruptly and was replaced by a new cohort of seedlings and sprouts, dominated by fast-growing and early- and mid-successional hardwoods. Because forest conditions in 1938 resulted from two centuries of prior land use, the hurricane of 1938 elicited a very different response from what we observed in our experiment.

Logging after the 1938 hurricane, which was the largest timber sal-

Figure 11.13. Salvage logging after the 1938 hurricane. Photograph from the Harvard Forest Archives.

vage operation in U.S. history, exacerbated the tendency toward dramatic environmental change and an increase in early-successional species. Logging involved cutting and removal of uprooted, damaged, and even many sound trees; burning of resulting slash; and extensive scarification of forest soils. These effects further reduced the density of the remaining canopies and the survival of damaged trees. Logging also greatly increased light penetration to the forest floor and altered the soil environment and seed-bed conditions. Under these open and highly disturbed conditions, the previous vegetation was largely replaced by early-successional and sprouting hardwoods. A legacy of this history may be seen in the even-aged stands of hardwoods that dominate the New England landscape today. In these forests, species such as paper birch and black birch are widely distributed and are not restricted to the mounds and pits as they are in the experimental blowdown. Widespread soil scarification, burning of logging slash and forest litter layers,

and open conditions in 1938 allowed these small, light-seeded species to achieve an unusual prominence and distribution.

We have strong evidence that these striking vegetation changes in 1938 were accompanied by soil and biogeochemical changes that differed from the results of the experiment. As shown by Jim Patric, for nearly five years after the 1938 hurricane the Connecticut River experienced a major increase in flow in comparison with rivers that received little damage to their watersheds. This effect is undoubtedly a consequence of the regional decrease in evapotranspiration due to the massive reduction in leaf area, resulting in an increase in soil moisture and, presumably, major alterations of biogeochemical processes. However, although many interpreted this effect as a natural response of the forest landscape to a natural disturbance, an alternative explanation is prompted from awareness of regional history, namely, the history of land use that led to the dominance by white pine and the extensive salvage logging. Indeed, it is quite likely that the response of increased river flow was an indirect consequence of human activity. The very young age of the 1938 forest, the dominance by white pine that was susceptible to blowdown and unable to sprout, the removal of leaf area, and extensive soil disruption by logging and burning all greatly reduced biotic control by the vegetation over soil and biogeochemical processes.

These results broaden the basis for decision making in forest management with regards to storm damage. The minimal change that we observed in soil, microenvironment, and important biogeochemical processes after the experiment suggests that natural hurricane disturbance is actually not highly disruptive to ecosystem integrity. In contrast, postdisturbance salvage logging may generate stronger ecosystem-level responses than the disturbance itself. If maintenance of biotic control over hydrology, soil environment, and nutrient fluxes is a priority of management, and if rapid forest recovery is sought, then leaving the site intact may be a preferred approach. This may be the approach taken, for example, in a municipal watershed where water quality is a primary objective or in natural areas or wildlife management areas in which natural conditions and structures such as standing and dead trees are objectives. Such an approach will also lead to the natural development of a diversified forest structure in which a range of classes and growth forms, including bent, misshapen, and broken trees, occurs along with quantities of downed wood.

In contrast, the desires to recoup value from windthrown timber and to influence the future composition and structure of the forest are major motivations for salvage logging. Salvage allows the elimination of leaning, bent, and broken trees and enables the promotion of low stump sprouts as opposed to stem sprouts. However, logging after extensive windstorm also increases soil disturbance and open conditions that will favor the establishment of early-successional species.

On one subject regarding salvage logging, namely, fire hazard, there are strong opinions but few data or little long-term study. Windstorms are often implicated as one of the catalysts of major forest fires in the New England landscape; indeed, a major motivation driving salvage, including the effort in 1938, is a desire to reduce fuel loading. However, there are essentially no studies of the changes in fuel loading after windstorms from New England forests, nor is there good historical research that links windstorm events and fire. Most of the support for salvage operations is actually predicated on the intuitive notion that the downed woody debris associated with windstorms necessarily enhances long-term fire hazard. Although blowdowns in conifer stands may produce a short-lived increase in hazard, data from the hurricane experiment suggest that fine fuels, which are the main fire concern, are unevenly distributed and highly transient because of the rapid decay of fine material and extended period over which the damaged trees die. As a consequence, overall fire hazard was only slightly increased in the experimental study and for a relatively short time. The decomposition of fine fuels and the rapid growth of new sprouts and understory plants quickly reduced the fire hazard.

Another approach to forest planning in a landscape subjected to infrequent tropical storms is to incorporate an understanding of hurricane disturbance regimes and forest response to winds into the harvesting regime in order to minimize damage and to increase long-term yields. Such considerations stimulated considerable research and discussion in the aftermath of the extreme losses suffered in 1938 and led to many studies, including the research by D. M. Smith. However, this burst of interest resulted in few significant changes in silvicultural practices across the region. Quite recently, however, the Metropolitan District Commission, the Massachusetts state agency responsible for metropolitan Boston's water supply, has taken this approach to heart in its management of the watershed of the Quabbin Reservoir, a 32,000-hectare area of forest and water. By cutting to decrease the average age of forests and to favor a diverse forest of wind-firm species, forest managers are seeking to minimize damage from the next storm, pathogen, or other disturbance. Although few data, including our own studies on the hurricane or hemlock woolly adelgid, suggest that these disturbances have negative effects on water quality, for areas in which logging is a management objective it should be quite feasible to direct silvicultural activities so as to mitigate other disturbances.

Many questions remain about the impacts of hurricanes on New England forests. In particular, we lack the ability to anticipate the effects of posthurricane management strategies on ecosystem function and vegetation development. Our experiment simulated the effects of intensive wind in the most common forest type on the central New England landscape today, and we found that at the scale of approximately 1 hectare, the forest has a remarkable ability to maintain internal functioning and resist major compositional changes. However, matched experimental treatments such as blowdown followed by fire, blowdown followed by salvage, and standard silvicultural treatments on comparable sites in hardwood and conifer forests could help to answer many remaining questions about the relationship between vegetation and environment, controls on regeneration after various types of disturbance, and the challenges of management in the face of the certainty of future hurricanes.

CHAPTER 12

Exploring the Process of Nitrogen Saturation

J. ABER, A. MAGILL, K. NADELHOFFER, J. MELILLO,
P. STEUDLER, P. MICKS, J. HENDRICKS, R. BOWDEN,
W. CURRIE, W. McDOWELL, and G. BERNTSON

Introduction and Historical Context

Plant growth in temperate zone forests is generally limited by the availability of nitrogen. Human activities may have played a major role in establishing this truism, and may play as large a role in its eventual contradiction.

Plants require more nitrogen (N) than any other element drawn from the soil. Nitrogen is generally absent from rocks and so is not released during weathering, the physical and chemical disintegration of minerals that provides calcium, magnesium, and many other essential elements. While N_2 gas constitutes 78 percent of the atmosphere, this form of nitrogen is not available to plants, so plants may be nitrogen limited while awash in an atmosphere rich in this element. Atmospheric N_2 must be converted to a reactive form either by microbial processes or by physical events in the atmosphere (lightning, for example) before it can be used for plant growth. Background inputs of nitrogen to native ecosystems are generally very low (1 to 2 kilograms nitrogen per hectare per year [$kg\,N{\cdot}ha^{-1}{\cdot}yr^{-1}$]) and can be more than overbalanced by losses through fire or by leaching of nitrate after disturbance, or by the continuous, slow loss of nitrogen in dissolved organic compounds to streams. This combination of a tenuous nitrogen balance, combined with relatively high rates of inputs of other elements to soils by weathering of geological substrates, causes nitrogen to be the most frequently limiting nutrient under natural conditions in recently glaciated areas such as New England.

Human use of the lands that now constitute the Harvard Forest has altered the nitrogen status of the landscape significantly. Early clearing and farming practices were extractive, with little in the way of fertilization or return of manure to pastures or plowlands. By the nineteenth century, pastures were enclosed and animals housed in barns, with manure returned to the fields. Nitrogen in grasses from extensive hayfields and pastures was concentrated into manures that were then spread over the adjacent but smaller areas of plowed cropland. Consequently, adjoining parcels of land were differentially depleted or enriched in nitrogen and other elements depending on their specific land-use history. Farm abandonment then resulted in the establishment of a patchwork of forest types across a landscape in which nitrogen cycling differed markedly as a consequence of this history, as well as inherent site factors (see Chapter 9).

One effect of this land-use history is that species-site relationships in the modern landscape can be very different from those seen in natural areas. For example, pine plantations established at the Harvard Forest soon after its founding were sited preferentially on the remaining open areas. These were mainly plowed lands that had become pastures near the end of the agricultural period and tended to be among the most productive and enriched farmlands. So while most natural pine forests occur on nutrient-poor and acid soils with low fertility, these plantations occupy some of the richest sites in the landscape. In contrast, old woodlots, dominated by hardwoods that had sprouted after repeated cutting and/or burning, are most likely depleted in nutrients relative to their presettlement conditions.

Regional urbanization and industrialization over the past century have altered the nitrogen balance at the Harvard Forest once again. Dramatic increases in the generation of the waste products of combustion regionally and farther upwind in the Great Lakes–Ohio River area have led to increased deposition of nitric and sulfuric acids, the major components of acid rain. Although sulfur emissions have been reduced over the past two decades through legislation, nitrogen emissions, which result more from automobile exhausts than industrial processes, remain high. Concern now focuses on the potential for "nitrogen saturation" or the availability of nitrogen in excess of biotic demand due to atmospheric deposition.

Background for the Chronic Nitrogen Experiment

The chronic nitrogen addition experiment was designed to test the long-term consequences of increased nitrogen deposition in a region in which nitrogen is historically and naturally a limiting nutrient. Understanding the experiment and the process of nitrogen saturation requires some basic background on the rates and forms in which nitrogen cycles through forest ecosystems.

The nitrogen cycle of a forest connects four pools of very unequal size (Figure 12.1). Over 98 percent of all nitrogen generally resides in organic forms in plants and soils. Nitrogen in these large pools turns over very slowly; any one atom of nitrogen may reside in these pools for

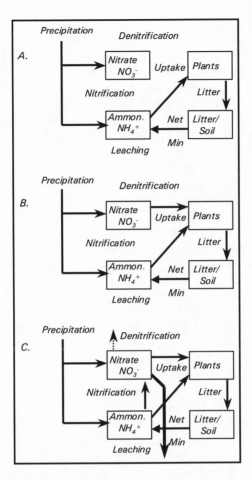

Figure 12.1. A simplified view of changes in the nitrogen (N) cycle in forest ecosystems in response to increased N deposition. A: Under N-limited conditions, N cycles mainly as ammonium due to competition for this form of N between plants (with mycorrhizal root symbionts) and free-living microbes. Net Min, net mineralization. B: Long-term N additions increase N availability, stimulating plant growth and partially alleviating N limitations. C: As deposited N enters the N cycle, further increases in availability occur, stimulating nitrification, nitrate cycling, and, eventually, leaching losses to groundwater.

decades to centuries. The remaining nitrogen is present in mineral form as ammonium and nitrate in soils with residence times measured in hours to weeks. Plants shed litter (dead plant parts such as leaves and roots) into the soil organic matter pool on an ongoing basis and contribute large woody debris like branches or whole stems after disturbance or tree death. Microbial decomposition of this dead organic matter releases ammonium, which can then be either reused by the microbes or taken up by plants. Under conditions of severe nitrogen limitation (Figure 12.1A), plants and their associated root symbionts compete effectively for ammonium against the free-living microbes that use ammonium as an energy source (nitrifiers) and produce nitrate (nitrification). Thus, in ecosystems where nitrogen is very limited, nitrogen cycles primarily as ammonium and there is no appreciable loss of mineral nitrogen from the system. Nitrogen can, however, be lost from ecosystems in dissolved organic form (the arrow showing leaching loss directly from litter/soil). Measured rates for this flux in most systems are low and do not appear to change with nitrogen status, so this flux can be important in the long-term nitrogen balance of forests in which nitrogen inputs are low and inorganic forms of nitrogen are retained.

The process of nitrogen saturation involves a number of integrated changes in biogeochemical cycling that are induced as increasing inputs of nitrogen from human sources gradually decrease nitrogen limitation in the system. Deposition of nitrogen from the atmosphere adds both nitrate and ammonium (Figure 12.1B). Plants and microbes can take up this added nitrogen, partially relieving nitrogen limitations. The increased availability of nitrogen stimulates increased plant growth and production of organic matter, resulting in larger inputs of litter, an acceleration of nitrogen cycling, and increased production of ammonium. As nitrogen inputs from the atmosphere accumulate, nitrogen limitations on microbes and plants are reduced, free-living nitrifiers in the soil have increasing access to ammonium, and the production of nitrate begins or increases (Figure 12.1C). Although both plants and microbes can use nitrate, production can eventually outstrip uptake potential, and nitrate will begin to leach to streams. Nitrate (NO_3^-) is a negatively charged ion. When it leaches from the system, it carries along positively charged ions such as calcium, potassium, and magnesium (preferentially), followed by aluminum and hydrogen, leading to acidification of soils and streams, increased aluminum concentrations in soil and stream water, and nutrient imbalances in soils and plants. All of these responses can stress forests, leading to declining growth rates and increased mortality.

At the initiation of the chronic nitrogen addition experiment, we published a set of hypotheses on the expected integrated response of nitrogen-limited forests to chronic nitrogen additions (Figure 12.2). A key feature of this set of hypotheses was that responses would be nonlinear; that different components and processes of the ecosystem would change at different rates and in different directions. A second key feature was that the initiation of nitrification would mark a critical step in the overall response. A third key prediction, and one we considered very unlikely to be realized, was that excess nitrogen availability would lead to reductions in forest growth, a phenomenon often termed *forest decline.*

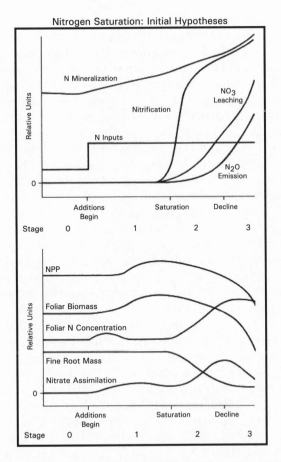

Figure 12.2. Initial set of hypotheses regarding the sequential responses of N-limited forests to chronic N additions. Key characteristics of this set of hypotheses are the nonlinear responses of several processes and the central role of nitrification. NPP, net primary production. Reprinted from Aber, Nadelhoffer et al. 1989.

Experimental Design

We selected a red pine plantation and an adjacent mixed hardwood (red oak, red maple, black birch) forest on the Prospect Hill tract for this study. The land now supporting the red pine stand was cleared and plowed during the eighteenth and nineteenth centuries and was in pasture at the time of acquisition by the Harvard Forest. This open field was planted to red pine in the 1920s. A distinct plow layer is visible in the mineral soil, and only a thin forest floor (organic soil horizon) has developed since the stand was planted. In contrast, the mixed hardwood stand was never plowed, as evidenced by the natural appearance of the soil horizons and high stone content of the upper soil. This stand, however, was harvested and used for pasture until around 1850, when it was abandoned and allowed to reforest naturally. The resulting forest was logged again in the 1930s, and at least part of the stand was burned by an intense surface fire in 1957. The stand is very slow growing and has a thick, densely rooted forest floor, which suggests strong competition for nutrients.

Our first surprise in this experiment was the difference in nitrogen cycling rates at these two sites at the beginning of the experiment. Planted on enriched plowland, the red pine stand had an annual rate of nitrogen mineralization (supply of N through decomposition) of 92 kg $N \cdot ha^{-1} \cdot yr^{-1}$, much higher than normal for needle-leaved evergreen stands. In contrast, the heavily used but never-plowed hardwood stand was cycling only 84 kg $N \cdot ha^{-1} \cdot yr^{-1}$, at the low end of the range for temperate deciduous forests. Nitrification was 21 kg $N \cdot ha^{-1} \cdot yr^{-1}$ in the pines and undetectable in the hardwoods. The imprint of two centuries of land use is still evident in these two stands nearly 100 years after the end of agricultural management.

Four large (30-by-30-meter) plots were established in each stand. Single large plots were used rather than smaller replicate plots to reduce edge effects and to include full-sized trees in each plot, allowing us to examine feedbacks between plants and soils. Each plot was assigned one treatment (control, low N, high N, and N+S [sulfur]). The low nitrogen treatment is 50 kg $N \cdot ha^{-1} \cdot yr^{-1}$ and the high nitrogen treatment is 150 kg $N \cdot ha^{-1} \cdot yr^{-1}$, added as a concentrated solution of NH_4NO_3 in six equal monthly doses applied to the soil surface with a backpack sprayer between May and September. The N+S treatment included the low nitrogen addition plus 74 kg $S \cdot ha^{-1} \cdot yr^{-1}$ as SO_4^{-2} and was intended to mimic the mid-1980s ratio (albeit at higher dosages) of nitrogen to sulfur in deposition over New England. In general, results from the N+S plots were indistinguishable from those from the low-nitrogen plot and will not be discussed further here.

One of the major goals of the chronic nitrogen experiment was to determine the fate of nitrogen added to the two stands through time: did it end up in the soil, the plants, or the microbes, or was it leached from the system? Although net budgets (measurements of changes in storage within different parts of the forest over time) provide partial insights to this important question, we also chose to add distinctive isotopes of nitrogen that can be traced through the measured pools to provide more precise measurements of nitrogen dynamics. In 1991 and 1992 nitrogen fertilizer additions in the low-nitrogen plots contained slightly elevated (natural abundance levels) of ^{15}N. At the same time, we added very small amounts of highly enriched ^{15}N to the control plots. This procedure allowed us to compare measurements between these two stands without adding sufficient nitrogen to induce a fertilizer response in the control plots. We added ^{15}N as NH_4NO_3 but split the labeling so that $^{15}NH_4NO_3$ was added to one half of each plot and $NH_4{}^{15}NO_3$ was added to the other half.

Process Measurements

We began by measuring processes that were relevant to the hypotheses in Figure 12.2. Nitrogen mineralization (the conversion of organic nitrogen to mineral nitrogen) and nitrification (the conversion of ammonium to nitrate) were measured by on-site soil incubation. We also measured nitrogen leaching losses into deeper soil layers using tension cup lysimeters, which collect soil water below the rooting zone. Concentrations of ammonium and nitrate in lysimeter samples were multiplied by monthly hydrologic fluxes derived from the PnET-II model (see Chapter 17). Net fluxes of important gases (nitrous oxide $[N_2O]$, methane $[CH_4]$, and CO_2) between the soil and the atmosphere were measured using chambers positioned on the soil surface. In experiment 1, below, the soda-lime trap technique was used to measure CO_2 efflux. Aboveground net primary production (ANPP, the production of new plant biomass) was calculated as the sum of litterfall and stem growth. Nitrogen concentration in tree foliage was measured on samples collected by shooting leaves from the canopies of trees with a shotgun in midsummer, and concentrations in fine roots have been measured intermittently.

Summary of Initial Results

In general, the two forests responded to the ongoing chronic additions of nitrogen as hypothesized, but on very different timescales. Nitrification increased quickly in both pine plots and eventually in the hardwood high-nitrogen plot as well. Nitrate leaching into lower soil horizons increased almost immediately in the pine high-nitrogen plot and was occasionally elevated in the low-nitrogen plot. In contrast, significant leaching in the hardwood stand did not occur in the high-nitrogen plot until the eighth year of treatment (Figure 12.3). In parallel with trends in nitrate loss, the concentrations of nitrogen in leaves and fine roots have increased dramatically in the pine stand, whereas increases in the hardwood stand have been less pronounced (Table 12.1). All of these results suggest that both stands are progressing toward nitrogen saturation but that the hardwood stand was more nitrogen limited initially and has responded more slowly to nitrogen additions than the red pine stand.

There were also surprises in the results. Nitrous oxide efflux from the soil, which we hypothesized would increase with nitrogen addi-

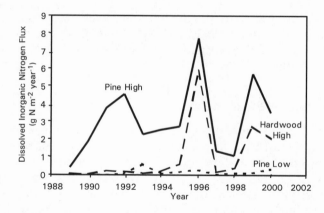

Figure 12.3. Pattern of nitrate loss from the chronic N plots at the Harvard Forest. Lines for high and low N additions in the pines and high additions in the hardwood stand are shown. Values for other stands are combined and are not different from zero.

tions (Figure 12.2), remained nearly undetectable. In contrast, there was an immediate and major decrease in the rate of methane consumption in both the pine and hardwood nitrogen-amended stands. Methane consumption in the high-nitrogen plots was approximately half that of the control stands in year two and has continued to decrease over time (Figure 12.4). The potential for microbes to switch from methane to ammonium as substrates for metabolism depending on relative abundances of the two molecules appears to underlie this result. Methane is a potent greenhouse gas, approximately twenty times as effective as CO_2 in trap-

Table 12.1. Differences in Nitrogen Concentrations in Foliage and Fine Roots in the Chronic Nitrogen Experiment

Stand	Control (%)	Low N (%)	High N (%)
		Treatment	
Pine			
Foliage	1.0	1.5	1.9
Fine roots			
Forest floor	1.6	2.2	2.0
Mineral horizon	1.1	1.8	1.6
Hardwood			
Foliage (oak)	2.4	2.6	3.1
Fine roots			
Forest floor	1.2	1.5	1.6
Mineral horizon	0.8	0.9	1.2

Note: Forest floor measurements include roots from the Oa and Oe horizons; mineral horizon measurements include roots from 0 to 10 centimeters. All root and foliar samples were collected in 1999. Foliar measurements from 1999.

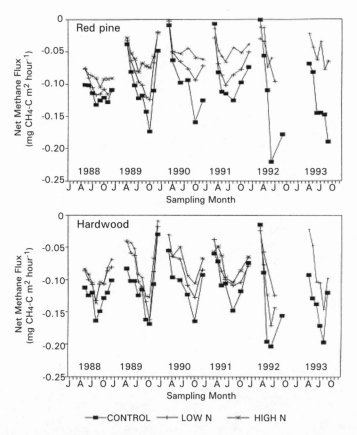

Figure 12.4. Changes in net methane flux in the red pine (top) and hardwood (bottom) stands in response to the first six years of chronic N additions. Modified from Castro, Steudler et al. 1995, by permission of the American Geophysical Union (copyright 1995, American Geophysical Union).

ping long-wave radiation. Increased nitrogen deposition over large parts of North America and Europe could actually be contributing to the measured rapid increase in atmospheric concentrations of methane.

We were surprised as well by a measurable decline in productivity in the red pine stands receiving nitrogen additions (Figure 12.5). While we had hypothesized that such an outcome was a feasible long-term concern on the basis of the literature on forest decline in both North America and Europe (Figure 12.2), we did not expect this response to occur this quickly. The decline in growth occurs despite an increase in foliar nitrogen concentration and is accompanied by a reduced foliar retention time and severe reduction in total needle mass (Figure 12.6).

It could be argued that the red pine stand is old and nearly mature

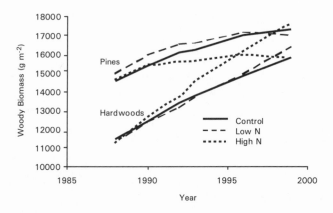

Figure 12.5. Comparison of changes in total woody biomass over time in the red pine and hardwood plots as a consequence of chronic N additions. Total accumulation in the high-N pine plot is significantly lower than in the control and low-N plots, while in the hardwood stand accumulation is highest in the high-N plot.

Figure 12.6. Foliar biomass has been greatly reduced in the pine stand receiving high-N additions (left) compared with the control (right). Photographs by D. R. Foster.

and therefore is being pushed to overmaturity by excess fertilization. However, similar results in a spruce-fir stand on Mount Ascutney, Vermont, and across a long transect of spruce-fir stands from New York to Maine, combined with results from Europe, suggest that high levels of nitrogen deposition may indeed endanger forest health in needle-leaved evergreen stands. The decline in tree growth that we have recorded is accompanied by decreasing ratios of magnesium to nitrogen and/or calcium to aluminum in foliage in the pine stand, at Mount Ascutney, and along the regional nitrogen deposition transect. The ratios of these elements are used as indicators of forest health, and our results suggest that nutritional imbalances involving elements other than nitrogen may be induced by excess nitrogen deposition. Measurements of photosynthesis and stress response at both the Harvard Forest and Mount Ascutney sites provide process-level explanations for declining wood production.

Perhaps the most surprising result is the very high potential for nitrogen retention in both forest ecosystems and the fact that most of the nitrogen retained was incorporated rapidly into soils without passing through plant biomass. This was shown by both mass budget calculations (direct measurement of changes in pool size and flux) and by the fraction of added ^{15}N recovered in each pool (Table 12.2). This result has been confirmed in a number of other studies in both the United States and Europe. We had anticipated that plants, being nitrogen limited, would be the strongest sink for added nitrogen and that this capacity would soon be exceeded by the high rates of nitrogen addition. Instead, total retention was much higher than we expected, and the most significant retention processes apparently occur in the soil. Efficient, long-term retention of added nitrogen is a critical result, as all of the negative

Table 12.2. Comparison of Two Methods for Estimating the Distribution of Added Nitrogen in Low-Nitrogen Plots

Sinks (%)	Pine Low		Hardwood Low	
	Budget	^{15}N	Budget	^{15}N
Foliage	12.7	9.6	5.3	5.9
Wood	2.9	1.4	12.7	4.4
Roots	9.6	8.7	11.6	13.7
Total plant	25.2	19.7	29.6	24.0
Soil (0–20 cm)	—	49	—	71.9
Soil (0–60 cm)	74.8	(80.3)	70.4	(76.0)
Total	100	68.7	100	95.9
		(100)		(100)

Note: Values are percentage of total nitrogen retained in each pool. The "Budget" column uses data collected as described in Magill et al. 2000. The "^{15}N" data represent percentage recovery by compartment after two years of natural abundance-level ^{15}N additions to control and low-nitrogen plots (Nadelhoffer, Downs, and Fry, 1999). The (est. 0–60 cm) values assume that the unrecovered ^{15}N resides in the soil compartment in all cases but could not be fully detected because of low concentrations and sampling limited to 20 centimeters' depth.

effects of nitrogen saturation are linked to elevated rates of nitrification and nitrate leaching that occur when nitrogen is no longer limiting (Figure 12.1). Consequently, if forests have a greater-than-expected potential to retain and store nitrogen before nitrate leaching begins, then the damage to forests and waters predicted to occur in the wake of nitrogen saturation is reduced, or at least delayed.

Mechanisms for Nitrogen Retention: A Series of Miniexperiments
Overview

We need to understand how added nitrogen is incorporated into soils and what the capacity of this mechanism is (how long will this protective function continue?). After the first three years of the experiment, it was clear that the soil was the biggest nitrogen sink. At that time (about 1991), new experiments tracing the short-term (twenty-four-hour) dynamics of ^{15}N in soils suggested that gross rates of uptake and release of nitrogen by microbes were many times faster than previously thought. Also, new insights into the possible role of dissolved organic carbon (DOC) in forest soils were emerging. Combining these perspectives, we hypothesized that DOC was an important substrate and energy source for microbes as they incorporated (and immobilized) the additional nitrogen into microbial biomass. This was something of a radical proposal in that basically all previous evidence indicated that soil microbes are carbon (or energy)–limited rather than nitrogen-limited.

We devised three experiments to test this new hypothesis that nitrogen incorporation was due to microbial immobilization using DOC as an additional energy source. The first looked for changes in the efflux of CO_2 from microbial respiration in response to nitrogen additions; the second examined the effect of nitrogen additions on concentrations of DOC and associated nitrogen (DON) in soil water; and the third measured nitrate dynamics using ^{15}N tracer-pool dilution techniques. Because our nitrogen additions are made directly to the soil surface, we also examined the results of a long-term leaf litter decomposition experiment for evidence of increased biological immobilization.

Experiment 1: Soil CO_2 Flux versus Nitrogen Incorporation

If soil microbes were limited by nitrogen, we hypothesized that the addition of mineral nitrogen would result in both the disappearance of that nitrogen and an increase in the flux of CO_2 from the soil as the microbes increased their metabolic activity and respiration in order to produce more biomass. To test this, we measured CO_2 efflux from soils to the atmosphere and the rate of disappearance of the added nitrogen in both pine and hardwood plots.

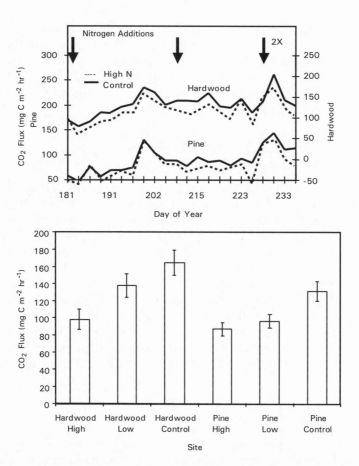

Figure 12.7. Comparison of measured rates of CO_2 efflux from the control and N-treated stands. Top: Changes immediately after N additions at doses equivalent to chronic N additions. 2X indicates timing of a double-high N dose. These data show no significant increase in CO_2 efflux after N additions (Micks 1994, unpublished thesis). Bottom: Mean soil CO_2 flux rates by treatment, June to August 2001, showing a strong reduction with the addition of nitrogen. Modified from C. Arabia and E. Davidson (unpublished).

The study revealed that total CO_2 flux might actually be lower in the nitrogen-amended stands and that respiration in both control and amended stands changed in response to moisture and temperature, but that there was no detectable increase in CO_2 efflux after nitrogen additions (Figure 12.7, top). The added nitrogen disappeared from the soil over a period of one to two weeks, and a very simple model of microbial dynamics suggested that CO_2 fluxes during this period would have increased severalfold if microbial biomass production were the primary process by which mineral nitrogen was converted to organic nitrogen.

More recent and more detailed measurements of soil respiration have indeed shown that CO_2 flux from soils has decreased in the high-nitrogen plots (Figure 12.7, bottom). There are several possible explanations for this, but none of them would support the hypothesis that nitrogen limitations were suppressing microbial metabolism in these stands before nitrogen additions began.

Experiment 2: Effects of Nitrogen Additions on DOC and DON Flux

A second line of evidence that could support the hypothesis of microbial incorporation driven by metabolism of DOC would be a reduction in measurable soil DOC. We measured the leaching losses of DOC from the forest floor to the mineral soil as an indicator of total DOC availability. The mean values for this DOC flux were somewhat higher, rather than lower, in the nitrogen-amended stands, but the differences were not statistically significant. There was, however, a significant increase in DON flux and a narrowing of the DOC:DON ratio in the water passing from the forest floor into the mineral soil. The size of the increase in DON flux (5 to 6 kg N·ha^{-1}·yr^{-1} in the high-nitrogen stands) is too small to account for a large part of the added nitrogen, but it may suggest a more important immobilization mechanism, as discussed below.

Experiment 3: Pool Dilution Studies and Abiotic ^{15}N Incorporation

We employed ^{15}N pool dilution techniques to examine rates of microbial uptake and release of nitrate. This technique involves injecting very small amounts of NO_3 that is highly enriched with ^{15}N into soils and then determining the form in which this nitrogen occurs at different time intervals up to twenty-four hours after addition. For example, if a small amount of enriched nitrate is added to a soil, then at time zero a given fraction of the extractable nitrate pool consists of ^{15}N. The gross rate of nitrogen uptake or immobilization of that nitrate is equivalent to the rate at which the extractable ^{15}N disappears through time. The rate of nitrate production is determined by the rate at which the remaining $^{15}NO_3$ pool is diluted by new $^{14}NO_3$. Different sampling times (often fifteen to thirty minutes and twenty-four hours) are used to identify and separate the potential roles of "fast" and "slow" immobilization processes. Rapid disappearance of ^{15}N (that is, a fast process) is often interpreted as being due to chemical or abiotic reactions, while disappearance between this first sample and twenty-four hours is generally attributed to microbial uptake.

We applied this technique using $^{15}NO_3$ in the pine and hardwood control and high-nitrogen plots, using several sampling periods and a

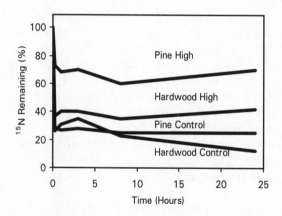

Figure 12.8. Gross nitrate immobilization in the control and high-N plots for the pine and hardwood stands as measured by ^{15}N pool dilution. Modified from Berntson and Aber 2000, 153, with permission from Elsevier Science (copyright 2000).

unique statistical method to determine more accurately the dynamics of the fast and slow processes. Surprisingly, our results suggest that the fast process is the only significant mechanism for nitrate incorporation in all but the hardwood control stand (Figure 12.8). In the pine stand, there was no significant slow immobilization, but fast immobilization was greater in the control plot than in the high-nitrogen plot. In the hardwoods, fast immobilization was similar between control and high-nitrogen plots, but the control plot showed additional slow nitrate immobilization while the high-nitrogen plot did not.

It is notable that total (fast plus slow) nitrate immobilization in each of the four plots measured was highly correlated with the concentration of nitrate in soil water below the rooting zone. This suggests that the processes measured by our modified pool dilution method could be important in nitrogen retention. The results also clearly indicate that fast immobilization is an important process in nitrogen retention. Much more work needs to be done here to determine whether "fast" and "slow" equate to abiotic and biotic reactions. If they do, then much of the efficient retention of nitrate in the high-nitrogen plots could be dominated by abiotic chemical reactions in the soil about which we know very little indeed.

Experiment 4: Carbon and Nitrogen Dynamics of Nitrogen-Amended Leaf Litter

To look at the role of the litter layer in nitrogen retention, we conducted two experiments on the mass loss and the nitrogen dynamics of litter. The first was a laboratory experiment in which NO_3 and NH_4 were added to freshly collected leaf litter in amounts equivalent to the field nitrogen applications. As an innovation, we also examined DOC and DON fluxes from this material in order to provide complete carbon and nitrogen balances and to link to the field DOC and DON measurements described above. The second was a long-term field study of leaf litter decay under the different nitrogen-addition treatments.

In the lab study, the addition of inorganic nitrogen produced small increases in rates of leaf litter decay and substantial increases in nitrogen immobilization. What was surprising was the large fraction of mass loss (5 to 35 percent) that occurred through the leaching of DOC, rather than as CO_2 fluxes from microbial respiration. This DOC flux has been largely ignored in decomposition studies and suggests that serious errors in the calculation of ecosystem carbon balances can occur when litter mass "disappearance" is equated with microbial respiration. These results also emphasize the importance of knowing the fate of DOC leached from litter. Is it then respired within the forest floor (in which case it would be equivalent to respiration from the litter itself), or is it eventually stabilized through chemical reactions?

Nitrogen immobilization into litter in this lab experiment did increase with nitrogen additions and accounted for a significant fraction of total added nitrogen. Using measured litterfall rates, and extending the results of the lab experiment to the point of humus production, the nitrogen sink associated with increased nitrogen immobilization in litter is estimated at 11 and 17 kg $N \cdot ha^{-1} \cdot yr^{-1}$ in the hardwood and pine high-nitrogen plots, respectively, or 7 to 11 percent of the total addition. Whether or not this immobilization occurs through microbial activity cannot be determined from these results.

The long-term field study confirms an increase in nitrogen retention in litter during decay. In this case, rates of decomposition in the nitrogen-amended plots were reduced in the later stages of the process (Figure 12.9), particularly through reduced rates of lignin decomposition. Nitrogen immobilization into this litter continued in the high-nitrogen plots, even though mass loss was reduced. At the end of the experiment, the concentration of nitrogen in litter was similar across treatments, but the total mass of litter remaining was higher in the high-nitrogen plots, so total nitrogen retained was also higher. Calculations of the total additional amount of nitrogen in the soil organic matter produced at the end of the entire decay sequence also suggest a net retention by this process of around 15 kg $N \cdot ha^{-1} \cdot yr^{-1}$, similar to the estimates obtained in the lab study. Results from both litter studies are consistent with those obtained from the ^{15}N additions in the field, in which 11 percent of the total ^{15}N added to the low-nitrogen plots was recovered in the litter layer.

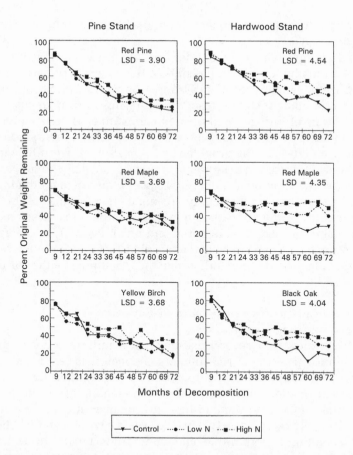

Figure 12.9. Effects of long-term chronic N additions on decay rates of different foliar litters in the field. N concentrations in older litter material did not differ between treatments, but these differences in total mass remaining in later stages of decay leads to increased total N content and higher estimated N incorporation and retention. Modified from Magill and Aber 1998, 304: fig. 1, with kind permission from Kluwer Academic Publishers (copyright Kluwer Academic Publishers 1998).

Summarizing the Experiments

CONSTRAINING THE IMMOBILIZATION PROCESSES

All four of these experiments contradict the hypothesis from which they were generated. Soil CO_2 efflux did not increase, DOC concentrations did not decrease, most of the immobilization appeared to occur very rapidly, and nitrogen concentrations in older litter increased in the absence of mass loss (or apparent microbial processing). All of these suggest that immobilization is not occurring through microbial uptake and biomass production as traditionally viewed. What other processes might drive this rapid movement of mineral nitrogen into soil organic matter?

In a 1998 review paper, we described two processes that fit within the constraints imposed by the results of the four experiments: abiotic immobilization and mycorrhizal assimilation without biomass production. Abiotic immobilization results from chemical reactions between mineral forms of nitrogen and organic compounds in soils and so requires no energy and generates no CO_2. Mycorrhizal assimilation into proteins would use carbon provided by the host plant. The efficiency of conversion of this carbon to proteins is much higher than the conversion of litter carbon to microbial biomass because of the cost of producing the enzyme systems required to decay organic matter. Because of this and the slightly lower ratio of carbon to nitrogen in proteins versus microbial biomass, less carbon is required and much less CO_2 is generated. The question remains as to what the fate of nitrogen assimilated by mycorrhizae might be. Is there other evidence to support either of these processes?

ABIOTIC INCORPORATION OF INORGANIC NITROGEN
INTO SOIL ORGANIC MATTER

It has been known for several decades that significant rates of direct chemical incorporation of NH^4 into organic materials can occur under the right conditions. Before the invention of chemical fertilizers, nitrogen-rich soil amendments were produced by combining peat moss with ammonium under conditions of high pH, temperature, and pressure. Several studies over the past two to three years have shown that chemical incorporation of both nitrate and ammonium can also occur under ambient conditions in lab incubations. One study estimates that abiotic immobilization accounts for 6 to 90 percent of total immobilization, with an average of about 40 percent across a very wide range of ecosystems. Two others showed that most nitrate immobilization at the Harvard Forest occurred in the first few minutes of ^{15}N additions, a result generally associated with chemical rather than biological immobilization, and that sterilization of soils did not alter this pattern. It is particularly surprising that these fast reactions occur for both nitrate and ammonium. There are no known or generally accepted mechanisms for abiotic incorporation of nitrate.

We have proposed the concept of the "DOM conveyor" as a possible mechanism for the rapid, abiotic immobilization of mineral nitrogen. This derives from the observation that increases in DON are a measurable if short-term response to the addition of mineral nitrogen to soil solutions. It is generally assumed that pools of DOC or dissolved organic matter (DOM) in the soil solution represent an equilibrium between very rapid rates of sorption and desorption from/to the solid soil phase. If there are chemical reactions by which mineral nitrogen can be incor-

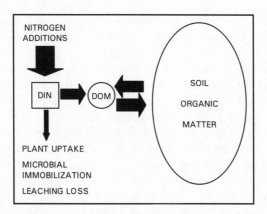

Figure 12.10. Diagrammatic presentation of a hypothesized mechanism for the incorporation of mineral N into soil organic matter. Pools of dissolved organic matter (DOM) are small but exchange rapidly with the solid phase of soil organic matter through sorption/desorption reactions. If chemical reactions occur that incorporate dissolved inorganic nitrogen (DIN) into DOM, then resorption of this DOM into the solid phase would provide a "conveyor belt" effect, loading added N into soil organic matter.

porated into DOM before resorption, then a rapid "conveyor belt" effect would be established by which relatively large quantities of nitrogen could be transported into soil organic matter (Figure 12.10), although the enrichment of the DON pool would be small at any one time. To test this, we need to establish both that DOC is exchanged rapidly with soil organics (a pool dilution–like experiment would do this), and that reactions between DOM and mineral nitrogen occur.

MYCORRHIZAL ASSIMILATION

If we listed components of forests in order by their importance in ecosystem processes divided by how much we know about them, there can be no doubt that mycorrhizae would be right up at the top. Mycorrhizae, the symbiotic combination of plant roots with soil fungi, are ubiquitous in forest soils and can compose a large fraction of the total fungal biomass and total microbial biomass. Mycorrhizae are longer-lived than free-living microbes, act as an extension of root systems, and are generally thought of as a mechanism by which plants increase the efficiency of nutrient uptake, but many uncertainties about processes and controls in this interaction remain.

We do know that the degree of mycorrhizal infection of roots and the species composition of the fungi involved change with nitrogen status, but we know almost nothing about the physiological processes of mycorrhizae in the symbiotic state in the field. This is especially true when compared against our knowledge of other critical processes such as photosynthesis. There are very few methods for the measurement of, for example, the simultaneous uptake of nitrogen and release of CO_2 by mycorrhizae in the field. This is one of the most important areas for future work in forest ecosystem studies. Without this knowledge, we can say only that mycorrhizae might assimilate mineral nitrogen with a low respirational cost, or they might not. Speculations on the evolutionary adaptability of such a process are premature.

Although direct studies of carbon and nitrogen processing by mycorrhizae in the field have not been undertaken, innovative isotopic approaches are beginning to provide some insights into changes in the role of mycorrhizae over nitrogen availability gradients. For example, studies on the fractionation of ^{15}N among soil organic matter, extractable nitrogen, roots, and foliage show that fractionation varies with nitrogen status. A modeling analysis suggests that increased fractional retention of nitrogen by mycorrhizae is the process most consistent with measured results. Reduced retention by mycorrhizae at high nitrogen availability may then contribute to nitrate leaching.

Synthesis and Conclusions

Results from the chronic nitrogen experiment, especially when considered along with those from similar experiments conducted elsewhere, have resolved several qualitative hypotheses about the process of nitrogen saturation. A generalized pattern of forest response to the increased availability of nitrogen has emerged that includes the following: (1) increasing rates of nitrification and nitrate leaching; (2) increasing foliar nitrogen concentration and decreasing nitrogen:element ratios; (3) initially increasing and then possibly decreasing rates of soil respiration and nitrogen mineralization; and (4) initially increasing and then perhaps decreasing rates of total net primary production (Figure 12.1). Growth declines in needle-leaved evergreens both in experimental trials and at the high end of ambient nitrogen deposition gradients here and in other places around the world suggest that reduced forest growth and yield in these types of forests in response to nitrogen deposition is a real possibility.

We have also documented a remarkable capacity for uptake and retention of nitrogen in soils and demonstrated that understanding the mechanisms responsible for this retention is the key to developing predictions of the timing of nitrogen saturation. By working on sites where the history of past land-use and land-cover changes are well-known, we have also recognized the critical role that the legacies of this history play in determining nitrogen status of forests and conditioning ecosystem response to nitrogen additions. Species composition also plays a role here. Broad-leaved deciduous forests appear to have a greater capacity to retain added nitrogen and to delay the onset of nitrogen saturation.

Why this is true remains unclear, although different physiological capacities to fix carbon, take up nitrogen, and produce biomass with higher nitrogen concentrations are logical candidates.

To put quantitative substance on the qualitative skeleton of our understanding of nitrogen saturation is more challenging. Negative effects of nitrogen saturation all relate to production and leaching of nitrate, so the key policy question relating to nitrogen saturation is the timing, magnitude, and spatial extent of excess nitrogen availability and the associated increases in nitrate leaching, cation loss, and soil and water acidification. Previous research has developed quantitative relationships regarding nitrogen availability and net primary production and also between litter quality and decomposition rates. The largest remaining unknown involves the greatest sink for added nitrogen in forest ecosystems, the organic and mineral soils pools. Consequently, although nitrogen deposition and saturation appears to pose a potential threat to forest ecosystems, we cannot yet judge the significance of that threat because of a lack of understanding of nitrogen cycling, both biotic and abiotic, belowground.

CHAPTER 13

Soil Warming
A Major Consequence of Global Climate Change

**J. MELILLO, P. STEUDLER, J. ABER, K. NEWKIRK,
H. LUX, F. BOWLES, C. CATRICALA, A. MAGILL,
T. AHRENS, S. MORRISSEAU, E. BURROWS,
and K. NADELHOFFER**

Rationale and Overview

Current models of climate change predict that as human activity increases the amounts of greenhouse gases such as CO_2, methane (CH_4), and nitrous oxide (N_2O) in the atmosphere, global mean temperature will rise over the twenty-first century, perhaps by as much as 1.4° to 5.8°C. Rising sea level, alterations of agricultural production, and gradual changes in the distributional ranges of animals and plants are among the many potential ecological consequences of temperature increases in this range. Some of the most important effects would be changes in a suite of soil processes that influence ecosystem function but also have critical linkages to atmospheric and global processes. Temperature-dependent belowground processes that are of particular interest include plant root respiration, organic matter decomposition, CH_4 production and oxidation, nitrogen mineralization, nitrification and denitrification, and phosphorus availability. These processes are involved in the fluxes of greenhouse gases and/or in ecosystem production. Understanding the direction, magnitude, and duration of their responses to temperature increases is important. Among the key questions concerning the effect of climate change on the soil environment are the following:

- What are the types and magnitudes of ecosystem responses?
- Are responses transient or persistent?
- Are there important feedbacks between these responses and global climate change?

There are many approaches to estimating ecological responses to global change, including simulation modeling, microcosm or closed-system experiments, and reductionist studies on individual organisms.

One of the most promising for studying the responses of belowground processes is carefully controlled and long-term manipulation of the soil temperature environment. Such studies seek to manipulate one component of intact ecosystems—the soil—generally using buried heating cables to maintain the temperature at a certain level above ambient conditions.

Soil-warming experiments in a number of ecosystems have shown a range of responses to increased temperatures, including increased fine root mortality in a yellow birch stand in New England and increased decomposition, nutrient availability, and foliar nutrient concentrations in an Alaskan black spruce stand. The warming of cores containing both plants and soil from a wet coastal area of the tundra of northern Alaska resulted in reduced ecosystem carbon storage, whereas the warming of an adjacent wet sedge area caused increases in concentrations of inorganic nitrogen, phosphorus absorption, and plant growth. Overall, these studies reinforce the interpretation that elevated soil temperatures from global warming have the potential to dramatically alter the carbon, nitrogen, and phosphorus cycles of ecosystems.

As a result of feedbacks among soil and ecosystem and atmospheric processes, soil warming may also have global effects, but the net effect of these interactions on a global scale is unclear. On one hand, faster decomposition of soil organic matter may enhance global warming substantially by increasing the release of CO_2 from soil to the atmosphere. On the other, this increase in organic matter decomposition may increase the availability of nitrogen for plant growth, increasing carbon storage in plants and litter.

To explore this complex set of interactions within the soils, we initiated a soil-warming experiment at the Harvard Forest in 1991. The purpose of this experiment was to determine the response of soil processes in a typical mixed deciduous forest to elevated soil temperatures, with special emphasis on soil processes that could substantially alter ecosystem function, atmospheric chemistry, and global climate. We were interested in considering warming as a stress that, like nitrogen deposition or logging and wind disturbance, would initiate forest ecosystem responses, but also sought to link those responses with broader processes. The initial ten years of the experiments generated some very interesting insights on both.

Experimental Design

The soil-warming experiment was established on the Prospect Hill tract in an even-aged, mixed deciduous forest dominated by black oak, red maple, paper birch, and striped maple that was quite similar to the forests in the other experiments. The soil profile and historical records indicate that the site underwent historical uses that are typical

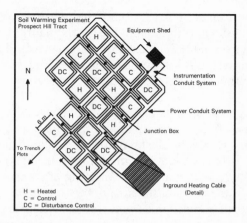

Figure 13.1. Design and layout of the soil-warming experiment at the Harvard Forest. The experiment had six replicates of three treatments: control plots where no manipulation occurred, disturbance-control plots where heating cables were installed but not turned on, and heated plots where buried heating cables maintained a 5°C temperature differential above the control plots through continuous monitoring by buried thermistors and a data logger. Modified from Peterjohn et al. 1994, with permission from the Ecological Society of America.

for the region. After deforestation the area was pastured, though not plowed, until the late 1890s, when it was abandoned and reforested naturally. Since that time, the regrowing forest was cut for firewood once.

In April 1991, we established eighteen 6–by–6–meter plots assigned to one of three treatments: (1) heated plots in which the average soil temperature at 5 centimeters was elevated 5°C above ambient using buried heating cables; (2) disturbance control plots that had buried heating cables identical to heated plots but that received no electric power; and (3) control plots that were left in their natural state (Figures 13.1 and 13.2). In the heated and disturbance control plots, heating cables were buried at a depth of 10 centimeters in rows spaced 20 centimeters apart. Electrical current was supplied only to the cables in the six heated plots. Application of the current was controlled by computers connected to an array of thermistors buried at a depth of 5 centimeters in all of the plots. These served to maintain the 5°C temperature difference between the heated plots and the others. The technology for heating soils is very well-developed and has many applications. In fact, the cables that we used are similar in design to those used in Mile High Stadium, where the Denver Broncos play football. Nonetheless, to develop an effective system to ensure closely controlled temperatures, we had to undertake considerable engineering and field experiments. Among the results from a study that we undertook on Cape Cod before establishing the soil-warming experiment at the Harvard Forest, we learned that soil temperatures

Figure 13.2. Several snow-free heated plots in late winter. In addition to raising soil temperatures and altering ecosystem processes, soil warming lengthened the growing season and decreased the duration of snow cover. Photograph by H. Lux.

were consistently lower near the edges of a heated area. Therefore, to avoid places where the temperature difference is less than 5°C, we made all of our measurements in a 5-by-5-meter area in the center of each plot.

Over the course of the study, we measured trace gas fluxes, various indices of nitrogen availability, and soil water content. Taken together, these measurements allowed us to quantify key biogeochemical responses of this forest ecosystem to soil warming. The three gases that we chose to measure—CO_2, CH_4, and N_2O—can accumulate in the Earth's atmosphere and trap enough heat radiating from the surface to cause global warming. As discussed extensively in the previous chapter, nitrogen is thought to be the nutrient that most limits forest growth in New England. Soil water affects both the carbon and nitrogen cycles in terrestrial ecosystems.

Soil Responses to the Treatment
Soil Temperature

Our soil-heating approach successfully and consistently elevated soil temperatures 5°C above ambient soil temperature throughout the ten years of the experiment (Figure 13.3). Although most of our measurements of soil temperature came from the upper 20 centimeters of the

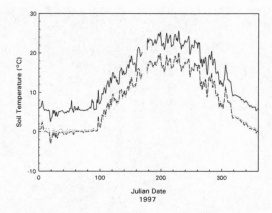

Figure 13.3. Daily soil temperatures in heated (solid line), disturbance control (broken line), and control (dotted line) plots over the course of one year (1997). The array of heating cables was capable of maintaining a sustained 5°C increase in heated plots over a wide range of temperatures and environmental conditions.

soil profile, we were able to verify that heating does affect soil temperatures to a depth of at least 1 meter. We also confirmed that the soil disturbance associated with the installation of heating cables had no effect on soil temperatures and only minor and variable effects on soil moisture.

Interannual Variability in Soil Respiration

Carbon dioxide flux from soils, which reflects respiration of both plant roots and soil microorganisms, exhibited substantial interannual variability across treatments. The highest annual emission rate of CO_2 was 1,090 grams per square meter from the heated plots during 1994 (Table 13.1). Interannual variability in climate parameters, temperature, and precipitation, all of which influence decomposition rates, appear to be responsible for these year-to-year variations. A warm and extremely dry period between May and August 1995 resulted in the lowest absolute fluxes for each of the treatments for the field measurement period, April through November.

Q_{10} Relationships between Temperature and CO_2

Over the course of our study, we observed a strong exponential relationship between soil temperature and CO_2 flux for each treatment (Figure 13.4). Initially, the Q_{10} values, which represent the change in respiration associated with a 10°C change in temperature, were very similar for all treatments. However, through time the Q_{10} value for the heated plots declined (mean of 2.4), while the annual Q_{10} values for the

Table 13.1. Cumulative Yearly CO_2 Flux by Treatment

Year	Control	Distrubance Control	Heated
1991	480	550	770
1992	660	730	870
1993	730	790	970
1994	900	870	1,090
1995	590	570	720
1996	640	610	740
1997	730	710	780
1998	890	880	940
1999	760	780	770
2000	720	730	780

Note: 1991 is a partial year (July through November). Values are grams carbon per square meter per year.

control and disturbance control plots remained similar (means of 2.8 and 3.0, respectively) (Figure 13.5). The Q_{10} values derived from these measurements on the three treatments bracket the median value of 2.4 calculated from an extensive review of the literature.

For each of the treatments, we observed substantial year-to-year variability in Q_{10} values. Annual Q_{10} values for the control plots ranged from 2.0 to 3.7, whereas those for the disturbance control plots ranged from 2.1 to 4.0 and for the heated plots from 1.7 to 3.4. The lowest Q_{10} values for all of the plots occurred during the very dry late spring and summer of 1995. There is a strong linear relationship between summer precipitation for the months of June, July, and August and Q_{10} values for the ten full years of soil respiration measurements (Figure 13.6).

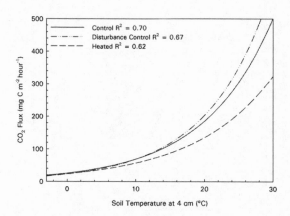

Figure 13.4. CO_2 flux as a function of soil temperature for each treatment over the period 1991–2000. Heating altered the exponential relationship of soil CO_2 flux with temperature, showing less sensitivity over time.

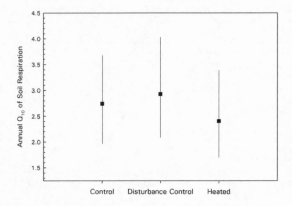

Figure 13.5. The mean and range of Q_{10} values calculated for soil respiration in the soil-warming experiment over the ten years of the study.

CO_2 Fluxes: Patterns Change through Time

Soil warming dramatically increased the CO_2 flux from soils, but this response changed over time (Figure 13.7). In this part of the chapter, we report results from the disturbance control plots as the reference plots, since they are most analogous in treatment to the heated plots. Over the first six years, soil respiration increased on average by about 28 percent. By year ten there was no measurable difference between the CO_2 emissions from the heated and disturbance control plots.

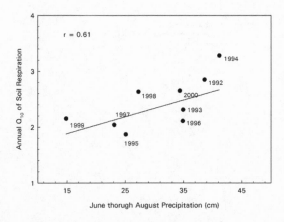

Figure 13.6. The relationship between Q_{10} values calculated for soil respiration each year from 1992 to 2000 for all three treatments and the precipitation for June, July, and August for these years in the soil-warming experiment.

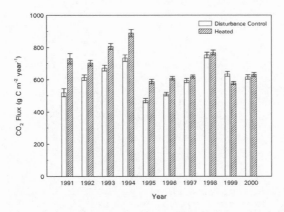

Figure 13.7. Soil respiration by treatment and year on the soil-warming experiment for the field measurement period of April through November. Soil respiration was measured using syringes to draw a time series of gas samples from closed-top chambers (static chamber technique). Samples were analyzed in the lab using an infrared gas analyzer, and rates were calculated from changes in concentration observed over time. Error bars represent the standard error of the mean (n = 6 plots) for each treatment. Based on Melillo et al. 2002.

This lack of a difference likely reflects the fact that warming had exhausted an easily decomposed carbon pool.

To estimate the size of this easily decomposed carbon pool we had to partition the two components of soil respiration—microbial respiration and root respiration. We carried out a second field experiment adjacent to the soil-warming experiment that involved trenched plots to exclude roots and a soil-warming treatment, in a full-factorial design in which there were three plots per treatment. We estimated that root respiration is about 20 percent of the total soil respiration, with microbial respiration accounting for the remaining 80 percent (Figure 13.8).

Combining the soil respiration data from the warming study with the data from the partitioning study, we estimate that carbon loss stimulated by warming for the entire ten-year period was 944 grams per square meter. This amounts to about 11 percent of the soil carbon found in the top 60 centimeters of the soil profile.

These results underscore the importance of long-term experiments. The large initial response, if extrapolated, would produce a large overestimate of the potential efflux of CO_2 from soils in a warmer environment.

Evidence for Exhausting the Labile Soil Organic Matter Pool

The overall decrease in the Q_{10} values in the heated plots over ten years suggests that there has been a reduction in the size of the car-

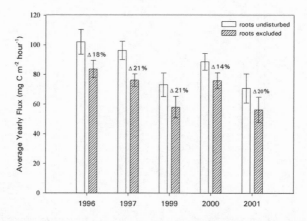

Figure 13.8. Average yearly fluxes at the trenched-plot experiment. Data represent average yearly CO_2 fluxes from unheated plots only, as measured by the static chamber technique. Percent values represent the contribution of roots to total soil respiration. Error bars represent the standard error of the mean (n = 3 plots) for each treatment. The same relative patterns were observed in the treatments that included heating and trenching. Modified with permission from Melillo et al., *Science* 298:2173–2176 (copyright 2002, American Association for the Advancement of Science).

bon pool that is readily available for decomposition. The narrowing of the differences between soil respiration rates in the heated and disturbance control plots is consistent with this interpretation. Thus, available evidence suggests that the CO_2 flux response to increased global temperatures in temperate forest soils will be short-lived unless carbon inputs are also increased. Inputs could be increased if temperature increases (or concomitant environmental changes associated with global climate change) elevate the production of leaf or root inputs to the soil organic pool.

Effects of Treatment on Methane Fluxes

Methane fluxes represent the net result of the activities of both CH_4-producing and CH_4-consuming soil microorganisms. Between 1991 and 1995, the average net uptake rate of CH_4 from the atmosphere in the heated plots was about 20 percent higher than in the control and disturbance control plots (Figure 13.9). Interestingly, we found that water-filled pore space, an index of soil water content, is a good predictor of CH_4 flux (Figure 13.10). Therefore, as soil temperatures increase, soil water content decreases, CH_4 uptake in the soil increases, and concentration of CH_4 in the atmosphere declines. This response further complicates the global change interpretation, because two major greenhouse gases (CO_2 and CH_4) respond in different directions to soil warming.

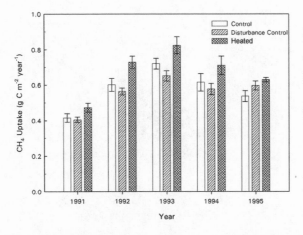

Figure 13.9. Methane uptake by treatment and year in the soil-warming experiment for the field measurement period of April through November. Uptake rates in the heated plots were generally 20 percent higher than those observed in the disturbance control plots. Error bars represent the standard error of the mean ($n = 6$ plots) for each treatment.

Effects of Treatments on Soil Nitrogen Fluxes

Nitrogen is a key limiting nutrient in forest ecosystems of New England (see Chapter 12). We therefore focused on three aspects of the nitrogen cycle:

- Net nitrogen mineralization and nitrification
- Concentration of inorganic nitrogen in soil water
- Nitrous oxide fluxes between the soil and the atmosphere

Net nitrogen mineralization and net nitrification were measured for the first eight years of the study. In this part of the chapter, we again report results from the disturbance control plots as the reference plots, since they are most analogous in treatment to the heated plots. Warming increased net nitrogen mineralization throughout this period (Figure 13.11). Overall, net nitrification rates were low in all plots and years, and in no year was it more than about 5 percent of net nitrogen mineralization (Figure 13.12). We observed neither large amounts of nitrogen (inorganic or organic) moving through the soil profile below the rooting zone nor large N_2O fluxes from the soil in any treatment (Figures 13.13 and 13.14). Presumably, this is due to the fact that these soils, like others at the Harvard Forest, have an ammonium economy with low rates of nitrification, low concentrations of nitrate, and very efficient immobilization of mineralized nitrogen. With little nitrate being produced, there is minimal leaching of inorganic nitrogen. Nitrate is also the primary form of nitrogen used by denitrifying organisms, the producers of N_2O, and so the lack of nitrate production leads to low N_2O production.

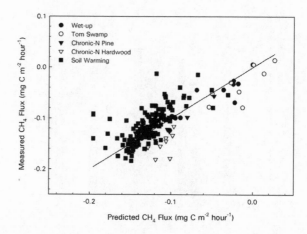

Figure 13.10. Relationship between the observed CH_4 fluxes on a number of field and experimental sites and those predicted with a simple regression model: CH_4 consumption = [0.00175 × (% water-filled pore space) − 0.1957], R^2 = 0.74. Modified from Castro, Melillo et al. 1995, with permission from NRC Research Press.

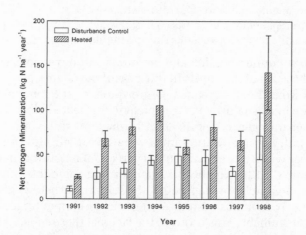

Figure 13.11. Annual rates of net N mineralization for the organic horizon plus the top 10 centimeters of mineral soil for the disturbance control and heated plots in the soil-warming experiment. Soil cores were incubated in situ and extracted in the lab using 2N KCl (buried bag technique). Error bars represent the standard error of the mean (*n* = 6 plots) for each treatment. Modified with permission from Melillo et al., *Science* 298:2173–2176 (copyright 2002, American Association for the Advancement of Science).

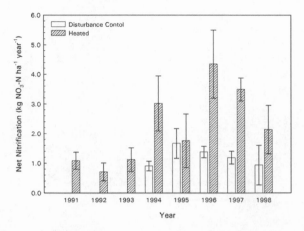

Figure 13.12. Annual rates of net nitrification for the organic horizon plus the top 10 centimeters of mineral soil for the disturbance control and heated plots in the soil-warming experiment. Nitrification was consistently a small contribution to net mineralization in all of the soil-warming plots. Error bars represent the standard error of the mean ($n = 6$ plots) for each treatment.

Since we measured minimal leaching or gaseous losses of nitrogen associated with this accelerated nitrogen mineralization rate, the "excess" inorganic nitrogen released through mineralization appears to have remained in the system. This result, which parallels the observations in the chronic nitrogen experiment (Chapter 12), raises an important set of questions:

- What is the mechanism for the capture of the "excess" nitrogen, and what is its fate?
- Is it transferred to deeper portions of the soil profile and stored there?
- If yes, what is the storage mechanism?
- Is it taken up by trees and shrubs?
- If yes, is it stored in woody tissue?

Answers to these questions will give us insight into the diverse ways in which warming and nitrogen deposition will affect carbon storage in terrestrial ecosystems. The observations and questions reflect the way in which identical fundamental biological processes can emerge as central issues in studies like chronic nitrogen and soil warming that initially were quite independent.

Global-Scale Consequences of Accelerated Carbon and Nitrogen Cycles in a Warmer World: Some Preliminary Evidence

Our results indicate that warming generally increases soil respiration. A number of scientists have argued that in many ecosystems the

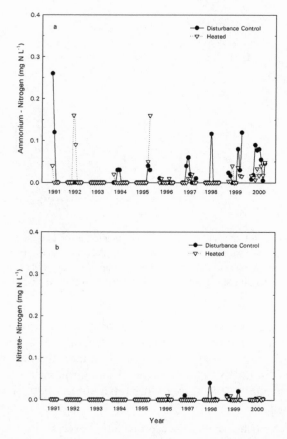

Figure 13.13. Soil solution ammonium (a) and nitrate (b) collected from tension lysimeters placed below the rooting zone at 50 centimeters in the soil-warming experiment. Nitrogen is largely retained in this system, in all treatments. Note low maximum concentrations.

increase of soil respiration with temperature is steeper than any increase of net primary production stimulated by the temperature increase, so that the net effect of warming is to reduce carbon storage and increase atmospheric CO_2. But warming may not always lead to a reduction of carbon storage in terrestrial ecosystems. In nitrogen-limited forests, particularly in cold climates, warming may increase carbon storage through a two-part mechanism. First, warming may increase the decay rate of labile soil organic matter with a low C:N ratio. This may release nitrogen into the soil solution and some CO_2 into the atmosphere. Subsequently, trees may take up this nitrogen from the soil solution and store it in woody tissues with a high C:N ratio (Figure 13.15). Thus, the transfer of nitrogen from the soil to the trees may lead to an increased net carbon storage in the ecosystem.

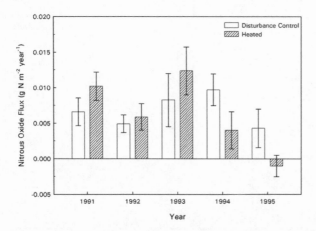

Figure 13.14. Nitrous oxide flux by treatment and year for the field measurement period of April through November in the soil-warming experiment. Nitrous oxide losses from the soil were consistently low. Error bars represent the standard error of the mean ($n = 6$ plots) for each treatment.

We have some early evidence that this nitrogen transfer may be happening in the heated plots. Our measurements of ^{15}N natural abundances in the soil-warming plots suggest that elevated temperatures have altered the pattern and rates of nitrogen cycling in the ecosystem. Nitrogen in fine roots, in soil microbes, in soil solutions, and in both the bulk and acid-insoluble (refractory) forest floor all show trends of in-

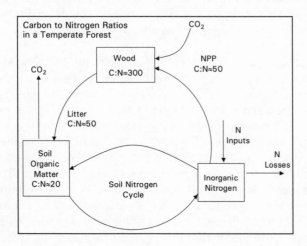

Figure 13.15. A model of the soil N cycle. Carbon-to-nitrogen mass ratios in net primary production (NPP), vegetation, litter, and soil organic matter for a temperate deciduous forest ecosystem.

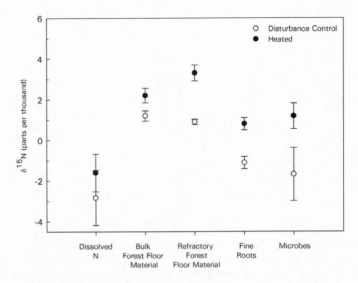

Figure 13.16. The δ^{15}N in fine roots, microbes, and soil N pools for the disturbance control and heated plots in 1992. Higher (more positive) δ^{15}N values indicate higher ratios of ^{15}N to ^{14}N for nitrogen in a sample. Trends suggest that N turnover from more recalcitrant pools is being accelerated in the warmed plots. Error bars represent the standard error of the mean (n = 3 plots) for each treatment. From Nadelhoffer et al., unpublished.

creasing ^{15}N abundances with warming (Figure 13.16). These patterns suggest that warming has increased turnover rates of relatively old, stable, ^{15}N-enriched soil organic matter, thereby transferring nitrogen from slow to rapidly cycling pools. This could lead to greater fluxes of nitrogen from soil to woody vegetation and to increased carbon sequestration, as noted above.

How much carbon might be sequestered at the ecosystem level through this "nitrogen transfer" mechanism? We can begin to address this question by using the results of our long-term nitrogen fertilization study at the Harvard Forest (Chapter 12). In that study we have been adding almost 5 grams of nitrogen per square meter per year for more than a decade to a hardwood forest stand very similar to the one we have been using for the soil-warming study. At the end of nine years we estimated that 12.7 percent of the total amount of nitrogen fertilizer added (approximately 41 grams of nitrogen per square meter) ended up in the woody tissue of the stand's trees. If we assume that 12.7 percent of the increased amount of nitrogen made available by warming ended up in woody tissue, we estimate that this would result in an additional 1,560 grams per square meter of carbon storage in the vegetation over the decade of warming. Combining this vegetative storage estimate with our estimate of soil carbon loss due to warming, we calculate that for this forest ecosystem, warming could have stimulated at least 600 grams per square meter of carbon storage over the decade, thereby creating a negative feedback to the climate system. The magnitude of this type of negative feedback in a climate-changed world would depend on more than soil warming. Carbon storage in woody tissue would also be affected by other factors related to aspects of climate change, including the availability of water, the effects of increased temperature on both plant photosynthesis and aboveground plant respiration, and the atmospheric concentration of CO_2.

Despite these uncertainties, there is some direct field evidence that supports this kind of negative-feedback mechanism. One example comes from a soil-warming study in a Norway spruce forest at Flakaliden in northern Sweden. Using a soil-warming design based on our Harvard Forest study, Linder and colleagues warmed large (10-by-10-meter) forest plots in a long-term study. After five years they found that there was a significant (more than 50 percent) increase in stem-wood growth of the trees on the heated plots relative to the controls.

The Harvard Forest warming study demonstrates the potential importance of understanding how microbial respiration in soils responds to global warming and how temperature-driven changes in the nitrogen cycle in forest ecosystems will affect their capacity to store carbon. If climate change leads to a net carbon loss from an ecosystem, the feedback to the climate system is positive, whereas if climate change leads to a net carbon gain for the ecosystem, the feedback to the climate system is negative. In addition, this study shows that warming can increase the capacity of soil microbes to oxidize the powerful greenhouse gas CH_4, thus leading to a negative feedback to the climate system. The possibility that warming can lead to both positive and negative feedbacks to the climate system suggests that it is essential that we accurately represent these feedback mechanisms in integrated global systems models designed to project possible future climates.

CHAPTER 14

Comparisons between Physical Disturbances and Novel Stresses

D. FOSTER, J. ABER, R. BOWDEN, J. MELILLO, and **F. BAZZAZ**

The three experimental treatments—hurricane pulldown, chronic nitrogen addition, and soil warming—simulate and enable us to begin to contrast the effects of important processes that have exerted and will continue to exert significant impacts on the forests of New England. Because we seek to understand long-term forest dynamics and work closely with land-management, conservation, and environmental organizations that develop forest and environmental policies, we are keenly interested in evaluating the long-term effects of these disturbances and stresses on forest ecosystem structure and function.

The results of our experiments yield surprises and a paradox. The different treatments generate strikingly different physical effects, ranging from apparent catastrophic disturbance in a jumble of uprooted trees to seemingly healthy ecosystems subjected to altered nitrogen and temperature regimes. And yet the long-term functional differences among forests subjected to these different treatments run in direct opposition to these first appearances. Functionally, nitrogen deposition and soil warming move forest ecosystem function in novel directions that contrast strongly with the intact forest in adjacent controls or the physically disturbed forest in the hurricane experiment. In contrast to the long-term changes initiated from chronic additions of nitrogen, recovery of key forest processes after the hurricane simulation was remarkably rapid. The unseen changes generated by nitrogen deposition and soil warming have important ramifications on local, regional, and global scales for our New England forest; however, the hurricane is very much "business as usual."

Despite the dramatic appearance of the hurricane blowdown, the major ecological effect was a vertical reorganization of the canopy and foliage from more than 20 meters in height to 1 to 5 meters above the ground surface. The combination of high survival and sprouting rates of damaged trees and rapid understory plant growth resulted in little change in leaf area and soil microenvironment. Most important for a range of ecosystem processes, soil temperature and moisture did not change detectably, as the prolific growth of plants shaded the ground and apparently maintained active uptake of moisture and nutrients. Diverse morphological responses of the surviving trees (for example, epicormic branching and stem, basal, and root sprouts) resulted in rapid redevelopment of leaf area and continuity in composition and foliage cover. As damaged trees have died off gradually, the growth of saplings and understory plants has provided additional leaf area. Because of the high rates of survival and vegetative reproduction, floristic composition has changed less than might be expected based on successional theory and on previous studies of the 1938 storm. Specifically, instead of a rapid influx of early-successional and weedy species, the site continues to be dominated by species characteristic of mature forests.

From an ecosystem perspective, the blowdown experiment is highly instructive. Despite massive structural alteration of the forest, net energy and nutrient processes remained largely intact. Productivity, as measured by litterfall, declined after the disturbance but recovered to 71 percent of predisturbance rates within four years. The similarity of nitrogen cycling, soil respiration rates, and nitrous oxide effluxes in the control and blowdown plots indicates that changes in nutrient availability were minimal. Continuous plant production and cover provided a high degree of biotic control over critical microclimatic factors and important ecosystem processes.

In contrast to damage in the blowdown experiment, the chronic nitrogen and soil-warming experiments simulate novel environmental stresses to which no visible, large-scale, integrated community responses occur. Tree growth and litterfall, for example, have changed slowly; vegetation composition and structure have not been altered. However, despite the absence of obvious community responses and every visible indication of ongoing health of the forest, measurable changes in ecosystem function are occurring (Table 14.1). In the chronic nitrogen addition plots, nitrate leaching is being induced, major changes in trace gas balances have occurred, and forest productivity has

Table 14.1. Changes in Ecosystem Processes Caused by Experimental Blowdown, Chronic Nitrogen Addition, and Soil Warming (percent)

Ecosystem Process	Blowdown	Chronic Nitrogen Addition	Soil Warming
Mineralization	+15.9 (1)	+138 (6)	+50 (2)
Methane uptake	−2.4 (1)	−36 (2, 3, 6)	+20 (2)
Soil respiration	+6.2 (1)	0	+76 (1)
Nitrate leaching	—	0	0

Source: Foster, Aber et al. 1997.
Note: The chronic nitrogen results compare the hardwood control plot with high-nitrogen plots. Numbers in parentheses refer to the year posttreatment for which data are shown.

declined. In the soil-warming plots, soil CO_2 balances have become negative, nitrogen cycling has increased dramatically, and nitrogen losses (as nitrate) are increasing. In time, altered chemical or physical environments caused by these novel stresses create altered nitrogen concentrations and altered rates of carbon and nitrogen cycling, which in turn alter ecosystem productivity. Although we cannot predict what future course these changes will follow because there is no historical analogue for these experiments, we expect that ecosystem function will be disrupted more in the long run by these novel disturbances than by physical disturbances because none of the major species present has evolved in an environment that contains these stresses. The type of plant response mechanisms seen in the hurricane experiment apparently do not exist for the stresses induced by the chronic nitrogen or soil-warming experiments.

Which of these treatments, physical disturbance or climatic and chemical stress, is actually most disruptive to the integrity of the community and most likely to lead to long-term changes in ecosystem function? Whereas the blowdown site appears severely disturbed, internal processes have not been altered significantly, and the stand is on a path to recovery of structure and function in keeping with the cyclic pattern of disturbance and development of this forest type. By contrast, the chronic nitrogen and soil-warming plots are visually intact and apparently healthy (with the exception of the decline in the pine stands receiving high nitrogen), yet the subtler measures of ecosystem function suggest serious imbalances with possible future implications for community structure, internal ecosystem processes, and exchanges with the global environment.

Reinterpreting the Role of the 1938 Hurricane

The understanding of temperate forest response to tropical storms is strongly influenced by studies of the 1938 hurricane. However, a comparison of our hurricane manipulation with the 1938 hurricane illustrates important differences and highlights critical issues that argue for a reevaluation of some interpretations based on the 1938 event.

The 1938 hurricane produced dramatic changes in forest composition that involved major increases in early-successional species, in striking contrast to the blowdown experiment. The 1938 hurricane also generated major changes in regional hydrology (see Figure 11.3). For a five-year period after the hurricane, stream flow increased greatly from the damaged watersheds, such as the Connecticut and Merrimack Rivers, as compared with the undamaged Androscoggin River, a result that has been interpreted historically as a natural consequence of hurricane-induced reduction in evapotranspiration. The lack of major compositional change, the relatively unimportant role of successional species, and the absence of a change in soil moisture in our experimental hurricane plot encouraged us to reanalyze the information and interpretation of the effect of the 1938 hurricane on the New England landscape.

The 1938 storm severely damaged a landscape in which white pine, a species that is highly susceptible to windthrow, covered extensive areas that had been abandoned from agriculture in the previous century (see Chapter 4). The storm was followed by the single largest timber-salvage operation in U.S. history, as more than 4.5 million workdays were expended in harvesting more than 1.5 billion board feet of timber—approximately 40 percent of the damaged timber across the New England states (see Figure 11.13). Salvage activity was associated with road development, soil scarification, and burning of residual slash. Loggers commonly cut broken, leaning, uprooted, and even undamaged stems. Consequently, given the novel forest composition of the region and the geographical extent of the salvage effort, the 1938 hurricane may be interpreted more accurately as a regional logging operation of a cultural landscape than as a natural disturbance. Logging-associated tree mortality, removal of leaf area and biomass, and disruption to the forest floor resulted in substantial modification of the understory and soil environment and prevented the rapid recovery observed in the blowdown experiment. In addition, loss of biotic control over ecosystem processes resulted in major changes in hydrology and, presumably, biogeochemistry.

In the simulated hurricane blowdown, by contrast, increases in resource availability were used quickly by survivors—sprouts and seedlings of many species adapted to natural disturbance. This rapid utilization of resources led to minimal change in vegetation composition and rapid recovery of critical ecosystem processes. The chaotic appearance of the recovering community results directly from a wide range of regenerative responses by the species in the system.

These results may have relevance for land managers in the face of many natural disturbances. As has been shown through many studies of fire in forest ecosystems and a recent investigation of forest mortality from the hemlock woolly adelgid, human attempts to salvage or minimize forest damage may often lead to more severe and long-lasting effects than the disturbance itself. If environmental integrity and ecosystem function, measured by biotic control of nutrient cycling, transpiration, etc., are major objectives, then allowing the disturbance to take its course and leaving the damaged ecosystem intact may be the preferred strategy.

CHAPTER 15

The DIRT Experiment
Litter and Root Influences on Forest Soil Organic Matter Stocks and Function

K. NADELHOFFER, R. BOONE, R. BOWDEN,
J. CANARY, J. KAYE, P. MICKS, A. RICCA,
W. McDOWELL, and J. AITKENHEAD

"Oh, I'm hoping for a thousand years at least."
—FRANCIS HOLE's reply when asked how long his soil experiment
should be maintained

Rationale and Overview

Organic matter is a key component of forest soils. Important properties such as moisture-holding capacity, aeration, and nutrient retention are strongly influenced by, and typically increase with, the amount of soil organic matter present. Organic matter in forests is a major reservoir for nutrients and carbon that fuel microbial processes and support complex communities of soil and forest floor organisms. Because annual inputs of limiting nutrients like nitrogen are low relative to annual demand (see Chapters 3, 10, and 12), plants depend mainly on nutrients released from decomposing organic matter to meet their nutritional requirements. Therefore, the amount and the "quality" (that is, the decomposability and relative amounts of nitrogen and different carbon compounds) of soil organic matter may strongly influence tree growth and forest dynamics. In turn, inputs of fine litter from aboveground (leaves, twigs, seeds, etc.) and belowground (mostly fine roots) determine the amount and quality of organic matter and nutrients in forest soils.

At the Harvard Forest, where the long history of land use has produced changes in soil organic matter and nutrient content, as well as major shifts in forest composition, many questions arise concerning feedbacks between the plants and soils. In particular, we need to understand the rate of incorporation of organic matter into soils, the rate at which soils impoverished in organic matter recover their stocks of carbon and nutrients, and the relative importance of belowground versus aboveground inputs of organics in these processes.

Recognition of the importance of feedbacks from plants in determining soil nutrient dynamics and carbon storage has led to a large number of studies of the rate and byproducts of decomposition of different types of litter. In many such studies, known amounts of plant material are placed in a fine mesh bag in or on the soil surface in forests. The decomposition processes are then typically followed for two to five years, or until 20 to 40 percent of the original litter material remains. Such studies have yielded much information about the roles of litter nutrient content and carbon quality in controlling the relatively rapid cycling of nutrients through the litter layer.

For example, litters with high concentrations of soluble carbohydrates and cellulose, such as leaves of sugar maple or ash, decay faster and both immobilize and mineralize nutrients earlier in the decay sequence than do litters with high concentrations of lignin and other complex polyphenolic compounds, such as leaves of oak or beech. Also, litters with relatively high nutrient concentrations tend to decompose quickly and to release nutrients (which may then be available to plants) more rapidly than do litters with low initial nutrient concentrations. Because individual plant species and different plant tissues often differ in litter chemistry, litter inputs to soils from the various species within a forest ecosystem partially regulate the rates at which nutrients become available to plants. In this way, forest composition may exert a strong influence on soil characteristics, ecosystem processes, and site productivity.

Far less is known about the longer-term fate of aboveground and belowground plant litter and its role in determining soil organic matter content and function over timescales ranging from decades to centuries. Because humus (well-decomposed litter) typically contains most of the nutrients and at least half of the carbon in forest ecosystems, this lack of understanding as to how plant processes influence humus formation represents a critical gap in knowledge about forest ecosystem function. To address this, we established a long-term study of the factors controlling soil organic matter formation: the DIRT (Detritus Input Removal and Transfer) project. The goal of the DIRT project is to assess how rates and sources of plant litter inputs control the accumulation and dynamics of organic matter and nutrients in forest soils over decadal timescales.

Our project is inspired by the work of Professor Francis D. Hole at the University of Wisconsin Arboretum. In contrast to many arboreta, the Wisconsin arboretum is much more than a horticultural collection. Established in the 1930s, its mission has been to re-create and manage a variety of ecosystems representative of those that confronted European settlers on their arrival to the midwestern United States in the nineteenth century. In the early 1950s, plant ecologist and arboretum Director John Curtis challenged a young Dr. Hole to design a long-term study of soil formation within the arboretum. Curtis's idea was that the restoration of plant communities required much more information about soil-forming processes and plant-soil interactions than was available at the time. Francis Hole, recognizing that the university had made a commitment to sustaining the arboretum as a long-term research site, devised an elegant and powerful experiment to meet this challenge and to address these fundamental but very practical questions about soil processes.

Dr. Hole located his study in native oak forests and prairies. His design called for a series of simple but sustained long-term manipulations of plant inputs to soils, coupled with periodic sampling to assess long-term changes in soil structure and properties. Treatments included altering the inputs of aboveground litter such as leaves and twigs and, in grasslands, the inputs of roots to soils belowground. Experimental treatments at Dr. Hole's plots in the Wisconsin arboretum were started in 1956 in two native forests and a restored prairie. They have been maintained for more than four decades through the ongoing efforts of Dr. Hole, arboretum staff, students, and community volunteers. We were allowed to sample at the Wisconsin forest plots in 1984 and again, in grassland and forest plots, in 1997. Results from the arboretum plots provide us with valuable long-term information against which the effects of the first decade of our experimental treatments at the Harvard Forest can be compared.

Controls on organic matter accumulation in soils has been a core theme of the National Science Foundation's LTER program since its inception. As part of this large program, we modified Dr. Hole's experimental design and established the DIRT project as a long-term intersite experiment, comparing the Harvard Forest with a nutrient-rich maple forest in Pennsylvania (Allegheny College Bousson Environmental Research Reserve) and a temperate rain forest in Oregon (H. J. Andrews Experimental Forest, U.S. Forest Service). Our hope is to develop additional linkages to similar experiments located across climate, vegetation, and soil texture gradients. This will allow an assessment of the importance of a range of physical and biological factors in controlling the accumulation of soil organic matter.

Experimental Design

Treatments in the DIRT experiment consist of chronically altered aboveground and belowground inputs of plant materials to permanent plots in a mid-successional oak-maple-birch forest in the Tom Swamp tract in close proximity to the experimental hurricane (see Figure 2.8). The manipulations, which are modified from Francis Hole's design (Figure 15.1) were started in the fall of 1990 and are as follows:

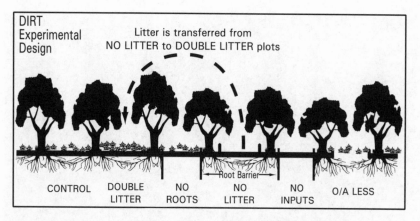

Figure 15.1. A conceptual diagram of the long-term Detritus Input Removal and Transfer (DIRT) experiment on the Tom Swamp tract of the Harvard Forest. Through a simple set of manipulations that can be carried out relatively easily over many decades, the project can assess many fundamental processes involved in the incorporation and dynamics of organic matter in soils. The surface organic soil horizon (Oea) is shown in black.

Treatment	Manipulation
CONTROL	Normal litter inputs
DOUBLE LITTER	Twice aboveground litter inputs
NO ROOTS	Roots excluded from plots by lined trenches
NO LITTER	Aboveground litter excluded from plots
NO INPUTS	No aboveground litter and no roots
IMPOVERISHED SOIL (O/A-LESS)	Organic and A horizons replaced with B horizon soil, normal inputs thereafter

Each of the 3-by-3-meter plots is located beneath an intact forest canopy and is replicated three times (Figure 15.2). The plots are placed between trees so that no stems are rooted in them, and the ground vegetation is removed as needed by clipping and occasional herbicide applications. Aboveground litter of leaves, twigs, and other fallen material is collected and excluded from NO LITTER plots with a thin mesh fabric. This collected material is then added to the DOUBLE LITTER plots in order to augment normal inputs. Root growth is prevented in the NO ROOTS treatment by excavating 1-meter-deep trenches around the plots, lining them with plastic barriers, and then back-filling the trenches with soil. The NO INPUTS treatment is a combination of the NO LITTER and NO ROOTS treatments. In the IMPOVERISHED or O/A-LESS treatment, soils were experimentally impoverished of organic matter by removing the forest floor and the upper 15 centimeters of mineral soil and replacing these with deeper, less organic-rich B horizon soil from an adjacent pit. The O/A-

Figure 15.2. A NO-LITTER plot in the DIRT experiment. In these plots, all aboveground litter (leaves, twigs, fruits, etc.) is removed and placed on an adjoining DOUBLE-LITTER plot. Low, coarse fencing keeps leaves from blowing onto the plot in fall. Instrumentation, including the lines for automated temperature probes, CO_2 flux measuring ring, and lysimeters, emerge through the soil surface. Photograph by J. Gipe.

LESS treatment does not involve ongoing manipulations beyond this initial treatment. This experimental impoverishment is intended to allow us to estimate (1) the fraction of total litter inputs (aboveground plus belowground) that is eventually transferred from litter to soil organic matter and (2) the amount of time that is required for organic-poor soils to recover to predisturbance conditions.

Our field measurements allow us to link quantitatively the changes in soil properties and processes to the amounts of carbon (energy) and nutrients entering the soils in organic matter. The value of this information will increase greatly as the manipulations continue over the next decades or centuries. In the field, we measure a number of parameters, including CO_2 fluxes from the forest floor, soil moisture, and soil temperature. In addition, we collect soil solutions (water and leachate) from beneath the forest floor using zero tension lysimeters and, at a depth of 30 to 40 centimeters, suction lysimeters. These water solutions are analyzed for ammonium, nitrate, phosphate, dissolved organic carbon (DOC), and dissolved organic nitrogen (DON). We also collect samples periodically from the forest floor and mineral soil (0 to 10 centimeters and 10 to 15 centimeters deep) in order to track changes in organic matter and nutrient contents. Soils were sampled from the plots at the beginning of the experiment in 1990 and again after one (1991), five (1995), and ten (2000) years. In the future we plan to sample once per decade. Forest floor and soil samples are assayed for total soil organic matter, carbon, nitrogen, and nutrient contents and for standard soil properties (acidity, cation exchange capacity, base saturation, texture). We have also measured CO_2 release and net nitrogen mineralization and nitrification under constant temperature and moisture conditions and gross nitrogen fluxes using ^{15}N pool dilution methods on the samples collected in Years 1 and 5. As with the field measures, laboratory results are used to quantify the effects of plant litter inputs on biological processes and carbon and nitrogen dynamics.

The biotic communities of the soils were analyzed in subsamples from our Year 5 collections. Microbes (bacteria and fungi) and microfauna (protozoa and nematodes) were counted and classified into functional categories (protozoa as flagellates, ciliates, and amoebae; nematodes by feeding type) rather than species to assess the effects of litter and root inputs on forest floor and soil biota.

Initial Results

Although the DIRT project addresses long-term questions about soil organic matter formation, plant-soil interactions, and nutrient cycling, results from the first few years of the experiment proved useful for addressing unanswered questions about important ecosystem processes. Processes investigated in the initial years of the study were fine root production, temperature sensitivities of rhizosphere (fine roots and closely associated microbes) respiration versus bulk soil respiration, and shifts in belowground community structure.

Partitioning Soil CO_2 Flux

As discussed in preceding chapters, measuring fine root production, decomposition, and respiration are among the most problematic issues in ecosystem studies. Because of the nature of our experimental design, in which some plots have roots whereas others are trenched to exclude roots, and some have new litter and others do not, we can use field measures of soil respiration during the first year after the start of manipulations along with mass balances to estimate these processes. Our mass balance approach indicated that live root respiration, production of aboveground fine litter (leaf, twig, and other fine litter), and fine root detritus each constituted about one-third of carbon inputs to soil in this stand. This suggests that fine root and leaf litter

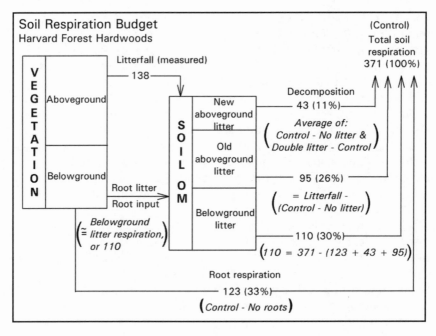

Figure 15.3. Soil respiration budget for a hardwood forest (based on the DIRT experiment) at the Harvard Forest. Italics show how live root respiration, decomposition of new (previous fall) and old aboveground litter, and decomposition of belowground (mostly fine root) litter were estimated from seasonal CO_2 fluxes occurring under different treatments. Respiration from the decomposition of aboveground and belowground litter are assumed equal to annual inputs. Flux measurements were made during the first full year after the start of treatments (see text). Numbers are fluxes (grams carbon per square meter per year) and percentages of total soil respiration for each component. OM, organic matter. Modified from Bowden, Nadelhoffer et al. 1993, with permission from NRC Research Press.

production are approximately equal in this forest type. Importantly, this finding narrows the range of uncertainty in estimating fine root production and suggests a method that can be applied elsewhere.

These conclusions can be drawn directly from differences in soil CO_2 efflux in Year 1 of the treatments (Figure 15.3). Soil respiration on the unaltered control plots is 371 grams of carbon as CO_2 per square meter per year (g $C \cdot m^{-2} \cdot yr^{-1}$). We assume that total respiration from the decomposition of leaf litter is equal to the annual contribution of leaf litter carbon (measured as 138 g $C \cdot m^{-2} \cdot yr^{-1}$). Of this, the amount decayed in the first year is equal to the mean of the differences in soil respiration between the CONTROL and NO LITTER plots and the CONTROL and DOUBLE LITTER plots (43 g $C \cdot m^{-2} \cdot yr^{-1}$). This leaves 95 g $C \cdot m^{-2} \cdot yr^{-1}$ as the amount of CO_2 generated by the decay of older aboveground litter. Live root respiration was estimated as the difference between CO_2 flux in the

CONTROL plot and the NO ROOTS plot (123 g C·m⁻²·yr⁻¹). The remaining soil respiration (371 − 138 − 123 = 110) is assumed to come from the decomposition of root litter, which is assumed equal to fine root production.

Temperature Regulation of Rhizosphere versus Bulk Soil Respiration

Soil respiration is a critical process in global as well as local biogeochemical cycles. Models of the global carbon cycle used to predict ecosystem-atmosphere interactions under global warming are sensitive to variations in the relationship between soil respiration and temperature. Broadscale simulation models, however, typically use a single exponential function (Q_{10}) to predict releases of CO_2 to the atmosphere from soil respiration (see Chapters 12 and 13). In the soil-warming experiment we learned that this Q_{10} function actually varied with changes in carbon quality and nitrogen availability. Analogous results emerge from the DIRT plots. Comparisons of soil respiration on treated plots in Year 4 showed that respiration by fine roots and associated rhizosphere organisms responds more to temperature than does bulk soil respiration (Figure 15.4). The Q_{10} value (increase in respiration for each 10°C increase in temperature) for the roots and rhizosphere (4.6) was significantly greater ($P < .05$) than the Q_{10} values for both the untreated controls (3.5) and the treatments without roots (NO ROOTS = 2.5, NO INPUTS = 2.3). Q_{10} values changed little with either addition or exclusion of leaf

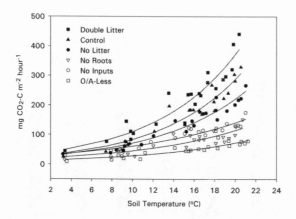

Figure 15.4. Relationship by treatment between mean daily soil CO_2 flux and soil temperature (5 centimeters soil depth) from 16 June 1994 through 14 June 1995. An exponential function of the form $y = \beta_0 e^{\beta_1 T}$ was fitted to the data, where y = flux, β_0 and β_1 are fitted constants, and T = temperature. Modified by permission from *Nature* (Boone, Nadelhoffer et al. 1998), copyright 1998, Macmillan Publishers Ltd.

litter. The findings suggest that soil respiration should be most sensitive to temperature in systems in which roots contribute a large portion of total soil CO_2 efflux. This finding has important implications for global carbon cycling models (compare with Chapter 13).

Litter Effects on Dissolved Organic Carbon

Soil solutions were collected after each rain event during the growing season of Years 4 (1994) and 7 (1997). In Year 4, there were no significant differences in DOC concentration between treatments, and within-treatment variance was large (Figure 15.5). However, by Year 7, DOC concentrations were significantly higher in the solutions collected below the forest floor from DOUBLE LITTER plots and were significantly lower in O/A-LESS plots. Overall results for DOC concentrations were DOUBLE LITTER > CONTROL = NO LITTER = NO ROOTS > NO INPUTS > O/A-LESS. There were no significant differences in DOC concentrations between treatments in the soil solution collected from the mineral horizon in either year. Results from the forest floor lysimeters suggest that changes in organic matter availability cause changes in the organic chemistry of forest soil solutions within less than a decade. In contrast, we infer from the mineral soil data that DOC losses from the mineral soil to groundwater are relatively insensitive to changes in forest floor organic matter dynamics. However, differences in DOC inputs to mineral soils as controlled by amounts of litter inputs are likely to influence carbon accumulation in mineral horizons.

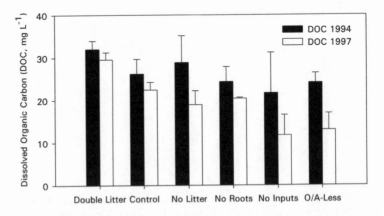

Figure 15.5. Mean concentrations of dissolved organic carbon (DOC) (±1 standard error) in solutions collected beneath the forest floor in 1994 and 1997. From J. A. Aitkenhead and W. H. McDowell, unpublished data.

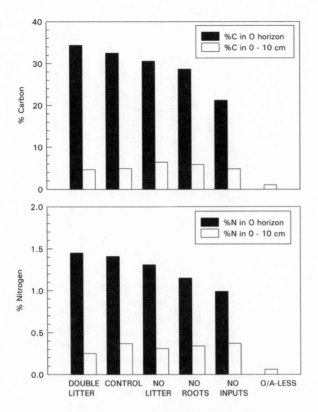

Figure 15.6. Percents of carbon and nitrogen in the forest floor (O horizons) and upper 10 centimeters of mineral soil after five years of litter and root manipulations on the DIRT plots. Bars show means ($n = 9$).

Cumulative Effects on Soil Properties

OVERVIEW

Changes in processes observed in the field during the initial years of manipulations were consistent with changes in both properties and processes after five years of treatment. The concentrations of carbon and nitrogen in the forest floor increased or decreased with increases or decreases in aboveground litter and root inputs (Figure 15.6). Mineral soils, however, did not show similar trends. Mineral soils respond less and/or more slowly to manipulations of plant inputs because most inputs occur directly to the forest floors. Also, organic matter in the mineral soil is likely more stable than is the organic material in the forest floor because of increasing structural complexity and physical protection by association with mineral particles.

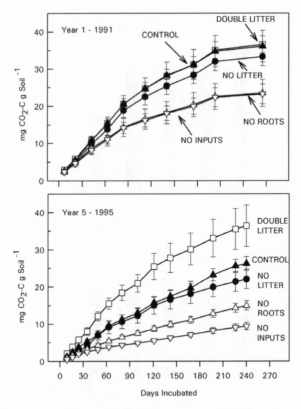

Figure 15.7. Cumulative respiration of forest floor materials (Oea horizons) collected from the DIRT experiment one year (top) and five years (bottom) after the start of manipulations in 1990. Samples were incubated at 22°C and −66 kPa moisture. Symbols show means and standard errors ($n = 9$).

LABORATORY INCUBATIONS

A comparison of laboratory incubations of forest floor samples collected during the first and fifth years of treatment was consistent with field measures. These trials, in which samples are held at constant temperature and humidity, show large effects on organic matter quality and microbial processes (Figure 15.7). Doubling the aboveground litter inputs increased six-month laboratory respiration by about 40 percent relative to respiration of samples from plots with normal (CONTROL) litter inputs. Preventing the ingrowth of roots (NO ROOTS) on plots decreased respiration by 43 percent relative to controls. Exclusion of aboveground litter and root inputs for five years (NO INPUTS) decreased cumulative respiration by almost two-thirds relative to controls. These patterns are consistent with field results (see Figure 15.3) indicating that litter inputs from aboveground and from roots are approximately equal. Respiration of

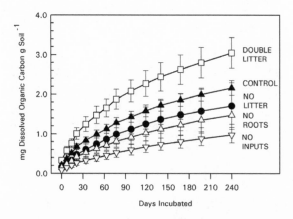

Figure 15.8. Cumulative dissolved organic carbon (DOC) release from forest floor materials (Oea horizons) collected from the DIRT plots five years after the start of manipulations in 1990. Samples were incubated at 22°C and −66 kPa moisture. Symbols show means and standard errors (n = 3 to 4).

samples from the NO LITTER plots, however, was not reduced as much as would be expected given the large increase in respiration after doubling litter inputs. Curiously, doubling litter inputs increased respiration much more than exclusion of litter inputs decreased respiration. This suggests that additional litter inputs might stimulate or enhance decomposition of existing more recalcitrant organic matter. Patterns of DOC release from incubations in response to five years of treatment were similar to those of respiration (Figure 15.8). Moreover, cumulative DOC release was about one-tenth of the release of CO_2 and showed overall patterns similar to those obtained in the field from lysimeters (see Figure 15.5).

Net nitrogen mineralization (release of plant-available nitrogen as measured in sequential leaching from incubated soils) under laboratory conditions was also influenced by five years of plot manipulations (Figure 15.9, top). However, differences in mineralization among incubations of samples from treated plots were not as consistent or as pronounced as were differences in respiration. Cumulative nitrogen release from DOUBLE LITTER, CONTROL, and NO LITTER incubations was similar. This could indicate that the source of most mineralized nitrogen is from leaf litter more than five years old. However, excluding root ingrowth from the plots, whether alone (NO ROOTS) or in combination with litter exclusions (NO INPUTS), decreased laboratory nitrogen mineralization. This suggests that root turnover, root exudation, or both processes contribute strongly over short timescales to mineralization.

The absence of roots, while decreasing net nitrogen mineralization

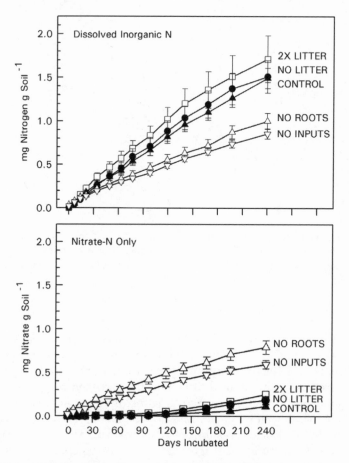

Figure 15.9. Cumulative dissolved inorganic nitrogen (NH$_4$ + NO$_3$) (top) and nitrate-nitrogen leached from incubations (bottom) (22°C, −66 kPa moisture tension) of forest floor samples (Oea horizons) collected from DIRT plots after five years of manipulations. Symbols show means and standard errors (*n* = 9).

overall (Figure 15.9, top), increased net nitrification (Figure 15.9, bottom); nitrate-nitrogen constituted more than half of the nitrogen released from NO ROOTS and NO INPUTS soils whereas nitrate-nitrogen release from soils collected from plots with roots was essentially zero until after three months of incubation. We speculate that the absence of roots and competition from mycorrhizal hyphae has allowed free-living microbes, including nitrifiers, to increase in the forest floor and that this activity carried over to laboratory incubations. The lack of response in net nitrogen mineralization to variations in aboveground litter suggests that microbial immobilization exerts strong control over soil nitrogen dynamics.

Effects on Soil Communities

Forest soils are generally dominated by fungi rather than bacteria. This is true in the Harvard Forest as well, where total fungal-to-bacterial biomass ratios averaged 200 across all treatments and horizons. Mean ratios for mineral soils (114) were significantly lower than for organic soils (305). The lowest ratios were found in the O/A-LESS mineral soils, where total fungal-to-bacterial biomass ratios averaged 20. This suggests that the fungal-to-bacterial ratios decline with increasing recalcitrance of soil carbon or total accumulation of soil organic matter.

Total fungal biomass was much greater than total bacterial biomass in forest floors (Figure 15.10) under all treatments. Total fungal biomass varied with leaf litter input (highest values in DOUBLE LITTER and the lowest in NO LITTER and NO INPUTS plots), but not in the absence of roots (NO ROOTS). Total bacterial biomass varied inversely with fungal biomass except in DOUBLE LITTER plots, where both fungal and bacterial biomasses were high. Active biomass of both fungi and bacteria were remarkably similar across treatments in forest floors. Given the strong effects of manipulations on mineralization and respiration, neither total nor active bacterial population size is a good predictor of soil processes. Active fungal biomass did not differ among treatments in forest floors.

Fungal biomass was also greater than bacterial biomass in mineral

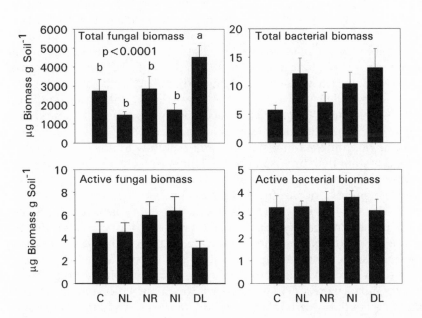

Figure 15.10. Fungal and bacterial biomass in forest floors (O horizons) in Year 5 of the DIRT manipulations. C, CONTROL; NL, NO LITTER; NR, NO ROOTS; NI, NO INPUTS; and DL, DOUBLE LITTER.

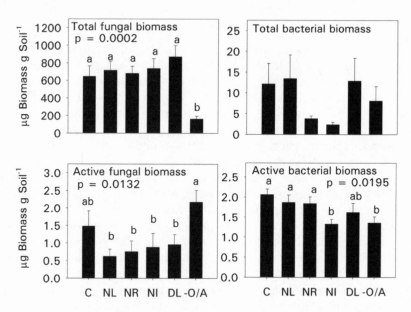

Figure 15.11. Fungal and bacterial biomass in mineral soils (0 to 10 centimeters) in Year 5 of the DIRT manipulations. C, CONTROL; NL, NO LITTER; NR, NO ROOTS; NI, NO INPUTS; DL, DOUBLE LITTER; and −O/A, O/A−LESS.

soils (Figure 15.11) and was lowest in O/A-LESS plots, while active fungal biomass was greatest in this same treatment. In contrast, patterns of active bacterial biomass in mineral soils followed patterns of easily degradable organic matter: the O/A-LESS plots had the lowest carbon and nitrogen contents, followed by the NO INPUTS soils (Figure 15.6), both of which had low active bacterial biomass (Figure 15.10).

Protozoan densities were extremely variable, with few significant differences among treatments or relationships to microbial abundance (Figure 15.12). Although protozoa are grazers of both bacteria and fungi, patterns of protozoa numbers appeared to follow trends in total fungal biomass rather than trends in bacterial biomass. This might be expected in these soils where fungal biomass dominates the microbial community. In organic horizons, fungal biomass was greatest in the DOUBLE LITTER and lowest in the NO LITTER treatments; total protozoan numbers were higher in the DOUBLE LITTER treatment than in either the NO LITTER or NO INPUTS treatments, but were greatest in the NO ROOTS treatment. This did not correspond to patterns of microbial abundance, and none of these trends was statistically significant. In mineral soils, protozoan numbers were low in the O/A-LESS treatment, corresponding to low total fungi and low carbon content, but other patterns in protozoa abundances did not match microbial abundance across treatments. Nematode abundance was variable in organic soils but greatest in DOUBLE LIT-

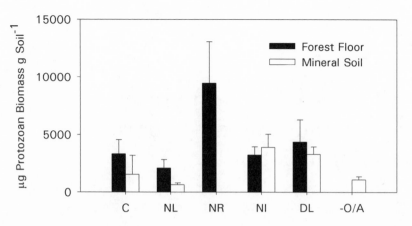

Figure 15.12. Protozoan populations after five years of manipulations in the DIRT experiment. C, CONTROL; NL, NO LITTER; NR, NO ROOTS; NI, NO INPUTS; DL, DOUBLE LITTER; and −O/A, O/A− LESS.

TER and NO ROOTS treatments, closely matching patterns of protozoa. In mineral soils, numbers of nematodes per gram of soil were very low but were lowest in the O/A-LESS and in NO ROOTS treatments. Again, patterns matched those of total protozoa abundance. Nematodes graze on fungi and bacteria as do protozoa, and they can also graze on protozoa.

Summary

Our manipulations of litter and root inputs to forest soils are designed to (1) quantify the proportions of aboveground litter and root inputs that become stored as organic matter with long residence times; (2) quantify how organic matter formation influences soil properties such as nutrient and water retention; and (3) characterize how the nutrient-supplying capacities of soils are influenced by plant litter and root inputs. These goals will require decades of manipulations to be achieved. We have, however, used results from the first years of the experiment to address important questions about forest ecosystem function. Thus, although the overarching goals are long term, there are important short-term benefits as well. This is a key to sustaining the interest necessary for justifying the continued maintenance of the plots. Another important feature of long-term experiments is that the manipulations themselves be simple and require a minimum of effort to maintain. This is the case for the DIRT plots, which require only several days of activity to remove and add litter annually to subsets of the plots. More effort is required to establish the plots and to retrench plots from which roots are excluded (every eight to twelve years). Once established, however, these plots are soon on their way to achieving Professor Hole's goal.

CHAPTER 16

Experimental Approaches to Understanding Forest Regeneration

S. CATOVSKY, R. CRABTREE, T. SIPE, G. CARLTON,
S. BASSOW, L. GEORGE, and F. BAZZAZ

Introduction: Disturbance, Heterogeneity, and Regeneration

Major changes in forests are initiated by disturbance. The previous chapters have presented both historical and novel forms of disturbances that are part of the New England landscape. Each of these will induce changes in the quantity and spatial distribution of resources required by plants as well as substantially alter vegetation structure and composition. Consequently, a single disturbance may initiate a highly complex chain of events that affect processes within a forest at many levels of organization and over long periods of time. Changes in species composition resulting from disturbance and succession may include important feedbacks between site and species characteristics, altering ecosystem function (Figure 16.1).

In order to explore the ecological mechanisms underlying the changes witnessed across New England, we have used a combination of field and glasshouse experiments to understand this dynamic interplay between species and site. A total of six different experiments are presented here that form the core of our Harvard Forest research. They address different aspects of environmental heterogeneity and change but share a common approach, which includes the use of tightly controlled experiments with a range of species drawn from the same genus. The species chosen for each experiment represent a range of successional characteristics. For example, with maples we use a very tolerant canopy tree (sugar maple), a canopy species of intermediate tolerance (for example, red maple), and a very shade-tolerant subcanopy tree (for example, striped maple). Variation in life-history strategies among the species allows examination of differences in responses to experimental environments while minimizing other evolutionary constraints.

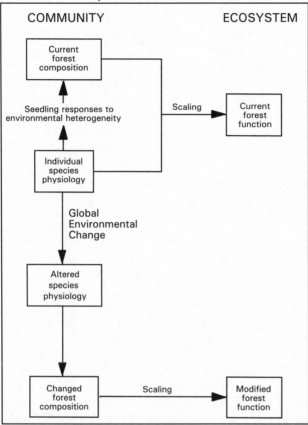

Forest Dynamics Research Framework

Figure 16.1. Research framework for studying forest dynamics, highlighting approaches for incorporating heterogeneity into the understanding of current forest composition and future forest responses to global environmental change, and for scaling from forest community structure to ecosystem function.

The experiments include the following:

- Plant response to experimental gap size and placement in the field
- Seedling response to temporal patterns of light availability in gaps
- Plant response to resource congruence in gaps
- Seedling responses to hurricane disturbance—consideration of multiple resources
- The influence of herb–tree seedling interactions and forest spatial structure
- Seedling regeneration and global environmental change

Experiment 1: Plant Response to Gap Size and Placement
Rationale

Ecological theory has suggested that fine-scale variation in environment and resources may enable a greater diversity of plant species to coexist in what are otherwise rather homogeneous forest conditions. Our research using experimentally created gaps in the forest canopy sought to test what has become known in the ecological literature as the *gap partitioning hypothesis*—that different species are adapted to different subsets of the entire range of environmental conditions to be found in a forest.

Experimental Design

We created three small (75 square meters) and three large (300 square meters) gaps in October 1987 in a 4-hectare stand of mixed oak, birch, and red maple on the Prospect Hill tract. In this study, which preceded our development of hurricane manipulations, we produced gaps by felling selected canopy trees with a chain saw and then cutting and removing all logs and branches manually with minimal disturbance to the soil. To minimize additional (albeit natural) heterogeneity, we cut back the understory vegetation to the ground each year. Each gap was broadly elliptical, with its long axis oriented in an east-west direction. Experimental plots were placed in the center of each gap, and in the northwest, northeast, southwest, and southeast corners. Equivalent plots in the same spatial arrangement were created as controls in adjoining areas of intact forest. Environmental monitoring stations were established at each plot to measure daily and seasonal variation in light (photon flux density, PFD), air and soil temperature, relative humidity, and wind speed. Seedlings of three maple species were planted, and physiological and demographic responses to the gap-understory continuum were measured for two years after gap creation.

Results

Light varied most among the resources measured, through both the day and the season, and between and within gaps, as would be predicted from the movement of the sun across the sky and the geometry of experimental gaps. In the northern hemisphere, the sun tracks across the southern part of the sky and thus exposes the north side of a gap, and the adjoining understories, to greater direct light. In contrast, the south side of the gap, although open to the sky, receives no direct light. Consequently, the number of days that direct beam radiation reached the forest floor decreased from north to south across the gaps, and in small gaps direct light fell only in the north half of each gap (Figure 16.2).

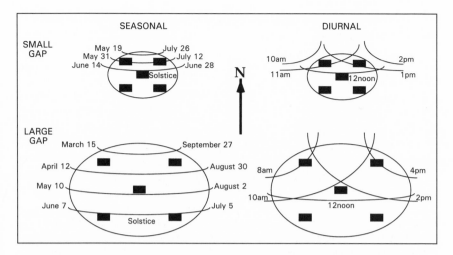

Figure 16.2. Seasonal and diurnal movement of direct beam radiation (direct sunlight) arcs in small and large gaps, based on field measurements and geometric calculations. Seasonal changes are shown with solar noon arcs, and diurnal changes are shown close to the summer solstice (hours are solar time). Modified from Bazzaz and Wayne 1994, 359–60, with permission from Elsevier Science (copyright 1994).

Superimposed on this north-south light gradient there occurred a strong east-west gradient in the diurnal light environment. As the sun rises in the east, west sides of gaps received morning sun, whereas east sides received beam radiation in the afternoon (Figure 16.2). Around the summer solstice, when the sun is highest in the sky, small gaps received direct light for only a little more than two hours each day, in contrast to large gaps, which received more than eight hours of direct light. These diurnal and seasonal patterns in light availability were observed only on clear days. On overcast days, the centers of gaps received marginally more diffuse light than the sides, and there were no differences among the four gap corners. Over the entire growing season there was significant spatial variation in light availability according to the following ranking: NW > NE > Center > SW > SE (Figure 16.3).

Species responses differed among large gaps, small gaps, and understory plots. Red maple seedlings showed a large increase in leaf area and in the rate of CO_2 assimilation with increasing gap size (Figure 16.4a and 16.4b). This flexibility in response gave red maple the highest survival overall and the largest increases in growth from the forest understory to large gaps (Figure 16.4c and 16.4d).

In contrast, striped maple showed a preference for small canopy gaps, as it grew better there than in large gaps or the understory. It also showed a modest increase in leaf area and a very substantial decline in

total radiation is received as intense light for a short period of time or diffuse light over the course of a whole day.

Experimental Design

We tested the effect of variable timing and intensity of radiation by simulating gap conditions and understory conditions and comparing seedling growth. Gap conditions were established using plywood walls (1.8 meters high) oriented north-south to control the timing and duration of direct radiation. The diffuse light in the understory was simulated using shade cloth of different thickness. Seedlings of birch were grown in each environment.

Seedlings of both early-successional gray birch and mid-successional yellow birch consistently grew larger in the shadehouses than in simulated gaps receiving equivalent amounts of light but in varying intensity (Figure 16.6). Gray birch was more sensitive to light availability in gaps than was yellow birch, while there was little difference between species' sensitivities in shadehouses.

The physiological mechanisms generating these response patterns are poorly understood, although a number of hypotheses have been proposed, including the following:

- Inefficient use of light: At levels higher than approximately 800 micromoles per square meter per second, light becomes saturating; any additional light intensity does not improve a plant's photosynthetic perfor-

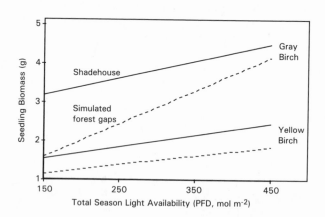

Figure 16.6. Growth responses of gray birch and yellow birch to differing diurnal distributions of light availability (photon flux density, PFD) after one growing season in simulated gaps (dotted lines) and shadehouses (continuous lines). Lines represent results of linear regressions fit to all data points. Modified from Wayne and Bazzaz 1993b, with permission from the Ecological Society of America.

mance. By concentrating a seedling's available light regime into a few hours in the middle of the day, the amount of excess light is increased.

- Trade-offs in biochemical processes within leaves: We might expect that seedlings receiving a relatively uniform distribution of light could adjust the photosynthetic machinery in their leaves to this rather narrow range of light levels, while seedlings receiving a more fluctuating light pattern would be forced into a more generalist and less efficient leaf biochemistry in order to cope with both high and low light. We did find that seedlings in simulated gaps showed greater reductions in leaf area ratio and increases in root weight ratio and leaf nitrogen and chlorophyll concentrations with increasing light than did seedlings in shadehouses. These adjustments to high light may have reduced seedling performance in low light, thereby decreasing their overall performance relative to plants grown in more uniform light conditions.
- Lack of congruence in resource availability: Plant responses to a particular resource are contingent on prevailing local environmental conditions, so the temporal pattern of resource availability could critically determine a plant's performance. We investigate this option in the next experiment.

Experiment 3: Plant Response to Resource Congruence in Gaps

Rationale

Plant responses to a particular resource are contingent on the availability of other resources. If high light coincides with abundant soil moisture (that is, resources are *congruent*), then plants will make better use of that light than if water availability is low. For seedlings in a canopy gap, where environmental conditions show distinct daily and seasonal patterns, resource congruency could play a major role in determining plant performance. These effects may be particularly marked between the west and east sides of a gap, which respectively receive morning and afternoon sun.

In particular, we hypothesized that seedling performance on the west side of a gap would be greater than on the east because western seedlings receive direct sunlight in the morning when other conditions are more favorable. In the hours before noon, vapor pressure deficit and temperatures are lower, imposing less stress on leaf function than afternoon conditions.

Experimental Design

We tested this hypothesis by placing birch seedlings on the east and west sides of simulated gaps so that they would receive direct light at different times of the day. A water treatment was also included, as diurnal changes in plant water status and moisture stress were expected to drive responses.

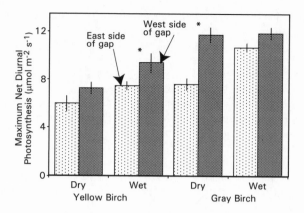

Figure 16.7. Maximum net photosynthetic rates measured during daily time courses of leaf-level gas exchange for gray birch and yellow birch seedlings grown on the east (afternoon sun) and west (morning sun) sides of simulated forest gaps and exposed to dry or wet soil treatments. Means ($\pm$ standard errors) are pooled across blocks, with significant post hoc comparisons ($p < .05$) between east and west treatments denoted with an asterisk. Data from Wayne and Bazzaz 1993a.

Results

In the second year of the experiment, the rate of carbon uptake by leaves was indeed higher for plants on the west side of gaps, which received morning sun, than for plants on the east side, which were not in full sun until after noon. These differences, however, varied with species and water availability (Figure 16.7). Gray birch seedlings grew larger on the west than on the east in dry soils, whereas yellow birch performed better on the west in well-watered soils. Gray birch has a high level of drought tolerance, and may only have experienced incongruent resources when water stress was particularly strong. In contrast, yellow birch is quite drought sensitive and may have suffered water deficits in all drought treatments regardless of the timing of light availability.

These differences in photosynthetic rates were not, however, reflected in the overall growth responses at the end of the experiment. In particular, seedling biomass did not differ significantly between morning and afternoon sun treatments. This result highlights the importance of whole-plant processes that may compensate for leaf-level photosynthetic differences. Differences between east and west sides will be evident only on clear, sunny days, and any effects may become masked when integrated with seedling growth on overcast days.

Experiment 4: Seedling Responses to Hurricane Disturbance—Consideration of Multiple Resources
Rationale

Hurricanes are the most important natural disturbance in New England forests. Wind damage generates canopy gaps such as those we have studied, but also much larger openings with greatly altered physical structure. As trees fall, they create a wide variety of microsites due to uprooting of trees (for example, pits and tip-up mounds) and the distribution of prostrate tree boles and crowns (see Chapter 11). This profound alteration affects belowground processes and resource availability, as well as the distribution of light.

To understand the role of forest structure on dynamics, we developed a series of studies investigating seedling regeneration patterns at the simulated hurricane experiment. These studies address changes in both aboveground and belowground conditions and related seedling responses. As well as re-creating a disturbance that is a critical determinant of forest structure and composition in this region, taking this multiple resource perspective allowed consideration of a broader range of potential opportunities for species' niche specialization. Thus, this experiment offered the potential for developing a better understanding of species coexistence and maintenance of diversity in New England forests.

Experimental Design

We conducted this work at the two experimental hurricane blowdowns in the Tom Swamp and Prospect Hill tracts (see Chapter 11). Environmental conditions and resource availability were characterized across the range of microsites created by the uprooted and fallen trees. The physical and environmental characteristics of these microsites were then related to patterns of seedling regeneration. Our assessment of the availability of various resources such as light, moisture, and nutrients revealed significant differences among microsites, and this heterogeneity had striking effects on patterns of seedling regeneration.

In these detailed studies we focused particularly on the regeneration of co-occurring birch species, as their differential performance on such microsites may play an important role in determining their contrasting roles in New England forests. Birch species rarely sprout after wind damage but instead rely on seedling establishment. The small size and wide dispersal capabilities of birch seeds mean that they can easily reach and potentially establish on a broad range of sites. The small size also means that energy reserves in each seed are small, and new seedlings must reach mineral soil in a short period of time. This generally limits successful establishment to areas of bare soil without surface organic matter.

Results

On the hurricane blowdowns, open microsites without vegetation and shade were most favorable for birch seedling establishment because these open areas created many small patches of exposed soil. A significant proportion of seeds also reached tip-up mounds, and the emergence of new seedlings was relatively high on these microsites, although it varied between years. Emergence was lower on the tops of tip-ups and especially in pits, where litter tended to collect, making it difficult for seedlings to establish. As well as differing in amounts of seed rain and seedling emergence, microsites had contrasting degrees and types of mortality (Figure 16.8). Open microsites had the lowest mortality, with snowshoe hare browsing as the main cause. Similarly, mound microsites had modest mortality, largely due to frost-heaving and physical displacement. Seedlings on tops of tip-ups had high mortality primarily due to low light levels. Almost no seedlings survived in pits, as the few seedlings that did emerge were eventually buried by litter.

Microsite type affected seedling growth as well as establishment. Mounds and tops of tip-ups supported the highest rates of seedling growth on account of high light availability and high nutrient availability, respectively (Figure 16.9). Within the tops, seedlings grew best at higher vertical positions, perhaps because light varied because of over-

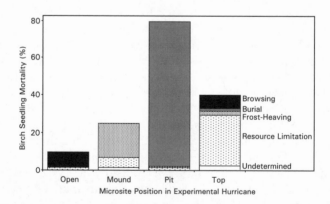

Figure 16.8. Mortality rate and cause of death (browsing, burial, frost-heaving, resource limitation, undetermined) for birch seedlings on different experimental microsites (see Figure 11.10 for diagram of microsites) over three growing seasons. Modified from Carlton and Bazzaz 1998a, with permission from the Ecological Society of America.

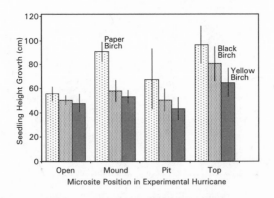

Figure 16.9. Mean height growth (± standard error) of paper birch, black birch, and yellow birch after three years of growth on different experimental microsites (see Figure 11.10 for diagram of microsites). Data from Carlton and Bazzaz 1998a.

hanging branches and residual understory vegetation. Paper birch responded most to the increased light and soil resource availability on mounds and tops, which might be expected from its early-successional status in New England forests. Seedlings were most responsive to increasing light availability at lower light levels and to increasing nitrogen availability at higher nutrient levels. Light responses were generally contingent on the level of soil nitrogen, with greater responsiveness at higher soil nitrogen availability. Nitrogen responses were less sensitive to prevailing light levels, except in very deep shade where all responses are severely limited.

Although seedling growth was enhanced on the mounds and tops of hurricane-created tip-ups, seedling establishment was very limited. Nevertheless, it appears that these growth responses could be important components of regeneration strategies for paper, black, and yellow birch as all three species maintain a seedling bank in undisturbed forests. After a hurricane, birch seedlings growing near a fallen tree's root mat will be raised up from the forest floor and exposed to high levels of light, substantially improving their chances of reaching the canopy. We included all the different components of seedling regeneration in projection matrix models to investigate the importance of hurricane events for the population dynamics of a number of forest species. These models showed that population growth rates of all species increase after a hurricanelike disturbance but that paper birch is particularly dependent on such events for long-term persistence in these forests (Figure 16.10). This analysis also revealed that regeneration in open microsites makes the largest contribution to population responses to a hurricane event, whereas regeneration on tip-up mounds themselves makes a much

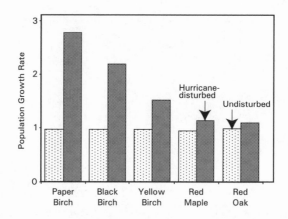

Figure 16.10. Population growth rates derived from the dominant eigenvalues of a projection matrix model of species' dynamics in undisturbed and hurricane-disturbed forests for five common temperate forest species. Data from Carlton 1993.

smaller contribution, primarily due to the small area covered by such microsites.

The work on seedling regeneration after a blowdown demonstrated that the complexity of forest responses to a large disturbance may be understood largely in terms of fluxes of the major resources in the forest ecosystem and seedling responses to these fluxes. Hurricanes increase heterogeneity in the availability of light, moisture, nitrogen, and space at the scale of individual seedlings by creating a wide variety of microsites at the forest floor. These microsites differ in both aboveground and belowground resource availability. Absolute differences in the levels of different resources, as well as congruency between different resources, may influence many stages of seedling regeneration, including dispersal, emergence, survival, and growth. Effects of disturbance on multiple environmental conditions may provide further resource axes, by which species may partition the environment.

Experiment 5: The Influence of Herb–Tree Seedling Interactions and Forest Spatial Structure
Rationale

Species differ in their ability to persist as seedlings and saplings in the shaded forest understory. For many species, development of a seedling bank forms a critical part of their regeneration strategies, as the size and status of seedlings in the understory may determine a species' response to a disturbance that will allow seedlings to reach the overstory.

Table 16.1. Environmental Conditions, Understory Vegetation Patterns, and Seedling Regeneration Dynamics between Contrasting Stand Types

	Stand Type	
Factor	Hemlock	Broad-leaved
Environment		
Light (mol day^{-1})	0.39*	0.94*
N (mg/kg soil)	90.6	112.8
Water (g g^{-1})	1.12	0.81
Soil pH	3.67	3.96
Herb/shrub species		
Density (m^{-2})	2.0*	28.6*
Species richness	8.3*	25.7*
Seedling demography		
Mean age (yr)	1.04*	1.53*
Emergence (m^{-2})	88.8*	19.6*
Survival (%)	28.1*	46.9*

Source: Unpublished data from S. Catovsky and F. A. Bazzaz.
*Significant difference.

Conditions in the understory can be both spatially and temporally heterogeneous due to site conditions, human disturbance, or differences in canopy trees overhead. One of the most striking patterns is the contrast between conditions under deciduous broad-leaved species (for example, red oak, red maple, and birch) and hemlock, an evergreen coniferous species. Hemlock-dominated stands have significantly less light at the forest floor than do broad-leaved tree stands, which leads to a sparse understory vegetation and higher seedling turnover and lower average age in the hemlock stand (Table 16.1).

Variation in the distribution and abundance of understory species may also influence seedling regeneration dynamics. Within temperate forests, herb and shrub species can form fairly stable patches that differ in species composition and density across different stand types. As all tree seedlings must pass through this herb and shrub layer before they reach the overstory, these understory plants may act as an ecological filter, determining the composition and spatial structure of the seedling bank and ultimately controlling future canopy composition. Ferns are one of the most important components of this ecological filter. They are widely distributed, have dense growth patterns, and tend to increase under conditions of increasing light.

Experimental Design

To investigate the influence of the major fern species (hayscented fern, *Dennstaedtia punctilobula;* and interrupted fern, *Osmunda claytoniana*) on forest regeneration, we removed ferns chemically (glyphosate application), pinned them back (shade-free treatment), or left them intact in a total of six plots. By comparing removal and shade-free plots, and by comparing shade-free and intact plots, it was possible for us to decouple competition for light from competition for water and nutrients. At each plot, the microenvironment and demographic responses of both naturally regenerating seedlings and planted seeds/seedlings were characterized.

Results

The two fern species did not differ in their influence on the understory microenvironment, but the presence of a full fern cover significantly reduced the levels of light reaching seedlings in the understory, from 3.4 percent to 1.1 percent of light reaching the top of the canopy (Figure 16.11a). Fern fronds also increased litter accumulations onto the

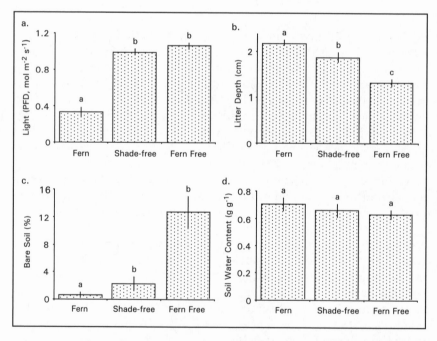

Figure 16.11. Microenvironmental characteristics of fern manipulation plots: light (photon flux density, PFD) (a), litter depth (b), percent exposed soil (c), and soil water content (d). Means ± standard error were calculated across all experimental sites ($n = 60$ plots). Within each panel, columns that share a letter are not significantly different ($p < .05$, Fisher's protected LSD) from one another. Modified from George and Bazzaz 1999a, with permission from the Ecological Society of America.

ground surface and decreased the degree of soil exposure, but they did not affect soil water content (Figures 16.11b, 16.11c, and 16.11d). Patterns of seedling emergence, survival, and growth were affected by the presence of a fern cover due to influences both on understory microenvironmental conditions and on the movement and activity of seed and seedling predators.

Emergence of yellow and black birch, white pine, and red oak seedlings was significantly reduced in fern plots, while red maple and white ash were unaffected (Figure 16.12a). Further analysis and experimental

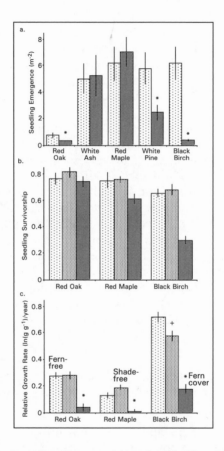

Figure 16.12. Variation in seedling emergence (a), survivorship (b), and relative growth rate (c) in experimental fern plots (fern-free, light shading; shade-free, moderate shading; and fern cover, dark shading) for different tree species. Species are shown in decreasing order of seed size. Panel a is from naturally established seedlings, while panels b and c are from planted seeds and seedlings, respectively. Bars represent means ($\pm$ standard error) pooled across six study sites. Symbols (*, +) designate bars that are significantly different from one another within a species ($p < .05$, Fisher's protected LSD). Data from George and Bazzaz 1999a and 1999b.

manipulation (for example, removal of litter, caging of the seedlings from herbivores) revealed that the factors leading to reduced emergence under ferns differed among species. Fewer white pines emerged under ferns mainly because of the reduced light levels. Birch emergence was also partially related to light levels but was primarily affected by the degree of litter accumulation in plots; the generally thick litter layer under ferns strongly inhibited emergence of very small-seeded birch. Red oak emergence was low in plots with ferns because of increased activity of small mammals that eat acorns and prefer the shelter and protection offered by fern-dominated areas. Again, because of small seed size, birch seedling survival was most sensitive to fern presence, whereas red maple survival was only marginally affected by ferns (Figure 16.12b). Red oak survival was not influenced overall by the presence of ferns, as the detrimental effect of lower resource availability under ferns was balanced by the protective advantage conferred by ferns against insect herbivory. Seedling growth in fern plots was consistently lower than in fern-free plots mainly as a result of reduced light availability (Figure 16.12c).

Ferns may have substantial effects on conditions in the forest understory, in terms of both microenvironmental characteristics (especially light and litter) and the activity of seed/seedling predators. All these factors may selectively influence seedling emergence, survival, and growth of different species, and thus affect forest composition. Projecting the results of this study forward suggests a strong selectivity of the fern layer and a strong role of this ecological filter. Red maple seedlings dominate in fern areas, whereas the contribution of birch and red oak to seedling bank composition increases in fern-free areas. The heterogeneous distribution and abundance of fern species across the forest may strongly influence the spatial structure of seedling regeneration. These spatial dynamics have important implications for neighborhood interactions between individuals and the genetic structure of tree populations.

Experiment 6: Seedling Regeneration and Global Environmental Change
Rationale

The responses to light, water, and nutrients described above may all be altered by the novel perturbations of elevated CO_2 and nitrogen deposition resulting from widespread human activity. As a result of fossil fuel combustion and extensive fertilizer production and use, forests across New England are experiencing high levels of nitrogen deposition (see Chapter 12). More than half the nitrogen currently deposited across New England is nitrate, whereas forest soils in this region are predominantly ammonium based. This switch in the nitrogen

species available to seedlings clearly has the potential to influence forest regeneration dramatically

Also as a result of the use of fossil fuels, CO_2 in the global atmosphere has risen from 280 to 360 microliters per liter over the past century. Carbon dioxide is a primary substrate for photosynthesis, and changes in atmospheric CO_2 concentrations are likely to affect plant function profoundly. Early work in this area demonstrated striking differences in tree species' responses to elevated CO_2, suggesting the possibility of substantial change in forest regeneration dynamics in the future. Elevated CO_2 may also influence plant responses to spatial and temporal heterogeneity in environmental conditions, leading in turn to alterations in the distribution and abundance of species across the landscape and to changes in forest species composition and spatial structure.

Experimental Design

To explore the implications of changing resource interactions further, we have run a series of experiments on the individual and combined effects of nitrogen (quantity and form) and CO_2 on regeneration and physiology. Our central hypothesis is that seedling responses depend on both species' characteristics (for example, leaf habit and life-history strategy) and environmental conditions (for example, light and soil resources) and that coniferous and broad-leaved species may respond differently due to fundamentally contrasting patterns of nutrient use.

Results

There were surprises in the results from these experiments. For example, the light environment exerted a strong effect on black birch response to the form of nitrogen addition. In contrast to predictions from the physiology of nitrogen metabolism, black birch seedlings showed a clear preference for ammonium over nitrate in high versus low light (Figure 16.13).

Although early-successional species tend to be characterized by greater flexibility in response to changing environmental conditions, our studies indicated that later-successional species showed the greatest growth enhancements in response to elevated CO_2 (Figure 16.14). More recent work at the Harvard Forest comparing eight coniferous and eight broad-leaved tree species suggests that leaf habit may alter the relationship between shade tolerance and CO_2 responsiveness. Growth enhancement in elevated CO_2 increased with increasing shade tolerance for coniferous seedlings, while CO_2 growth enhancement decreased with increasing shade tolerance for broad-leaved species seedlings.

We found that elevated CO_2 caused the greatest enhancement of

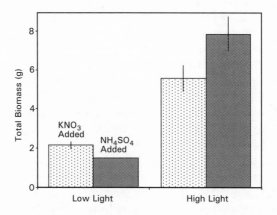

Figure 16.13. Black birch growth responses to the form of nitrogen addition (KNO_3 and NH_4SO_4) under low (8 percent photosynthetically active radiation [PAR]) and high (70 percent PAR) light treatments. Means ($\pm$ standard error) were pooled across block and nitrogen addition rates (2.5 and 5 grams per square meter per year). Data from Crabtree and Bazzaz 1993a.

growth in treatments with high nutrients and low light availability. The nutrient interaction was most noticeable in earlier-successional species such as gray birch and white ash, whereas the low light effect was particularly striking in mid-successional species, including red oak and yellow birch. The low-light, high-nutrient combination may produce the greatest CO_2 growth enhancements because it is the treatment in

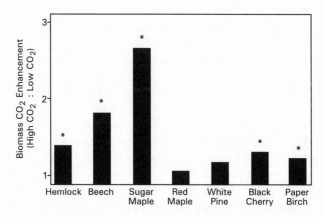

Figure 16.14. CO_2 enhancement ratios of total seedling biomass (biomass at 700 microliters per liter/biomass at 400 microliters per liter) after one growing season for seven co-occurring temperate forest species, ranked from left to right in order of decreasing shade tolerance. Asterisks designate significant differences between growth at ambient and elevated CO_2 for that species. Data from Bazzaz et al. 1990.

which plants undergo the least photosynthetic down-regulation after prolonged exposure to elevated CO_2. Down-regulation is thought to occur when carbohydrate supply (that is, photosynthesis) exceeds a plant's demand. Capacity to utilize carbohydrates is a function of the size and number of sinks within a plant. High nutrient availability may increase sink strength, while low light availability may reduce carbohydrate supply rates, and thus both factors may reduce the degree of down-regulation in response to elevated CO_2. The reasons for differences in species' sensitivity to down-regulation, however, are currently unclear. Importantly, we noticed one major consequence of down-regulation: substantial declines in CO_2 responsiveness were observed over three growing seasons for a number of temperate forest species, although the extent and rate of decline differed among species.

Although no consistent effects of CO_2 on species' niche breadths relative to drought stress were found, significant shifts in species' responses across moisture gradients did occur. Species that are more tolerant of drought, such as gray and paper birch, increased their representation in dry soil treatments, while drought-intolerant species, including yellow birch and red maple, were more responsive to CO_2 in well-watered treatments (Figure 16.15). These patterns may reinforce current distributions of these species in relation to soil moisture.

Synthesis and Extrapolation

Extrapolating from short-term seedling experiments to the longer-term effects of novel environmental perturbations on forest struc-

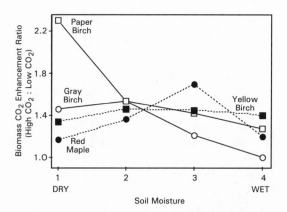

Figure 16.15. CO_2 enhancement ratios of total seedling biomass (biomass at 700 microliters per liter/biomass at 375 microliters per liter) after one growing season for two drought-tolerant (gray birch and paper birch) and two drought-intolerant (red maple and yellow birch) species under different water treatments. Data from Miao et al. 1992 and Catovsky and Bazzaz 1999a.

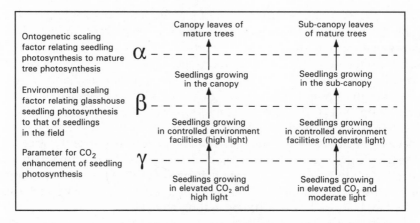

Figure 16.16. Approach for predicting future responses of mature canopy trees to atmospheric CO_2 levels based on scaling seedling photosynthetic responses to elevated CO_2. Based on Bassow 1995 and Bazzaz et al. 1996.

ture and function is one of the most challenging aspects of global change research. We have developed a scaling approach that allows us to estimate how mature tree photosynthesis may respond to elevated CO_2 based on seedling responses in controlled environment facilities (Figure 16.16).

We measured the α effect by growing seedlings on canopy access towers in the canopy and subcanopy strata of mature trees, so that microenvironmental conditions are identical, and measuring leaf photosynthetic characteristics. The β effect is calculated by comparing seedlings growing in a forest canopy with seedlings in controlled environment facilities, where similar microenvironmental conditions are simulated (high versus low light, matched temperature and vapor pressure deficit). The final factor, γ, is measured by comparing seedling photosynthetic responses with ambient and elevated CO_2. Our results suggest that such a scaling approach is feasible provided that the developmental scaling factor α is not too large. For species like red maple and birch, seedlings and mature trees did not differ greatly in photosynthetic responses, whereas red oak and yellow birch showed much larger photosynthetic differences. For these latter species, nonlinearities in photosynthetic responses between scaling components may limit the utility of such an approach.

As part of a second scaling approach, we incorporated our results on variation in CO_2 responsiveness as a function of species and resource availability into a spatially explicit temperate forest dynamics model (SORTIE) in conjunction with collaborators at Princeton to investigate the community- and ecosystem-level consequences of rising CO_2 levels.

The model is driven by canopy tree species effects on light environment at the forest floor and seedling growth and survival responses to this variation in light availability. By re-parameterizing the model, we showed that differences in species' responses to elevated CO_2 can lead to dramatic changes in future forest species composition. The model predicted particularly large increases in the contribution of red oak to community biomass (Figure 16.17), with a simultaneous decrease in species diversity. This alteration of community structure was also a primary determinant of ecosystem-level change in the future. The model predicted an additional 35 percent in basal area if species composition was allowed to change over time, compared with runs where composition was constrained.

The experiments outlined above demonstrate that substantial alteration of forest species composition could occur as a result of future environmental changes, in particular, elevated atmospheric CO_2 and nitrogen deposition. Novel environmental perturbations will interact with current spatial and temporal patterns of resource availability to influence forest regeneration dynamics. Species' responses to different resource combinations will determine which species come to dominate the future forest. In central New England, model results predict a noticeable increase in red oak abundance in response to elevated CO_2, with accompanying decline in overall species diversity. The exact nature of the change, however, will depend on the degree of future forest disturbance. Human activities, such as land-use management, pathogens, and the introduction of exotic species, may influence disturbance regimes and determine the kinds and rates of future changes.

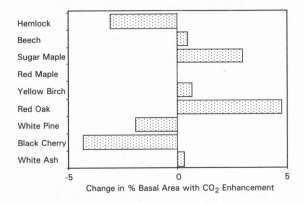

Figure 16.17. Projected relative change in species composition after 100 years of CO_2 enrichment. Species are ranked from top to bottom in order of decreasing shade tolerance (hemlock to white ash). Data from Bolker et al. 1995.

Synthesis and Extrapolation
Models, Remote Sensing, and Regional Analysis

J. ABER, W. CURRIE, M. CASTRO, M. MARTIN,
and **S. OLLINGER**

Introduction and Background

How relevant is the research we do at the Harvard Forest to the larger New England region we claim to represent? Most of our experiments are set in forests we consider "typical," but all show sensitivity to site conditions, species, or history—all three of which vary widely across the region. How can we extrapolate across time and space to make useful predictions into the future? "Regionalization" of results from individual sites or studies has been a core concern of the LTER program for many years. How do we contribute to this goal?

At the first All-Scientists' Meeting of the LTER program in 1991, we proposed an approach that would tie intensive, site-level research to predictions across New England and into the future (modified in Figure 17.1). This approach embodies the traditional steps in scientific research: data acquisition, synthesis of information into a set of working theories, and generation of hypotheses to be tested out of those theories. The figure, though, includes two loops: one for intensive site work and another for extensive or regional work. These two are linked through a box called "Models."

By "models" we mean computer models that represent ecosystem processes as a series of linked equations. In our approach, those equations come directly from statistical analyses applied to collected data. Models are formed by tying together the results of different field research projects expressed usually as regression equations between some environmental factor (for instance, light) and an ecosystems process (for instance, photosynthesis).

Several different models, operating over very different scales of time and space, have been developed as part of the Harvard Forest LTER program, but all of them share a common approach. Harvard Forest models tend to be strongly empirical, combining measured data with the cur-

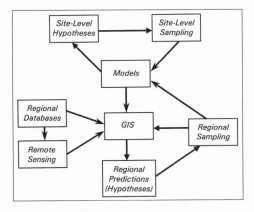

Figure 17.1. A conceptual approach to the integration of intensive site-level research and extensive regional-scale research with spatial data sets within a geographic information system (GIS). Note that models of ecosystem function are the integrating focus of this approach. Based on Aber, Driscoll et al. 1993.

rent understanding of the physiological processes involved. They are generally data-rich and well-tested, or validated, against additional measurements not used to derive the model. Our more complex models are not calibrated, in that the input parameters are not "tuned" in order to achieve better agreement between model predictions and measurements. Rather, discrepancies between predictions and observations are taken for what they are—indications of incomplete or erroneous understanding of the processes involved (Box 17.1).

To regionalize Harvard Forest results, then, we run these models derived from field research (the upper loop in Figure 17.1) using information on the regional variation in factors such as temperature, radiation, or precipitation that drive the processes. Both input data and model outputs are managed through a geographic information system (GIS), essentially a digital map of the region (several sample outputs are shown below). Maps of model predictions are essentially hypotheses stated at the regional scale. We also test these where the data exist, as we have for forest production and stream flow. Disagreements between predictions and measurements at this level also provide insights into how the models can be improved (Figure 17.1) or how good they might actually be at predicting our collective future.

From Simple to Complex Models of Ecosystem Processes

This process of model development and refinement can best be illustrated through several examples from the Harvard Forest. These

Box 17.1.

While the use of models is widespread in broadscale ecological predictions, such as biotic responses to global change, many ecologists remain skeptical of the modeling process and tend to disbelieve or discount insights provided by modeling exercises. In this way, ecology differs from many other scientific fields in which quantitative model predictions, and verification of those predictions, are central.

There may be a good reason for this general distrust of models in ecology: modeling projects and modeling papers are not universally held to a consistent, rigorous set of standards of full disclosure during peer review, especially as compared with data papers. We have proposed two achievable objectives that could help increase the value of the modeling process in ecological research: (1) establish a set of guidelines or standards by which papers presenting modeling results should be judged, and (2) increase clarity in the understanding of the difference between calibration and validation.

On the first point, all modeling papers should contain, at a minimum, the following sections, with the suggested content:

- Model Structure: The diagram or schematic should be complete with all components and connections shown. More important, the equation(s) used for each connection should be stated explicitly or clearly referenced, and citations should be given justifying that equation form. If the equation is theoretical or invented, this should be stated and justified on the grounds that no data were available on this process. This section of the paper should become a literature review of previous work on the processes modeled, thus ensuring that the modeler is aware of previous field and laboratory work. The modeling process and literature review may suggest an equation form not previously used in presenting empirical results, which can be a major contribution of the modeling process.
- Parameterization: *All* of the parameters used in the model should be listed (with units), and all values for those parameters given, along with references to the sources of those parameters. If the parameters

are derived by calibration, this should be clearly stated, the calibration method described, and the calibrated values given. If the model is mostly a theoretical construct used for identifying questions, this should be stated explicitly. However, whenever possible, models should include realistic, empirically based parameterizations that tie the model as closely as possible to experimental data.

- Validation: *No* modeling paper should be accepted without at least some attempt to compare model predictions against independent data sets, that is, data not used in any way in the derivation of the model's parameters. Ecology is data-rich and model-poor relative to other fields. There are very few aspects of ecology for which no validation data exist. Where this is the case, such as with predictions of large-scale phenomena for which experiments cannot be run, this should be explicitly stated by the authors. Even in such models there are often intermediate variables that are predicted by the model and for which independent experimental data can be found.

- Sensitivity Analysis: *Every* modeling paper should present the effects of altering model parameters or input variables on model predictions to give the reviewers some idea of model responsiveness to such changes. This also provides information on the importance of specifying each variable correctly. A greater degree of uncertainty can be tolerated in parameters to which the model is relatively insensitive. A second type of sensitivity analysis might be called the "null model" approach, stated as, "How does the predictive ability of the model compare with that of a simple multiple linear regression model?" Stated another way, what is the increase in predictive accuracy achieved by moving from a statistical model to one that includes knowledge of the processes in the system?

- Prediction: *Only* after the above standards have been addressed should the model be used to predict something. Perhaps the greatest disservice ecologists can provide comes from allowing poorly described and invalidated models to be used to predict the results of policy actions. This is equivalent to basing policy decisions on data we know to be seriously flawed. It also fosters the false impression that we know more than we do about the systems we study, which is then often in contradiction to what the experimental data suggest.

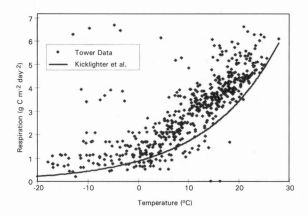

Figure 17.2. Relationship between soil temperature and soil CO_2 efflux for temperate deciduous forests: comparison of total ecosystem respiration measurements at the eddy covariance tower on Prospect Hill (data points) with values predicted by summary equations using data from several sites around the world (line; Kicklighter et al. 1994). The effects of temperature are similar in both relationships, whereas total respiration is higher from the eddy covariance data because of inclusion of aboveground (tree) respiration. Modified from Aber et al. 1996, 261: fig. 4, with permission from Springer-Verlag (copyright 1996).

Methane Consumption in Forest Soils

Measurements of methane consumption in soils at the Harvard Forest have been made in the chronic nitrogen plots under ambient conditions throughout the frost-free season over a six-year period. In addition, experimental water-exclusion and water-addition experiments have been carried out in subplots within and adjacent to the manipulated areas. Together, these data sets cover a range of conditions of soil moisture, temperature, and nitrogen availability and allow a multifactorial analysis of controls on this important process.

Field data show that methane consumption (1) increases with increasing temperature, (2) decreases as water-filled pore space (WFPS or soil water content) increases from 20 percent to 100 percent (Figure 17.4), and (3) is very sensitive to nitrogen availability in soils, having declined by 25 percent in the first year of the chronic nitrogen experiment, moving to a 64 percent reduction by Year 6 (Chapter 12).

A multiple factor model built to predict methane consumption based on these relationships has been applied globally at the 0.5° × 0.5° scale. Using this model, we estimated that soils consume about 100 teragrams per year at the global scale. This estimate is about twice the previous estimates of the strength of the global soil methane sink.

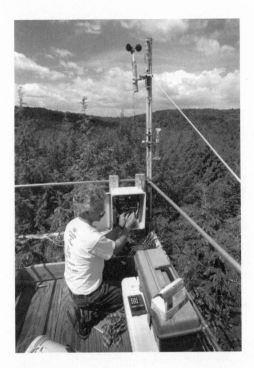

Figure 17.3. An eddy flux system that records exchanges of CO_2 and other material between the forest and the atmosphere. One of three such systems in operation at the Harvard Forest, this equipment is situated above an old-growth hemlock forest in the center of the Prospect Hill tract. Results are compared with detailed physiological and ecosystem results and are being used to develop a simple model of forest ecosystem function. The canopy walk-up tower used for this system also provides access to leaves, branches, and boles of the tree for detailed physiological measurements. Photograph by J. Gipe.

Whole-Canopy Gross Photosynthesis

The long-term, continuous eddy covariance gross and net carbon exchange data available for the Harvard Forest offer a unique opportunity to challenge whole-canopy models of forest photosynthesis. The daily summations of gross photosynthesis, or gross ecosystem exchange (GCE, Chapter 10), are accompanied by direct measurements of temperature, radiation, and other climatic parameters that affect these processes.

The PnET-Day model predicts daily gross and net photosynthesis using an empirical relationship between foliar nitrogen concentration and maximum potential photosynthesis (A_{max}), a similar relationship between photosynthesis and stomatal conductance (see Chapter 3), along

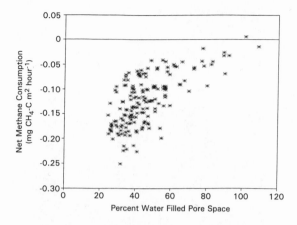

Figure 17.4. Effects of soil moisture (expressed as percent water-filled pore space) on net methane consumption at the Harvard Forest. As soil moisture increases, methane consumption in the soil declines and more methane is released to the atmosphere, where it can serve as an important greenhouse gas. Modified from Castro, Steudler et al. 1995, 6, with permission of the American Geophysical Union (copyright 1995, American Geophysical Union).

with standard functions for response to radiation intensity, temperature, and vapor pressure deficit in conjunction with daily climate data to estimate gross and net photosynthesis and potential evapotranspiration.

Comparing PnET-Day predictions of daily gross photosynthesis with values measured at the EMS tower is a good example of model testing or validation (Figure 17.5, top). The mean difference between predicted and observed is 0.13 grams carbon per square meter per day or 3.7 percent of the daily mean GCE of 3.45 grams carbon per square meter per day.

Models are also interesting when they fail. PnET-Day does not calculate water stress. During a significant period of drought in August of 1995, PnET-Day predictions were higher than observed GCE values by 1 to 2 grams carbon per square meter per day (Figure 17.5, bottom), suggesting that water stress was indeed important during this period and that the model needs to include this factor.

Litter Decomposition and DOC/DON Flux

The first field studies of litter decomposition at the Harvard Forest were carried out nearly twenty-five years ago. Overall, several thousand small bags constructed from fine-mesh materials and containing senescent plant tissues of various kinds have been placed out in the forest, retrieved after different periods of time, and analyzed for chemical

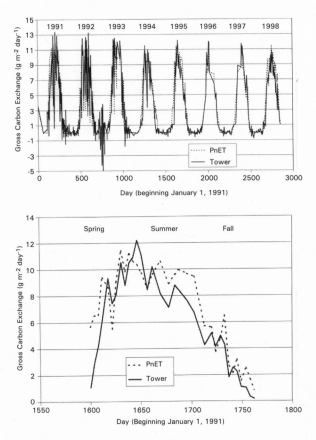

Figure 17.5. Comparison of gross carbon exchange estimates by the PnET-Day model with measurements at the eddy covariance tower for the period 1991–1998 (top) and expanded view of data from the summer of 1995 (bottom). Based on Aber et al. 1996.

content. Predictive equations describing both the rate of mass loss and the gain or loss of nitrogen during decomposition have been developed, using carbon quality or the chemical composition of the material as the driving variable (Figure 17.6). In general, decomposition rate decreases with increasing content of lignin (a complex phenolic compound that encrusts cell walls, making wood woody and oak leaves leathery) and increases with increasing content of materials that dissolve in hot water ("extractives," including sugars, starches, and amino acids). Cellulose, the primary constituent of cell walls and probably the most common biopolymer in nature, decays at intermediate rates unless "protected" by lignin. We now know (see Chapter 12) that not all mass loss from litter occurs through conversion to CO_2 by microbial decay. A significant fraction of carbon loss can occur in dissolved form as dissolved organic

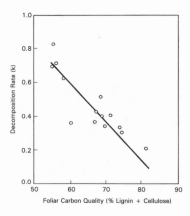

Figure 17.6. Summary relationship between the carbon quality of different types of foliage and their decomposition rates. The "k" value is an expression of the speed of decay. Lignin plus cellulose is the sum of these two carbon fractions (equivalent to 1 − extractives, or the soluble carbon content) in litter. Modified from Aber et al. 1990, with permission from NRC Research Press.

carbon, or DOC. Similarly, fluxes of DOC from the forest floor to the mineral soil and on to streams and groundwater have received increasing attention.

DocMod (Figure 17.7) is a model that predicts litter decay rates based on the carbon quality and nitrogen content of different materials, dividing plant materials into their extractive, cellulose, and lignin-cellulose components. These three materials decay at different rates, producing various amounts of CO_2, DOC, dissolved organic nitrogen (DON), and stabilized organic matter (humus) in the process. Decay rates are modified by temperature and moisture using actual evapotranspiration (AET) as a surrogate. Nitrogen dynamics are driven by decay rates, microbial carbon-use efficiencies for each class, and resulting nitrogen release or immobilization.

The model was linked to a GIS and applied to the White Mountain region of New Hampshire using spatial models for climate and forest production generated by other regional models (see regional PnET discussion below). The linked DocMod-PnET model predicted spatially explicit patterns in forest floor mass and nitrogen content as well as CO_2 and DOC and DON fluxes. Across elevational ecotones in vegetation communities in the region, the linked DocMod-PnET model accurately predicted increases in DOC and decreases in CO_2 efflux with increasing conifer content in forests, due to temperature gradients and increases in low-quality carbon content (lignin) in litter. DocMod was also found to be robust in a modeling intercomparison of blind predictions of litter

DocMod Model

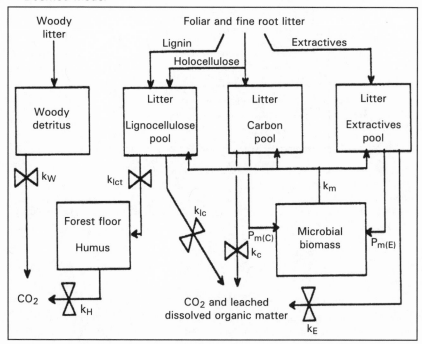

Figure 17.7. Structure of the DocMod model showing carbon fluxes. The variables k_W, k_H, k_{lc}, k_E, and k_c represent decay rates for woody detritus, humus, lignocellulose, extractives, and unprotected cellulose, respectively. The variable k_m represents turnover of microbial biomass, and k_{lct} indicates lignocellulose transfer, the movement of mass from the lignocellulose pool to the humus pool. $P_{m(C)}$ and $P_{m(E)}$ indicate production of microbial biomass from substrates in the unprotected cellulose and extractives pools. Modified from Currie and Aber 1997, with permission from the Ecological Society of America.

decomposition across four terrestrial ecosystems in the LIDET study (LTER Intersite Decomposition Experiment Team).

Ecosystem Models

Equations that summarize individual processes need to be linked together to capture the dynamics of whole ecosystems. For example, if photosynthesis in PnET-Day is linked to the concentration of nitrogen in foliage, then changes in the decomposition processes in a model like DocMod will change foliar nitrogen and total carbon gain for the forest. Conversely, changes in photosynthesis will alter the production of litter, which will alter decomposition rates.

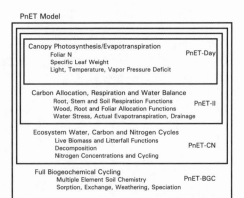

PnET Model

Figure 17.8. The nested structure of the PnET models showing the various ecosystem functions that are added with each layer of complexity.

The PnET Models

The PnET family of models are actually a nested set of algorithms of increasing complexity for predicting carbon, water, and nitrogen cycling in temperate and boreal forests (Figure 17.8). The simplest core model is PnET-Day (discussed above), which calculates gross and net photosynthesis. This is embedded within PnET-II, which adds a water balance routine, including water stress and carbon allocation and respiration algorithms. PnET-CN adds pools for woody biomass and soil organic matter and routines for litter production and decomposition, as well as adding nitrogen content to all pools and fluxes. The advantage of this approach is that the simpler versions require fewer input parameters to run. So, for example, if only daily photosynthesis values are of interest, only about one-third as much information is required to run the model, in comparison with the full PnET-CN version. PnET is an open-code project (code available at www.pnet.sr.unh.edu). Each of the versions of PnET has been applied at the Harvard Forest and also at the Hubbard Brook LTER site in New Hampshire.

PnET-II has been tested against data on monthly carbon balance as measured at the EMS tower and against annual foliar and wood production in the control plots of the chronic nitrogen amendment experiment (using measured foliar nitrogen values for the area around the tower and in the plot itself, respectively). In the chronic nitrogen plots, PnET-II predictions for wood and foliar production on these plots are within 10 percent of measured values (Table 17.1; again, input parameters have not been calibrated in any way). The relatively low productivity values for the hardwood control stand reflect the relatively low measured foliar nitrogen concentration (2.0 percent).

Table 17.1. Predicted (PnET-II) and Observed Values for Wood (Aboveground Plus Woody Roots) and Foliar Production at the Control Plots in the Chronic Nitrogen Stands

	Pine			Hardwood		
	Foliage	Wood	Total	Foliage	Wood	Total
PnET-II (no O_3)	307	341	648	294	527	821
PnET-II (with O_3)	280	301	581	284	430	714
Measured	316	310	626	300	475	775

Note: Units are grams of biomass per meter square per year. Both measured and predicted productivity data are for the period from 1988 to 1996. Estimates for net primary production with ozone effects use the algorithms presented in the regional applications section in the text.

As discussed below, ground-level ozone is a potentially important stressor for forests in central New England. Our PnET estimated net primary production (NPP) values have always been slightly above the measured values. Using the algorithms for ozone effects on photosynthesis described in the regional modeling section below as well as measured ozone concentrations at Ware, Massachusetts, we tested the predicted effect of ozone on NPP (Table 17.1). These predictions are now slightly below measured values. The area around Ware has among the highest ozone concentrations in New England, and values from this site may be too high for the nearby Harvard Forest site, which is at a higher elevation and is heavily forested.

PnET-CN incorporates the long-term effects of land-use and disturbance history on ecosystem function by adding long-term storage pools (wood and soil organic matter). As a result, this version requires an accurate description of site history and offers insights into the duration and nature of disturbance effects.

Two different tests of the PnET-CN model have yielded very different results. At Hubbard Brook, the model successfully captures the major features of the long stream nitrate record in the reference watershed (W6) once both variation in climate and all the biotic and abiotic disturbances known for the watershed are included (Figure 17.9, top). This is a fairly rigorous test as nitrate loss to streams is a small net flux that is the difference between many large gross fluxes (for example, mineralization and plant uptake).

In contrast, the model fails miserably when tested against the nitrate loss record for the high-nitrogen treatment in the hardwood stand of the chronic nitrogen experiment (Chapter 12). That forest captured and retained nearly all of the 100+ grams of nitrogen added per square meter over the first eight years. PnET-CN with monthly fertilizations during the growing season equivalent to those applied in the chronic nitrogen experiment predicted immediate and very large losses of nitrate (Figure 17.9, bottom).

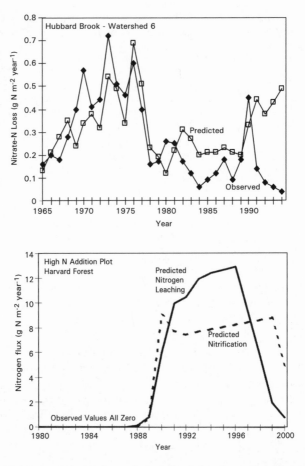

Figure 17.9. Predicted and observed annual nitrate flux below the rooting zone and into streams or groundwater as predicted by PnET-CN. Two examples are shown: the reference watershed (W6) at Hubbard Brook (top), and the high nitrogen addition plots in the hardwood stand at the Harvard Forest chronic nitrogen addition experiment (bottom). Measured values are all zero. Top panel modified from Aber et al. 1997, 69, with permission from Elsevier Science (copyright 1997); bottom panel based on Aber and Driscoll 1997.

We learn as much or more when models fail as when they succeed. This lack of agreement shows that there is a process of nitrogen retention operating in the chronic nitrogen plots that we do not understand and have not included in our models (see Chapter 12 for a discussion of several possibilities). Understanding and measuring these processes so that they can be described quantitatively and added to PnET would increase the completeness of the model and the accuracy of predictions at high rates of nitrogen application.

Using models as consistency checks on our understanding of ecosystems processes is a valuable application, but federal agencies generally fund models for their predictive capacities. Along with nitrogen deposition, increasing concentrations of ozone and CO_2 in the atmosphere are important components of the changing environment at the Harvard Forest. By adding relationships to PnET that capture current physiological understanding of the effects of these two gases on photosynthesis, and running the model for different land-use histories at the Harvard Forest (intensive agriculture) and Hubbard Brook (harvest only), we can predict the interactive effects of disturbance history and atmospheric chemistry (CO_2, ozone, and nitrogen deposition). These effects are compensatory rather than reinforcing (Figure 17.10). Both nitrogen deposi-

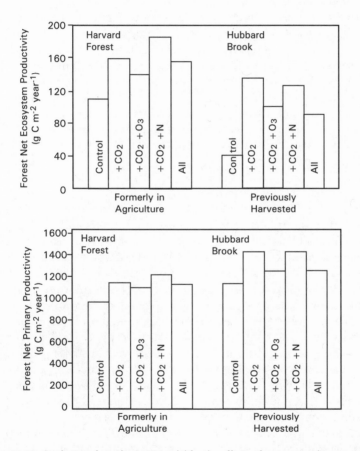

Figure 17.10. Predictions from the PnET model for the effects of various combinations of CO_2, ozone (O_3), and nitrogen deposition as conditioned by land-use history at the Harvard Forest (extensive agricultural use) and at Hubbard Brook (forest harvesting only) on total carbon balance (net ecosystem productivity) and total plant production (net primary productivity). Modified from Ollinger et al. 2002, with permission from Blackwell Science Ltd. (copyright 2002).

tion and increased CO_2 lead to increased forest production (NPP) and greater carbon storage (net ecosystem production, NEP), but these effects are more pronounced after nineteenth-century extractive agriculture at the Harvard Forest than in the harvest-only history at Hubbard Brook. In both cases, increased ozone offsets some of these gains.

The TRACE Model of ^{15}N Dynamics

PnET-CN could not predict or explain the continuing high rate of nitrogen retention in the chronic nitrogen plots. It also does not incorporate the information available from the ^{15}N tracer studies applied in these plots or make use of the more complete description of the decay process included in the DocMod model.

The TRACE model (Tracer Redistributions Among Compartments in Ecosystems) was developed as a tool for interpreting processes controlling ecosystem-level redistributions of ^{15}N at the Harvard Forest. It links plant processes of the PnET-CN model with soil submodels derived from the DocMod model described above (Figure 17.11). Plant and soil pools of carbon and nitrogen in TRACE are physically meaningful, in most cases designed to allow straightforward comparison with field or laboratory sampling methods. In one major structural change from PnET-CN, there are separate compartments for the forest floor and mineral soil horizons.

TRACE was first used to interpret mechanisms underlying plant-soil partitioning of ^{15}N over a two-year period following the additions of $^{15}NH_4^+$ and $^{15}NO_3^-$ tracers in the chronic nitrogen plots. In an iterative sequence of comparisons between model predictions and field data, alternative formulations of soil sinks for NH_4^+ and NO_3^- were tested to account for measured distributions of ^{15}N in soils and vegetation as observed in the field. Reasonable agreement between model predictions and field data required high gross rates of ^{15}N assimilation from inorganic nitrogen pools by detritus (and associated microorganisms) for both $^{15}NH_4^+$ and $^{15}NO_3^-$ additions to ambient (nonfertilized) and chronically fertilized plots. The modeled plant uptake of ^{15}N (verified by plant tissue concentrations of ^{15}N), followed by litter production, could not account for the high rates of ^{15}N incorporation into organic fractions in soils. This finding was significant as it indicated strong soil sinks for NO_3^- in deciduous and coniferous stands at the Harvard Forest, in agreement with some, but not all, findings elsewhere.

Perhaps of equal importance was something TRACE taught us when it failed. To capture the very high rates of immobilization into litter required to match measured ^{15}N redistributions, nitrogen assimilation into combined detrital-microbial pools had to be decoupled from the traditional concept of carbon bioavailability in litter pools. In other words, the very high rates of nitrogen immobilization and retention ob-

TRACE 2.2 Model

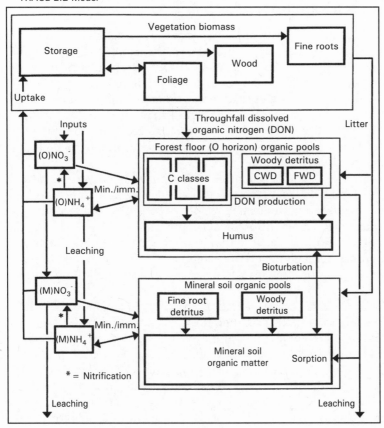

Figure 17.11. Schematic diagram of the hierarchical structure of pools and fluxes of nitrogen in the model TRACE 2.2. Plant uptake of nitrogen, detrital nitrogen dynamics, and nitrogen transformations are calculated separately in each soil layer. Pools of available nitrogen are separated by soil layer. O, organic horizon; M, mineral soil; CWD, FWD, coarse and fine woody detritus; Min./imm., mineralization and assimilation. Inputs = NO_3^- and NH_4^+ in atmospheric deposition, fertilizer, and isotopic tracer additions. For clarity, not all fluxes are shown in detail. Modified from Currie and Nadelhoffer 1999, 7: fig. 1, with permission from Springer-Verlag (copyright 1999).

served in the chronic nitrogen plots cannot be modeled using traditional views of carbon-use efficiency and biomass production by free-living microbes. This indirectly supports the hypotheses in Chapter 12 that this immobilization occurs through either abiotic or nontraditional (for instance, mycorrhizal) microbial processes. This direct and not fully understood process for nitrogen immobilization was the most important pathway for nitrogen immobilization into soil organic matter.

Modeling the Effects of Environmental Change at the Regional Level

If models give reasonable results for intensively studied sites, how can we then extrapolate those results across whole regions? The concept is simple: specify all of the input parameters required by the model and it can be run anywhere. The reality, however, is more difficult and relates to model structure. The more complex the model, and the more input parameters required, the more difficult it will be to run at poorly studied sites or across broad regions. Said another way, the model can be run only for those locations where all the input parameters can be specified (or estimated).

We recognized this potential limitation when constructing the logic behind Figure 17.1 and designed the PnET models to be as simple as possible and to require as few inputs as possible. The goal was to capture most of the dynamic of forests in the region with the fewest possible parameters. Still, we need spatial information on climate and vegetation type across the region, and the accuracy of these spatial data sets are as important as the realism of the physiological functions within the model.

We have based our system for predicting climate in the New England region (Figure 17.12) on a digital elevation model (DEM), or an electronic map of regional topography. Simple regression models using data from more than 300 weather stations across the region predict mean temperature and precipitation as a function of latitude, longitude, and elevation yield estimates with mean errors of less than 1.5°C and 0.67 to 1.25 centimeters per month. These equations are then mapped back on to the DEM to yield spatial images of climate variables for the region (Figure 17.12).

It is interesting that the high-density data sets available for temperature and precipitation are not replicated for solar radiation. Even primary weather stations record only a crude "percent sunlight" value that approximates the fraction of any day with sunlight intensity above a certain value. Extensive solar energy monitoring networks initiated in the 1950s were discontinued in the 1970s. However, these older data sets can also be related to geographic location and used to derive seasonal and regional changes in solar energy input (Figure 17.13).

Mean monthly deposition of elements and ions in precipitation, averaged over several years, can be predicted from regional concentrations multiplied by predicted precipitation rates. For example, the concentrations of both nitrate and sulfate vary linearly with longitude across this region, reflecting the strong sources of these components of acid rain in the industrialized areas to the west. Particulate and gaseous components of nitrogen deposition ("dry deposition") vary linearly with latitude, reflecting the shorter residence time in the atmosphere and the

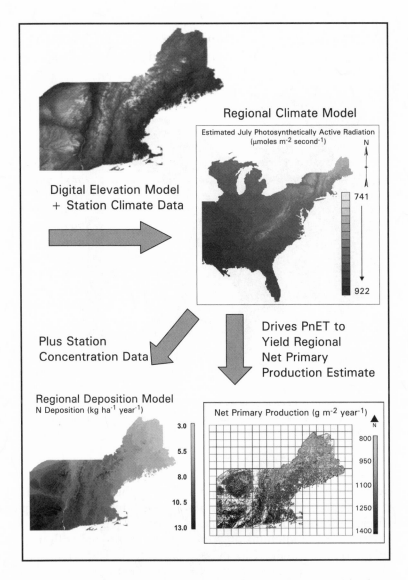

Figure 17.12. Schematic diagram of the combination of a digital elevation model with station-level climate and deposition data to generate spatial estimates of the physical and chemical climate for the New England region, which can then be used to drive the PnET model. Based on Ollinger et al. 1993 and 1995, and Aber, Ollinger et al. 1995.

predicted changes in temperature, precipitation, and atmospheric CO_2 concentration. Using an early climate change scenario in which doubled CO_2 results in changes of $+6°C$ and -15 percent precipitation, PnET-II predicts an increase in forest production (Figure 17.14, bottom) but a decrease in water yield, similar to values predicted for the Harvard Forest site alone. The more complete set of change scenarios analyzed with PnET-CN for Harvard Forest and Hubbard Brook has not yet been repeated at the regional scale.

The Link to Remote Sensing

Our regional modeling efforts have been hampered by the lack of spatial data on some of the most important indicators of ecosystem function. Two of the most critical of these are characteristics of foliage: foliar nitrogen concentration, which is the most sensitive parameter controlling canopy photosynthesis, and foliar lignin and cellulose concentrations, which control decomposition in the DocMod model. These are whole-canopy parameters, and remote sensing is the only practical approach to measuring whole-canopy characteristics over large continuous areas. Early recognition of this problem led us to explore the potential for acquisition of basic model input parameters by remote sensing.

Traditional broadband sensors such as AVHRR (Advanced Very High Resolution Radiometer) and LANDSAT-TM (Thematic Mapper) do not provide enough spectral resolution (enough channels of information at the right wavelengths) to detect subtle changes in canopy chemistry. Instead, we have used a prototype sensor developed by the National Aeronautics and Space Administration (NASA) called AVIRIS (Airborne Visible Infrared Imaging Spectrometer), which provides 220 channels of information in the visible and near-infrared portions of the electromagnetic spectrum with approximately 10-nanometer spectral resolution. The instrument flies aboard a NASA ER-2 aircraft at an altitude of 70,000 feet, providing a spatial resolution on the ground of approximately 20 meters.

The high spectral resolution of the surface reflectance data provided by AVIRIS allows the application of laboratory-level spectrophotometric techniques for extraction of information. We have used first and second difference spectra (rates of change in reflectance across wavelength) in multiple linear regressions to calibrate equations predicting foliar nitrogen and lignin concentrations for a number of forest stands (for example, Figure 17.15), with accuracy approaching that of field measurements. We have used the data from the Harvard Forest as input to the PnET-II model and predicted spatial patterns of net carbon exchange for the Prospect Hill tract (Figure 17.15). The spatial average of these values

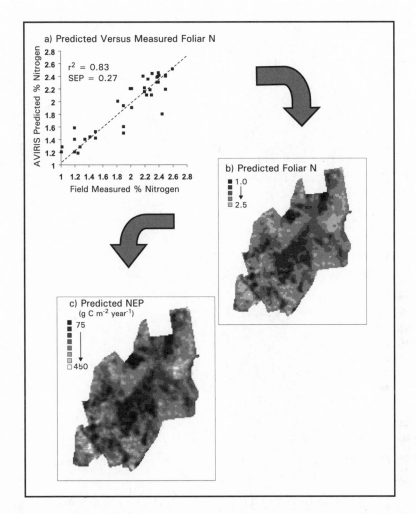

Figure 17.15. Use of high spectral resolution remote sensing (AVIRIS, airborne visible infrared imaging spectrometer) to derive whole-canopy foliar nitrogen concentration, used in turn as an input to the PnET-II model to estimate fine spatial scale patterns in net ecosystem production (NEP) for the Prospect Hill tract. a: A relationship between field-measured foliar nitrogen concentration and that predicted using AVIRIS; b: estimated foliar nitrogen for Prospect Hill; and c: PnET-II-estimated NEP. Modified from Martin and Aber 1997, with permission from the Ecological Society of America.

around the EMS tower is not inconsistent with the mean net ecosystem exchange estimates from the eddy covariance method.

AVIRIS data have also been valuable for refining maps of species distribution on Prospect Hill. Species vary in predictable and somewhat nonoverlapping ways with respect to the lignin and nitrogen concentra-

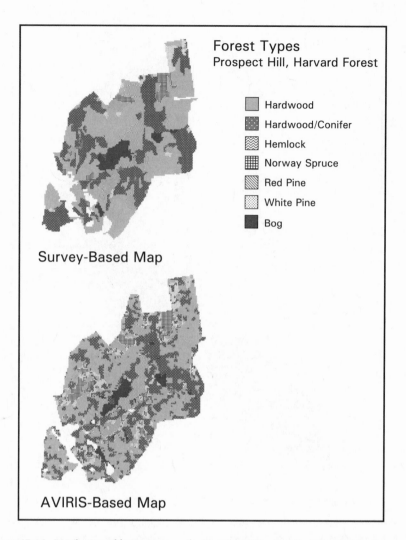

Figure 17.16. Distribution of forest types on the Prospect Hill tract of the Harvard Forest as mapped on the ground and as predicted from AVIRIS data. Note the identification of small areas of different species composition in the AVIRIS image that fall within larger areas mapped as large, heterogeneous units on the ground-based map. Modified from Martin et al. 1998, 252, with permission from Elsevier Science (copyright 1998).

tions in foliage. Using wavelengths that were appropriate for determining foliar chemistry, traditional supervised classification reproduced measured distributions of forest types and actually increased precision by delimiting pine and hardwood subcomponents within mixed stands (Figure 17.16). Separation at the species level was possible in pure stands (such as red pine versus white pine and spruce versus hemlock).

V

Lessons from the Forest and Its History

Through the preceding chapters we have outlined our understanding of the many changes that have occurred in the New England landscape over the past millennia and across a wide range of spatial and ecological scales. We have emphasized the role that changes in human and natural processes play in shaping the land and ecosystems in which we study and live today. One unifying theme has been a hallmark of Harvard Forest studies over the past century. Whether we examine forest structure and composition, wildlife dynamics, soil conditions, nutrient cycles, or energy flow, we have discovered that modern conditions and future responses can be interpreted only in light of ecosystem and landscape history. In conjunction with our ecological understanding, this historical perspective becomes invaluable as we work with conservationists, land managers, and policy makers to develop effective management strategies for the landscape of New England and other regions.

An understanding of landscape history enables us to extend the temporal perspective of long-term research and to document the range of variability of ecosystem conditions and processes in ecological rather than human time frames. From this perspective, we learn that most landscapes and natural ecosystems bear strong legacies of past events. In fact, as we anticipate future changes in our forests, we fully expect that these will be driven as much by recovery from historical processes as by responses to novel conditions.

In this final section, we leave the details of our specific studies and experiments to reflect on some of the broad ecological and environmental lessons that have emerged from our collaborative efforts. We discuss these lessons in terms of broad themes and with regard to current and future issues of environmental and societal concern. In each of these discussions, we consider the interplay among environmental change, natural disturbance, and human activity in shaping the past, current, and future states of the New England landscape.

We begin with the lessons that paleoecology, historical studies, and modern research in forest ecology yield regarding the basic organization of forest communities and assemblages of plants and animals. We then describe how we apply this temporal perspective to the conservation and restoration studies that have become an increasing focus of our work.

We then shift from forest communities and landscapes to consider forests as ecosystems and reflect on the insights that come from the "long lens of history" that we have applied in our research. As we examine the changes that have occurred in the physical environment, in the patterns of plants, animals, and forests, and in the chemical climate of carbon, nitrogen, and pollutants in the atmosphere, we glean many insights into current forest function. We focus specifically on carbon because of its importance in forests, in human life, and in global climate concerns. Carbon is also a unifying theme for many disparate studies because it links diverse ecosystem components and provides an important currency for discussions of contrasting ecosystems. Our work has brought many new approaches toward measuring and understanding carbon dynamics, and these, in turn, have forced ecosystem and atmospheric scientists and ecological modelers to embrace history as integral to their work and interpretations.

Carbon, human history, and environmental concern force us ultimately to expand our spatial scope beyond the forest and the region to consider the globe. How do the lessons from the Harvard Forest fit into the global picture and how do land-use history and ecosystem process play out at this scale? These varied perspectives on our work emerge by placing our Harvard Forest research within the context of a wide range of other studies. Although personal and reflective, these essays place the research in this book into a broader context and underscore common and unifying themes.

ing broad-leafed tree was common throughout much but not all of southern New England. Forests of chestnut and oak, oftentimes mixed with hickories, birches, and other species, were a natural and seemingly stable forest type across the region. Certainly they must have appeared that way to the nineteenth-century villagers in central Massachusetts or their Native American predecessors, both of whom relied heavily on the nuts and wood of these trees. And yet, chestnut did not become abundant in Massachusetts until approximately 1,500 years ago, when a subtle climate change initiated significant shifts in forest composition. In contrast, oak, which is frequently likened to and found with chestnut, has been dominant across the region for at least 9,000 years. Thus, although chestnut and red and white oak overlap to some extent in their habitat requirements and many ecological characteristics, their physiological tolerances, resource requirements, and migration rates are sufficiently different that through most of the postglacial period they differed substantially in their distribution and abundance. In fact, when examined over centuries, or generations of trees, as we have done at Chamberlain Swamp (see Chapter 6), there has been no long-term stability in the oak-chestnut association or other forest types that we encounter today or that were important historically.

Similar results from studies of fossil faunal remains suggest that animal assemblages have been equally dynamic and that individual species operate with striking independence. Our reconstruction of the history of changes in wildlife populations since European settlement highlights dramatic and largely individualistic changes in animal species distributions over just the past three centuries, driven by habitat and environmental change in addition to major human impacts including selective hunting. Thus, the individualistic quality and different rates of response of species to changing climate and disturbance suggest that community assemblages are typically rather transient and that "nonequilibrium" conditions may be widespread. These observations force us to anticipate the continuing reconfiguration of species assemblages and ongoing individualistic change in the future.

The relative independence of species is a major source of the resilience that our forests have shown in the face of many different forms of physical and biological disturbance over past millennia. When a selective process eliminates or transforms the ecological role of an important species, such as occurred in hemlock's response to a forest pest 5,000 years ago, slaughter of the passenger pigeon in the late 1800s, or chestnut blight in the early twentieth century, ecosystems do not collapse and catastrophic shifts in broad processes do not occur. Despite the apparent keystone or core role of some species, other species adjust in abundance, sometimes over lengthy periods, and a range of forest ecosystem processes exhibit great resilience. This is not to say that the loss of species does not have important ramifications, including effects that may appear to be devastating at the time. To the squirrels, bears, domestic animals, or rural New Englanders that relied on the nuts or wood of chestnut trees, or to the many people, industries, and wildlife that cherish hemlock forests, the arrival of the chestnut blight or the hemlock woolly adelgid has had major repercussions. But historical studies suggest that all species have evolved in a dynamic environment and under conditions of change and that each has its own particular biology and response to these changes. This independence and individualistic character of species confers great resilience to the forest overall. Tight linkages are relatively few, and so, though species can be traced through remarkable stretches of geological time, their assemblages are transient and innumerable. The loss of any one, or even many, brings change, but continuity is facilitated through readjustment. Though the squirrel populations of New England apparently crashed initially with the loss of chestnuts in the early 1900s, it would be difficult to conclude that from the large numbers of gray, red, or flying squirrels in the woods today.

However, when we discuss "continuity of change" in our landscape, it is essential that we acknowledge the relative nature of this term. In particular, our paleoecological records embrace thousands of years and may provide somewhat deceptive interpretations of the nature of change. In general, the shifts noted before European settlement are many times slower and much less frequent than those occurring today or over the past few centuries. In fact, some sites exhibit evidence of exceedingly slow rates of change in vegetation or terrestrial and aquatic ecosystem processes over the few thousand years before European settlement. For example, although pollen diagrams from Lily Pond in central Massachusetts depict a rapid shift 1,500 years ago from oak to chestnut, both of these dominant tree species showed essentially no substantial change in abundance for the thousand years preceding or following this transition. In northern Vermont, our pollen diagram from Levi Pond exhibits a 2,000-year history of gradual and exceedingly slow increase in spruce and no evidence of fire or other disturbance to the watershed. Otherwise, this remote site exhibits few dynamics, even after European settlement. In similar fashion, hemlock has dominated the forest in the center of the Prospect Hill tract on the Harvard Forest for the past 8,000 years despite many changes in associated species and a 1,000-year fluctuation due to insect attack.

Although disturbances such as fire and insect outbreaks occurred over time, in many instances the broad environment changed only gradually, and the forest underwent repeated sequences of recovery requiring 250 to 1,000 years after these isolated impacts. Thus, although change is a constant factor in our forests, the rates and nature of these changes have varied considerably through time. By all measures the rate and magnitude of change and introduction of novel impacts and stresses have accelerated tremendously over the past three centuries.

The Importance of Disturbance History

Many angles of investigation underscore the fact that the composition and stability of species assemblages over time are strongly influenced by disturbance, and in particular, its frequency, intensity, and geographic scale. Through paleoecological, historical, and modeling approaches, we have documented the long-term influence of fire, hurricanes, pathogens, and people on New England forests. In addition, we have seen that gradients in disturbance regimes influence vegetation patterns at site, landscape, and regional scales. Knowledge of the scales and manner in which disturbance processes operate assists greatly in the interpretation of stand and landscape patterns across New England. For example, there is a regional decline in hurricane frequency from the southeastern coast inland to the north and west across New England, and a local gradient of intense winds that is related to topography. Similarly, the influence of disturbances such as ice storms and pathogens varies geographically and topographically, although the effects of these types of disturbances on stand development and composition are often more difficult to document than broadscale events such as hurricanes. The frequency of disturbance required to maintain particular species or associations also varies considerably. For example, although moderately shade-tolerant species such as oak, pine, and chestnut are thought to require disturbance such as fire for their establishment, once established these species may persist for centuries with only relatively infrequent disturbance.

Because many plant species are relatively long-lived, significant lags may occur in vegetation response to shifting climate or disturbance regimes. This "ecological inertia" complicates interpretations of the factors that control community composition and structure and drive vegetation dynamics. A good example of this is the difficulty in teasing apart the causes of regional vegetation change in New England over the past 500 years. Most of these changes have been interpreted as driven by European land use for the simple reason that many studies compile vegetation descriptions at the time of settlement, compare these with modern data, and attribute the difference to human activity. In fact, broadscale changes such as the decline in hemlock and beech and the decrease in regional differences in forest composition across central New England appear to result from interactions among changing climate, land use, and possibly fire. Climate changes associated with the Little Ice Age extended from at least the 1400s to the mid-1800s and affected landscapes worldwide. In New England, this period overlapped with major cultural transitions, including shifts in Late Woodland Indian activity, disruption and near-eradication of Native American populations by European settlement, and major land-cover and land-use changes. Changes in forest composition that began in central Massachusetts just before European settlement included an increase in oak and bracken fern and a decrease in hemlock and beech; these shifts continued after Europeans arrived. Although many of the other historical changes that occurred in vegetation, such as an increase in early-successional and sprouting species (for example, birch and red maple), are reasonably attributed to cutting, clearing, and burning, others may represent long-term lags in vegetation response to previous climate change.

Reconstructive studies enable us to place the composition of modern forests on the reforested agricultural lands in a long-term perspective. In fact, despite the lapse of a century or more since the decline of widespread agriculture, there is no indication that forest composition is reverting to that which occurred before European settlement. It seems that 200 to 350 years of active use, in conjunction with ongoing disturbance and environmental change, has significantly, and perhaps irrevocably, altered the forests of New England. This appears to hold true not only for sites that were cleared of native vegetation, such as pastures and tilled fields, but also for areas in which the only impact was a brief episode of intense logging. Whereas 100 years is clearly inadequate for complete recovery, there is considerable evidence to suggest that forest composition, both locally and across the region, has been permanently altered by the activities of the past few centuries.

The pervasive effect of human activity is often difficult to perceive and appreciate. Under close scrutiny, many areas that are valued today because they support remnant "old-growth" stands show direct or indirect evidence of human disturbance, including early historical cutting, grazing, or maple sugaring or introduction of nonnative pathogens. In fact, all of the New England landscape, and essentially all areas of the globe, has to some extent been affected by human disturbance, either directly or indirectly. As a result, it is likely that the distribution and abundance of every species on the landscape, both rare and common, have, to different degrees, been altered by historical land-use practices. Although it is quite remarkable that forests have become reestablished across much of New England relatively soon after agricultural abandonment, we realize that the forest is different in terms of composition, structure, and function from that which occurred before European arrival. Some species that we suspect were widespread on particular types of sites before European arrival were largely eradicated by agriculture and have not successfully recolonized old fields. Other species are apparently much more widespread today than they were in the early historical and prehistoric periods, having taken advantage of the increase in open and disturbed conditions that have come hand in hand with historical land use. We conclude, therefore, that modern species distributions and community patterns not only result from variation in current resource conditions, but also in many instances reflect a wide range of historical factors that have persistent influence on the landscape. When we view the forested landscape today, we must recognize that we are observing species at different points in their response to or recovery from past disturbances. As a result, even in the absence of further disturbance, forest composition and structure will continue to change in the future in response to past human disturbance.

Finally, we must emphasize that the history of land use over the past 300 years has been sufficiently complex that we will never know many of the details of historical disturbances. Although many of the effects of this complex history are likely to be significant, they are still largely undocumented and perhaps will always remain so.

The Importance of Retrospective Approaches for Conservation and Restoration

The historical approach that we have used to understand the composition and function of the modern landscape is also critical for evaluating conservation objectives and for developing management approaches for achieving those objectives. Retrospective studies are necessary to document the range of communities and processes that have been important within a region or on a site of interest over time. In turn, this information may help in our attempts to select desired and practical conservation goals as well as effective management approaches to achieve them. We must emphasize, however, that because most systems are characterized by continual change over the previous hundreds or thousands of years, generally no static baseline conditions emerge as clear conservation targets from any reconstructive study. Although many studies have attempted to determine conditions just before European settlement as a baseline for comparison with modern conditions, the fact that the pre-European period was highly dynamic suggests that it is arbitrary and oftentimes inappropriate to consider such a point in time (for example, just before European contact) as representing "natural conditions" or as the sole legitimate context for conservation and restoration efforts.

Even in cases where it is possible to develop a detailed picture of the pre-European landscape, it is typically impossible to re-create former conditions precisely. Several centuries of human disturbance as well as altered climatic and disturbance regimes have transformed the landscape sufficiently such that it is usually not possible to "restore" some prior state, although it may be possible to manage for particular attributes (for example, disturbance regimes, specific vegetation structure, minimum population levels of rare species, etc.). For example, it has long been recognized that fires have strongly influenced vegetation dynamics on sand plains throughout the Northeast, and paleoecological investigations have confirmed that fires were important prehistorically on many of these sites. As a result of the widespread perception that modern sand plain communities differ from their prehistoric "analogs" largely as a result of twentieth-century fire suppression, many conservation and resource management agencies throughout the Northeast have implemented prescribed fire management programs in these communities. For several reasons, however, it is unlikely that simply reintroducing fire to these systems will result in the restoration of prior community states. In fact, most sand plain communities have experienced a wide range of disturbances in addition to twentieth-century fire suppression that have influenced their current composition and structure. On sites where species were largely eradicated by historical agriculture, new species have become established, and soil conditions have been changed (see, for example, Chapters 8 and 9), and it is highly unlikely that simply restoring fire will result in the reestablishment of historical assemblages. In addition, the landscape setting of modern sand plain communities differs substantially from that which occurred in the early historical period. Whereas thousands of hectares of sand plain vegetation formerly occupied many areas, today most stands are small and isolated, creating significant differences in processes that are sensitive to landscape patterns, including species dispersal and establishment after disturbance. Finally, management fires do not closely mimic pre-European burning regimes, which were undoubtedly quite varied and at least occasionally burned intensely and during the most flammable conditions over large areas. Because it is not practical to set such types of intense or large fires, other management approaches (for instance, mechanical treatment and soil scarification) may be necessary to regenerate species that require such disturbances. Thus, even with the reintroduction of some fire into these systems, it is unlikely that conservation managers can create conditions that closely approximate the pre-European landscape.

One of the paradoxes of New England's history of intensive land use is that it has created landscapes that are often attractive to us and harbor plants and animals that we value but that depend on continued human disturbance for their perpetuation. Thus, in New England, many aesthetically desirable and indeed characteristic scenes—including open fields, spreading fencerow maples, graceful old-field pine stands, heathlands, and sand plain grasslands—are products of a cultural history that is transient, changing rapidly, and frequently difficult or impossible to reproduce. To retain fields, heathlands, and grasslands, we would need to replicate or reintroduce the traditional methods of plowing, grazing, mowing, and burning that produced these landscapes in the first place. Similarly, though we could generate new old-field pine stands, it is not possible to retain or regenerate the existing forests on the same sites.

Such successional communities and landscapes are inherently transient.

Likewise, the plants and animals that thrive in these cultural settings are highly dependent on human disturbance for their maintenance and therefore represent some of the most vulnerable species in our region, and indeed in many parts of the United States and even globally. Shifts in agricultural activity, including reforestation in New England and intensification of agricultural use in many other regions, have resulted in a loss of the cultural practices that maintained traditional landscapes and their wildlife and plant assemblages. Regardless of the condition of the land at the time of European arrival, recognition that many current landscape characteristics are the result of cultural activity is a critical step in interpreting and conserving important landscape features. It is precisely this appreciation of history that has recently led some conservation organizations to promote the use of sheep and cattle grazing, mowing, and other practices in the management of many early-successional habitats. Similarly, such an understanding has led to the pragmatic decision to attempt to protect particular species such as grassland birds, which were once found abundantly in New England's agricultural landscape, in alternative cultural habitats today, such as airports, right-of-ways, military training grounds, and landfills. These peculiar artifacts of modern human activity replace some of the function that historical fields played in centuries past.

In this example and in many others, historical perspectives are useful in identifying the range of conditions and processes that have been important over time, in clarifying species response to disturbance, and in setting some constraints on what conservation goals and approaches may be desirable or practical on particular sites. However, although decisions about what we should manage for may be informed by ecological study, they are ultimately based on our cultural values. The extent to which cultural values influence conservation objectives is well-illustrated by a comparison of European and American approaches to conservation. In many parts of Europe, where a history of intense human activity has altered the landscape over millennia, the conservation value of cultural landscapes such as agricultural meadows, hedgerows, heathlands, and wood pastures is widely accepted and serves as a major motivation for management and financial expenditure. In fact, these cultural landscapes and the species they support are regarded as among the highest priorities for conservation precisely because they embody cultural history and identity while maintaining biological diversity. In contrast, and in large part as a result of the relatively brief history of intensive human disturbance in North America and the wilderness ethic that pervades American culture, conservation efforts in the United States typically emphasize protection of "natural systems," "natural dynamics," "wilderness," "wildlands," "native species," and similar concepts. This emphasis on naturalness causes human-modified and cultural land-scapes to receive less appreciation and conservation focus. When coupled with incomplete historical information, such perspectives may also lead conservationists into misinterpreting features of the modern landscape as "natural" and thus managing with the wrong tools.

Conservation Directions in New England: The Possibilities from History

The history of rapid change across New England has left many opportunities and challenges for land managers and conservationists. First and foremost, species abundances and habitats are changing dramatically, with many woodland species increasing in the vast maturing forest and openland birds, insects, and plants declining as forests and human development encroach on grasslands, thickets, and shrublands. The great variation in landscape setting and history in New England leads to contrasting approaches to conservation within the region. For example, in portions of northern New England (for instance, the so-called "northern forest lands") where historical agriculture was not widespread and where considerable expanses of primary (that is, previously cut, but not cleared) forest remain, there has been an emphasis on conservation of vast tracts of forestland and associated woodland species, including the reintroduction of native predators. In contrast, in southern and coastal New England, with an intensive agricultural history, large human population, and increasing development pressure, more attention has been focused on the maintenance of uncommon species that are largely dependent on agricultural or other cultural landscapes. Across New England, there is also a need to address increased utilization of resources ranging from wood products to wildlife itself. In fact, each of these conservation approaches is consistent with aspects of the region's history and may be appropriate in order to achieve differing objectives. However, our analyses suggest that each strategy is being frustrated by a lack of regional perspective and coordination and incomplete information, including a consideration of history.

For example, in our collaborations with local land trusts, environmental organizations, and government agencies in central Massachusetts, we have been surprised by the incomplete use of broadscale information—on land-use history, ownership patterns, modern disturbance, and activities of the diverse land-management organizations. We recognized that although a tremendous amount of conservation land existed in the region, there was little information pertaining to a simple series of important questions:

- Who owned the land and what was its conservation status?
- How effective was communication and coordination among the range of

organizations and agencies involved in land management and conservation?

- What was the current pattern of land use, especially logging, across the region on private and public lands?
- Were there possibilities for planning a more spatially coherent and ecologically effective strategy of ownership and management across the region?

The results of our inquiries have been surprising, and, perhaps equally surprising, they have had a modest effect on conservation planning. First, we determined that approximately 40 percent of the region's land was protected from development (but not harvesting) through legal restriction, agency mandate, or other encumbrance. Although striking, the GIS-based analysis also showed that the resulting pattern of protected lands was haphazard and collectively less effective than might have been achieved through design and coordination (Figure 18.1). The lack of coordination and haphazard design were the result of more than twenty-eight public agencies and private organizations and untold numbers of private citizens operating independently with individual mandates including drinking water protection, wildlife habitat protection, timber production, rare species protection, open space protection, research, and wildland protection.

Equally impressive was the intensity and pattern of forest harvesting (Figure 18.1, bottom). Approximately 1.5 percent of the forest landscape was selectively harvested annually in a spatially random pattern that was unrelated to physiography, land-use history, broad cover type, or distance from roads. The only correlate of harvesting pattern was ownership class, a result that yielded some surprises of its own. Indeed, among major land ownerships, the Quabbin Reservation, which is the water-supply land for Boston and the largest conservation property in southern New England, is harvested most intensively, followed by private land, and then the state forests.

Thus, although conservation had successfully protected a large percentage of the natural landscape from development, the resulting protected lands were disjointed, vulnerable to fragmentation, and managed in a fashion that provides most effectively for neither intensive natural resource removal nor extensive blocks of maturing interior forest. Instead, current management distributes early-to-mid-successional habitat and forest openings from logging throughout the region. An alternative approach is possible that considers history in an attempt to fashion a regional plan to address many conservation and societal objectives simultaneously.

Although forests in this region were historically fragmented by deforestation and agriculture, natural reforestation over the past century or more has generated a modern landscape in which more than 80 percent of the uplands are forested, including a few large continuous blocks (Figure 18.1, top). One major opportunity for land conservation would

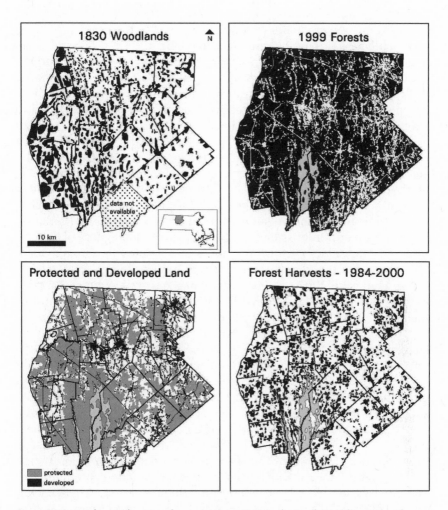

Figure 18.1. Landscape change and conservation activity in the North Quabbin region of central Massachusetts. Like most upland areas of the state, the region has been transformed from largely agricultural to heavily forested in the past 170 years (top figures). In 1995 our compilation of land-ownership records indicated that nearly 40 percent of the land area was protected from development (gray; bottom left). However, the spatial pattern of protection was haphazard because of the history of uncoordinated activities by diverse agencies and independent organizations. Across the region, forest harvesting is widespread and frequent and forms a spatially random pattern that is unrelated to protection status, distance from roads, or physical factors such as topography. Forest cover data for 1999 from MassGIS 2002.

therefore be to promote natural ecological processes and interior forest species by setting aside extensive areas, protected from future harvesting or other human activities. Though these forests will retain many of the legacies of past land use, including stone walls, cellar holes, soil plow horizons, and unique species assemblages, the forests will grow and age and will gradually produce many of the structures, processes, and characteristics of old-growth forest, including snags and dead wood, tip-up mounds, windthrows and active beaver meadows, and immense forest trees. This approach would be most successful in areas with extensive protected land under individual ownership and little fragmentation that might be effectively buffered by land subjected to logging or rural land use. An obvious choice for consideration for such an approach in central New England would be the watershed of the Quabbin Reservoir, a continuous 30,000-hectare area of land and water. As the largest conservation property in southern New England, the Quabbin is often heralded as "an accidental wilderness," and yet timber harvesting is more intensive here than elsewhere in the region. A wildland created from Quabbin lands would not be wilderness and certainly would retain cultural legacies, particularly given the history of agriculture, milling, and dams. It would, however, also provide a unique habitat and preserve common forest types, wildlife, and ecological processes that are typical of the New England uplands. Over time it could emerge as an unmatched reserve for wildlife and forest processes while fulfilling its role as a municipal watershed.

The current distribution of the other conservation lands in the region suggests that a more effective pattern might emerge from protection of a relatively small area of additional connecting lands. This not only would forge linkages among existing parcels, thereby facilitating corridors and movement for wildlife and humans, but also would generate much more extensive continuous blocks of conservation land, thereby improving their effective management. With large blocks of wildlands established, much of the remaining forestland could be devoted to timber management, through various coordinated approaches depending on forest type and diverse objectives. Large areas managed for timber could buffer intact unharvested blocks and would provide habitat for species dependent on openings, gaps, young forests, and landscape mosaics.

The region around the Harvard Forest is typical of the broad New England uplands, where extensive agricultural lands and "hot-spots" of high biodiversity are uncommon. Conservation for those attributes is best achieved where such species and features predominate, including such areas as the Connecticut River Valley and the diverse coastal landscape. Although the uplands are not high in diversity, they have a history that has produced a remarkable expanse of forest that may be best suited for the promotion of large populations of common woodland species and habitats.

In ecological and environmental science we are conditioned to work outside policy circles, and we typically exert little effect when we seek to enter into these. However, the information and considerations described above have had some modest effect. In central Massachusetts, the North Quabbin Regional Land Partnership, a loosely coordinated group composed of the diverse conservation organizations and agencies in the region, was formed in response to analyses that highlighted the obvious need for communication and cooperation. This group has launched some focused projects that address mutual needs, and, in turn, it has attracted attention from the state government and considerable attention and funding for land protection and education. Although many groups and constituent members may not endorse the regional scenario proposed above, there is broad agreement on the need for regional information, coordination, and action that are rooted in the history and modern characteristics of the land. The success of this partnership suggests that effective conservation of diverse values and landscapes throughout the New England region would benefit greatly from similar coordinated planning efforts.

Conclusion

The conflict between conservation values that emphasize limited human disturbance and a history of intensive land use that has altered community composition and function has given rise to significant conservation dilemmas. As woodland species have increased in abundance with reforestation, species that are characteristic of agricultural and disturbed landscapes have generally declined, including numerous species that are currently rare or very uncommon. Should we expend considerable effort and resources to maintain these uncommon species and their habitats, or should we allow them to decline and perhaps disappear from the region? Were these openland species native to the region, and if so, at what population levels and in what types of habitats? Whether or not openland species and assemblages occurred before European settlement, is there value in maintaining the cultural landscapes and associated species that have developed over time and are now part of our landscape heritage?

Conservation and natural resource managers must not only acknowledge the dynamic nature of the landscape and the influence of history on modern communities, but also find ways to incorporate landscape change into long-term planning. It is often inappropriate and futile to attempt to manage natural areas as static and unchanging. Instead, protecting a wide range of sites across physiographic boundaries may enable species to adjust to changing climate and disturbance regimes. Protection of multiple examples of communities of interest may also be desirable, as it is unlikely that all sites will experience catastrophic disturbance simultaneously. In addition, in developing conservation objectives, it may be worthwhile to emphasize protection of functioning systems, including varied and often complex disturbance regimes, rather than concentrating on attempts to maintain the distribution of individual species or current species assemblages.

CHAPTER 19

The Harvard Forest and Understanding the Global Carbon Budget

S. WOFSY

Why Make CO_2 Flux Measurements at Harvard Forest?

Harvard Forest today is very different from the forest that existed before European settlement; trees are smaller and younger, and the forest is composed of different species. There are more red oak, red maple, and paper birch (and other species typical on disturbed lands) and fewer hemlock, beech, yellow birch, and spruce. Chestnut have been eliminated by introduced disease. Harvard Forest today is more similar to the forests of the Middle Atlantic region and less like forests of colder climates in northern New England, as compared with the seventeenth century.

The evidence collected in previous chapters indicates that several hundred years would be required for Harvard Forest to attain the status of "old growth," and this is true for similar tracts throughout New England. We can imagine that if the forest remained undisturbed for such a long period, the processes of growth and decay, recruitment of seedlings, and death of old trees would attain an approximate balance. A large area might have to be considered, however, in order to find this "steady state," since we must encompass an ensemble of patches of different ages and disturbance history.

Only 100 to 150 years have passed since the agricultural era and not yet 70 years since the devastating hurricane of 1938. The climate today is warmer than the climate during the sixteenth century. Forested lands are virtually all manipulated and harvested in various ways to realize and enhance economic value. Cycles of repeated disturbance and succession, creating even-aged stands 50 to 100 years old, are typical of vast areas of forests in the United States today. "Equilibrium" forests, if they ever existed here, are unlikely to be seen again on any broad scale in central Massachusetts, or in most regions of North America.

Recent evidence indicates that the large areas of young and mid-successional forests in North America may represent a significant sink in

Table 19.1. Global CO_2 Budget (Pg C yr^{-1}), 1980–1990

Sources	
Fossil fuel + cement	5.3
Tropical deforestation	1–2
Total	6.3–7.3
Sinks	
Atmospheric accumulation	3.2
Ocean uptake	2.1
"Missing sink"	1–2
Total	6.3–7.3

Note: 2.1 Pg C = 1 ppm atmospheric CO_2 (Ciais et al. 1995).

the global budget for atmospheric CO_2. Atmospheric measurements of CO_2 consistently show an imbalance between global inputs of CO_2, from combustion of fossil fuel, and the accumulation of CO_2 in the atmosphere. For example, in the decade 1980–90, the "missing" (unaccounted) uptake of CO_2 amounted to approximately 1.5 gigatons of carbon per year (1 gigaton = 1 billion metric tons, or 10^{12} kilograms), equivalent to about 30 percent of the input from fossil fuel (Table 19.1). Even larger amounts are "missing" in the 1990s. Other data show that this uptake of CO_2 is associated with release of oxygen and with increase of the $^{13}C/^{12}C$ isotopic ratio in CO_2, both indicating that "missing" carbon has been converted into organic matter. From one year to the next, the unaccounted uptake varies by a factor of three or more, implying sensitivity to El Niño and other climatic fluctuations. Were sequestration of CO_2 by forests and other vegetation to cease, the annual increase of atmospheric CO_2 would be 50 to 100 percent greater than observed in recent decades.

Motivating Questions

Do the forests of the northeastern United States contribute significantly to sequestering atmospheric CO_2? The spatial patterns of CO_2 concentrations in the atmosphere suggest substantial areas of uptake in northern middle latitudes, with one study pointing toward the temperate forests of North America in particular. We would like to know whether this region is indeed a significant sink and to understand the factors regulating the amount taken up. Therefore, one objective of the studies at the Harvard Forest is to assess how long sequestration might continue, how much carbon might be stored, and for how long it may persist as organic matter. A related goal is to learn how to manage forests to optimize economic, aesthetic, and environmental benefits.

This interest provided strong motivation for initiating the long-term study of carbon fluxes at the Harvard Forest described in Chapter 10. We wanted to know whether the Harvard Forest is storing carbon at rates large enough to be consistent with the sink inferred in global carbon budgets. Would factors found to regulate carbon uptake at the Harvard Forest help elucidate causes of year-to-year variations observed in the global budget? Would we find that the ecosystem responds to environmental variations as predicted by models based on process studies (such as described in Chapter 17)?

Approach

New types of observations and a new strategy for long-term measurements were developed to answer these questions. The carbon budget for the entire ecosystem was measured accurately, for many years. The factors that regulate net carbon uptake had to be determined quantitatively. We needed additional information beyond that available from the two principal methods used before 1990 to estimate carbon sequestration:

- *Biometric surveys,* such as carried out in the U.S. Forest Service Forest Inventory Analysis (FIA), estimate net changes in aboveground wood volume by measuring the diameters of tree boles at thousands of locations across the United States. Long intervals (years to decades) between samples are required to allow time for significant change in volume, making it impossible to resolve responses of the ecosystem to seasonal or annual climatic anomalies. Changes in soil organic matter are not measured. The FIA is intended to assess current harvestable stocks of wood and therefore does not give the amount of timber removed over time from the stands nor account for reclassification of land to nonforest uses. Complex calculations are required to estimate carbon fluxes from the observed wood volumes, and the results are very difficult to check. Nevertheless, the FIA provides important regional and national information and a critical constraint on carbon fluxes and budgets in forests.
- *Ecological process studies,* such as measurements of rates for leaf photosynthesis or soil respiration, record the response of components of the forest system to environmental conditions over short time intervals. To assess net carbon exchanges, we have constructed computer models to aggregate these observations to the relevant spatial scales (whole ecosystems, landscapes) and time intervals (seasons, years). Before the eddy flux studies, there were no data to test these models on timescales from hours to years. Our observations test critical aspects of these models, providing a quantitative measure of effects of factors that influence net carbon balance on long timescales, such as climate change and forest succession.

Harvard Forest proved to be the ideal setting for the first long-term study using continuous direct flux measurements. The infrastructure and long-term ecological data base provided the means for us to develop new instrumentation and methods. The ongoing collection of ecological data and coordinated manipulations provided the foundation for testing and demonstrating the success of the approach.

The first step was to adapt for long-term, automated studies the eddy covariance method for measuring fluxes at the ecosystem scale (see Chapter 10). We wished to apply this technique, developed for short-term process studies by agricultural scientists, to studies in forests spanning many years. The next step was to show that the observations obtained using this technique could be aggregated to long time periods accurately; that is, that systematic errors could be controlled sufficiently to resolve the important features of ecosystem carbon sequestration. A series of papers published between 1993 and 1996 provided these demonstrations. We used the vast numbers of measurements and the repeated patterns of diurnal and seasonal variations to test and dissect the data set, establishing rigorous quantitative bounds for errors and biases. This analysis established the capability of eddy flux data to record rates of key carbon exchange processes for time intervals as long as a decade. These papers provided the impetus and conceptual foundation for several networks of tower flux measurement sites established in North America, Europe, and Japan since 1995. In 2001, there were more than fifty long-term flux sites in operation.

A New Way of Seeing the Forest

The long-term flux data set provides an extraordinarily rich view of ecosystem processes, with quantitative resolution spanning timescales from extremely fine grained (thirty minutes) to decadal. Examining these data is like looking at a huge photograph of a city, in which the focus remains sharp whether viewed from a great distance, taking in the whole landscape, or from very close, looking at the salad at the local pizza parlor.

Figure 19.1 shows the average rates of CO_2 sequestration and release for eight years of data, October 1991 to September 1999, representing each hour and day of a mean year. We have measured over and over again the response of the forest to environmental changes: daily variations of solar input and temperature, precipitation events, synoptic weather changes, and seasonal variations of temperature and day length. The waxing and waning of rates for respiration, associated with seasonal warming and cooling and wetting and drying, are clearly defined by data from thousands of nighttime observations. The data define both the mean conditions shown in the figure and the deviations from the mean caused by environmental variance.

Uptake of CO_2 appears dramatically in May, when the forest canopy develops. The date when this happens varies, with shifts of up to two weeks in the timing of leaf-out and the onset of photosynthesis. Varia-

Table 19.2. Carbon Sequestration in Forests: Expectations versus Results, Harvard Forest

Expectations	Results
Uptake of CO_2 (photosynthesis): limited by light, water, nutrients; *insensitive to T, short season* (near optimum)	Uptake of CO_2 (photosynthesis): *very sensitive to short season,* light and water; large increase in CO_2 uptake in warm years
Land cover change → small CO_2 uptake: reforestation after agriculture completed	Land cover change → big CO_2 uptake: reforestation after agriculture creates a long-lasting legacy, amplified by warming
Decomposition and respiration very sensitive to temperature → strong release of CO_2 with warming	Decomposition and respiration temperature limited in winter, but water limited in summer → *weak T ↔ respiration* response
Carbon storage in temperate forests expected to <u>cease after 50–100</u> years after reforestation and to decrease in response to climate warming.	**Carbon storage in temperate forests <u>persists for >100</u> years after reforestation and <u>increases</u> in response to climate warming.**

est, but the *following* years were low-uptake years. Some of the forest components were still affected by stress and mortality from the drought, and there was more-than-average readily decomposed organic matter on the forest floor. Thus, the effects of drought enhanced uptake in the near term and reduced uptake later.

Even in the Northeast, the immediate effects of drought may differ among sites, depending on soil depth and moisture-holding capacity. For example, in 1997 Xuhui Lee of Yale University observed net *release* of CO_2 at an oak-dominated site at the Great Mountain Forest in northwestern Connecticut, not far from the Harvard Forest. This site is on a ridge top and has much shallower soils, and leaf senescence occurred weeks early. Evidently, to assess regional or global effects of the climate anomalies of 1997, we would need to observe the responses of forests over major land areas, which could be obtained from remote sensing. We would also need observations of net CO_2 exchange from representative sites, which can be obtained from towers like the Harvard Forest EMS.

The results have provided a new way of viewing ecosystem processes. Formerly, the only way to study ecosystem carbon flows with fine resolution over long timescales and large spatial scales was to develop and exercise models that incorporate results of small-scale studies of ecosystem processes. Some of these models are very complex and attempt to compute carbon uptake and release in terms of diverse environmental, physiological, and ecological parameters. It was difficult or impossible to check these computations by direct measurements. The long-term flux data, when combined with measurements of the structure of the ecosystem (for example, leaf area index and nitrogen content of leaves), demography and growth of trees, and data on soil organic matter and respiration, provide powerful new ways to check the models and to develop new conceptual paradigms.

Table 19.2 summarizes the response of the Harvard Forest to several slowly varying environmental factors and contrasts them with expectations before the start of the eddy flux measurements. What was expected was rather different from what was observed:

- Over most of North America, climate has warmed over the past thirty years. Summertime temperatures are close to the optimum for photosynthesis, but rates for respiration should increase with temperature. It was therefore expected that climatic trends would be linked to decreasing uptake of CO_2. The opposite was observed, for two reasons. Summer temperatures have in fact not changed recently in New England, but warming has been observed in winter and spring, leading to a longer growing season. The strength of the ecosystem response to a longer growing season was stronger than expected.
- The forest is relatively old, seventy to ninety years, compared with many forests in the Northeast, and it was not anticipated that recovery and succession would currently sustain significant annual sequestration of carbon. It is clear, however, that the red oaks are currently overtopping and outgrowing red maples from upland areas through exactly the process documented here by Chad Oliver in the 1970s. Their larger stature and higher wood density lead to net carbon accumulation even though other factors, such as the leaf area index of the canopy, do not perceptibly change.
- Most ecosystem models represent soil hydrology and rooting depths in a very simplified manner. It was not anticipated that dry summer conditions would lead to seasonal transients with *more* uptake of carbon, nor was it recognized that episodic dry periods would play signal roles in driving the successional replacement of maples by oaks. Thus, the summer drought had the opposite sign and much larger delayed effects than expected.

A summary of what we have learned about the factors regulating monthly and annual net uptake of CO_2 at the Harvard Forest is presented in Figure 19.4. The legacies of the past control the size of the trees, the amount of organic matter in soils, the species composition of the forest, and successional trends. These are the primary drivers for net carbon uptake by the forest. The principal climatic effect has been the lengthening of the growing season over the past thirty years. Variations in snow cover have also been important, but the effects are complex and hard to define unambiguously. We looked for, but were unable to detect, any negative effects of air pollution or episodic droughts. We have been

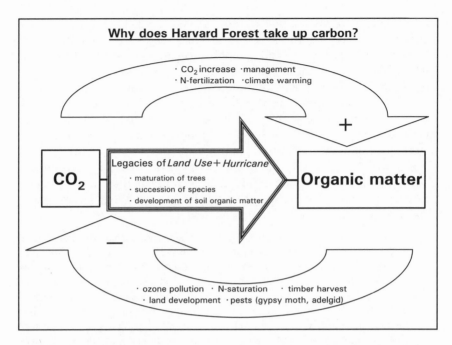

Figure 19.4. Controls on the uptake and release of CO_2 at the Harvard Forest. The major factor driving carbon dynamics is recovery from prior agricultural land use, logging, and disturbance by the 1938 hurricane. The rate of uptake is modulated by numerous environmental factors. Factors promoting sequestration or inhibiting decay of organic matter are placed in the upper arrow; others inhibiting sequestration or promoting the oxidation of organic matter are placed in the lower arrow.

able to quantify the effects of these factors, at least to first approximation, with a combination of eddy flux measurements and careful adaptation of more traditional ecological biometric studies.

A number of critical questions remain to be answered, especially in relation to factors that have changed little over the course of the study and to issues on the future course of the forest. Rising levels of CO_2 and inputs of nutrients from fossil fuel combustion are both expected to enhance forest growth, but these factors do not exhibit sufficient variance for us to detect effects in our experiment. Manipulation experiments addressed elsewhere in this volume are needed to understand these factors. We cannot confidently predict the future course of succession of the vegetation assemblage. We note the current dominance of red oak but a puzzling lack of recruitment of oak seedlings. We don't know what the future climate will be.

We would like to think that studies of the Harvard Forest, and the other sites in the AmeriFlux and Euroflux networks, can tell us what the forests of the United States and Europe are currently doing, and what they will do in the future, with respect to carbon sequestration. Can we design management strategies to enhance carbon uptake and simultaneously optimize economic return and environmental services (flood and erosion control, aesthetics, pollution uptake, moderation of microclimate)? These are the challenges of the future, to which the work represented in this book is dedicated.

CHAPTER 20

The Long Lens of History

J. ABER

Environmental Change and the Future Health of the Harvard Forest

Ecosystem studies at the Harvard Forest proceed from a unique view of the landscape through the long lens of history. This longer-term and historical view of ecosystem response to human use at the Harvard Forest did not begin with the formation of the LTER program or the work described in this book. In his classic essay on forests and social change titled "The View from John Sanderson's Farm," Hugh Raup, former director of the Harvard Forest, compares the rapid pace of economic development and the maturation of the New England economy with the relatively slow development to maturity of a forest. As Raup put it, "Foresters have always had trouble with time because the human mind produces changes in the uses of wood several times faster than trees can grow." He concluded by describing the Harvard Forest as a failure in its initial purpose, "a firm commitment to the ideal that good forest management could and should be economically self-supporting." The root cause of this failure was the inability of foresters and conservationists to envision rapid changes in technology and human values of the land and woods and to focus instead on the inherent functions of the land itself— a belief that "the stage was more important than the actors."

Raup's essay was published in 1966, at the beginning of an energetic era of environmental concern and just as the first classic papers in modern ecosystem studies were appearing in scientific journals. In a sense, Raup's descriptions of failure of purpose at the Harvard Forest were prey to the same rapidly moving social forces he described. He apparently could not envision the continued development and persistence of the Harvard Forest as a center for long-term environmental research: the setting for a series of interrelated experiments exemplifying modern (from the perspective of the early twenty-first century) interdisciplinary environmental science. In a sense, this failure of vision proves the very idea at the core of the essay: we cannot predict what holders of the land will do next and how that will be conditioned by forces in the larger world.

The past, however, does not change. At the Harvard Forest we are blessed with a historical record that allows some precision in the description of how the lands have been used, and also how natural disturbances such as hurricanes weave into this history. Surprisingly, the study of human impacts on ecosystems has only recently become an important part of what American ecologists do. Although classic papers on the development of communities and ecosystems over successional and even geological time frames have formed the basis of the science, the vast majority of studies occur in areas deemed to be "free" of human impact. This assumption has allowed the dismissal of land-use history as an important component of the responses measured in the present.

Over the past decade this perspective has changed. There may be several reasons for this. One must be the accumulation of images of the Earth from space, which constantly reinforce the reality that very few if any large regions of the Earth are predominantly "natural." Environmental awareness reinforces the idea that the study of "pristine" ecosystems, however that might be defined, may have declining relevance in a world driven by human activity. A second could be an increasing emphasis on interdisciplinary work, especially between the natural and social sciences. Even a brief introduction to anthropology or some aspects of history points to the recurring waves of human cultural concentration and dissipation that have swept over the inhabitable parts of the globe. A third is clearly the result of investigations like those described here that underscore the enduring legacy of past human activities on modern conditions.

The longer the legacy of past land use, the more critical it is to understand that history and the imprint it leaves in the soils and forest conditions that we sample today. Here, the Harvard Forest results have particular relevance. The detailed history of change over several centuries presented in this book (Chapter 4) is linked directly to the responses we measure in the long-term experiments. The tower-based measurements of carbon accumulation (Chapters 10 and 19) may vary from year to year in response to short-term changes in climate, but the long-term average is well above zero because of previous removals of carbon in wood and soils going back to the time of John Sanderson. Similarly, the hardwood stand in the chronic nitrogen experiment accumulates and retains huge amounts of added nitrogen because of soil and plant conditions resulting from the centuries-long history of the site. We are just beginning to understand the effects and longevity of ecological legacies from history. There is still much to be learned.

In his introduction to a volume titled *Humans as Components of Ecosystems*, published in 1993, the historian William Cronon captures the recent movement of ecology away from the enumeration of ecological museums (the remaining pristine sites) to the understanding of disturbed systems. He notes that understanding responses to disturbance is critical and that such responses often offer greater insights into the function of ecological systems (after all, every experiment is a disturbance designed to gain just such insight). He concludes that recent interactions of ecologists with environmental concerns have "encouraged ecologists to recognize that their field, like geology and evolutionary biology, is a genuinely *historical* science, with analytical problems and approaches that are unavoidably different from those of less time-bound fields like physics or chemistry."

How important is this temporal perspective? One could liken ecological research without the historical dimension to the plight of the residents of Flatland in the whimsical Victorian volume of the same name by Edwin Abbott. Limited to a life in two dimensions, the residents of Flatland can see only the perimeter of things, and the true shape of objects and beings must be intuited. A brief journey into three dimensions is a revelation to one Square as he perceives the outline of previously unknown dimensions.

So with a historical perspective, we may begin to understand what appears unexplainable in the present. Two ecological examples can be used. In Chapter 5 we described homogenization of tree species distribution across central Massachusetts due to human use of the landscape. What was ordered along climatic gradients and explainable in terms of environmental controls before the European presence has become disordered. Currently, there is no apparent relationship between forest composition and environmental variables at this spatial scale. We cannot reorder the distribution perfectly because we do not understand the history of every site, but if we could, with the perspective of time and history, a logical ordering would likely reappear. We wonder if the same is true for the distribution of soil nutrient content and cycling rates. These rates vary widely over distances of tens of meters and those differences are hard to explain. If the history of each site (each field in Sanderson's farm) were known, would we be able to predict the distribution of nutrients and their rates of cycling?

If our long lens into the past remains imperfect, the lens into the future is even murkier. Hugh Raup concludes that we cannot foresee human valuations or activities in the forests, and the rapid change in social forces relative to the rates of natural change in ecosystems is inescapable. Raup highlights this statement with a glance toward the immense colonial houses, built between 1820 and 1850, that line the main street in the town of Petersham. The owners were prosperous farmers and tradesmen living well in nineteenth-century New England and with every intention of staying for generations to come. They could not foresee the continental-scale revolutions in industry and transportation that would ultimately drive their open countryside back to forest. With the current population less than half of its nineteenth-century maximum, Petersham and its large houses testify to the immensity of social and landscape changes and our limited ability to anticipate them. Certainly, the farther into the future we try to project, the less likely we are to be right. But even over the short term, Raup gives us the example of the Erie Canal as a watershed event that changed land-use patterns in New England in a matter of decades, suggesting that change may come precipitously and in ways that cannot be anticipated.

With this caveat, we can look to the future of the New England forests and attempt to discern at least which forces and stresses might result in the most rapid shifts in distribution and function.

Critical Changes in Physical and Chemical Climate

Regionally and globally, human activity is altering the chemistry of the atmosphere. At the global level, the burning of fossil fuels and a net conversion of forests to agriculture cause a continuing increase in CO_2 concentrations. This is predicted to drive changes in physical climate, generally increasing temperature and altering precipitation. The changes in CO_2 are clear and measurable. Increases in temperature are becoming detectable. Changes in the more chaotic climate variables such as total precipitation, frequency and intensity of storms, and recurrence of global climatic shifts such as the El Niño Southern Oscillation continue to elude direct measurement, and predicted changes vary widely depending on the model used. Predictions for changes in both temperature and precipitation at the regional scale are especially variable between models.

Coincident with the release of carbon to the atmosphere is the release of nitrogen, as well as the conversion of atmospheric nitrogen to reactive oxides, especially in high-compression automobile engines. Both the reactive nitrogen oxides and partially burned hydrocarbons contribute to the formation of ground-level ozone ("smog").

Together, these changes in atmospheric chemistry and the resulting changes in climate represent the strongest potential threat to the health of New England's existing forests. However, the interactions of all of these major "stressors" may be surprising, and not all predicted responses are negative.

Carbon dioxide is the basic substrate for photosynthesis, and the very low concentration of this trace gas in the atmosphere severely limits carbon fixation in most plants. Increases in CO_2 are predicted to increase rates of photosynthesis and also to increase water-use efficiency (the ratio of carbon gained through photosynthesis to water lost through transpiration), thus reducing drought stress. The relative degree of change in photosynthesis and water-use efficiency depends in part on whether plants respond to higher CO_2 concentrations by reducing the rate of gas exchange with the atmosphere.

Nitrogen is the primary limiting nutrient in New England forests, as we have discussed in previous chapters. While nitrogen deposition may act as a fertilizer in this respect, Chapter 12 presents the concept of nitrogen saturation, by which excess nitrogen may lead to nutrient imbalances in plants and acidification of soils and streams. However, the chronic nitrogen experiment suggests that the amount of nitrogen required to reach saturation may be very large, and any increases in plant growth due to CO_2 enrichment would increase this capacity.

In contrast, there is no positive side to ozone. As an oxidant gas, ozone penetrates into foliage and damages cell membranes, which then require investments of energy to repair. This reduces net photosynthesis in a linear and predictable way and also reduces forest growth (Chapter 17). As entrance into leaves is required for damage, any reduction in this exchange rate due to increases in CO_2 could mitigate ozone damage. On the other hand, any reduction in forest growth due to ozone will reduce the forest's capacity to take up nitrogen and could speed the process of nitrogen saturation.

Most predictions for changes in precipitation call for increases globally, while the regional distribution of this increase varies widely between models. In some scenarios, critical areas of the Earth are predicted to see reduced rainfall. Although New England is a humid and generally well-watered region, drought can limit forest growth in certain places and times. Thus, we would expect increases or decreases in precipitation to increase or decrease forest growth.

Response to temperature remains a major unknown. Although photosynthesis and respiration are both highly sensitive to temperature and the optimal temperature for a plant at a given time can be described, we know relatively little about a plant's ability to acclimate—to create a different optimal temperature for net photosynthesis. This remains one of the major unknowns in predicting the response of forests to climate change.

The daily range in temperature (from daily maximum to minimum) also affects plants to the extent that it represents the dryness of the air. In less humid air, the difference between maximum and minimum temperature is larger. More humid air also means less transpiration and a higher water-use efficiency, and so less chance of drought stress. In general, minimum temperatures are predicted to increase more than maximum temperatures in a higher CO_2 atmosphere.

We are currently extending the PnET model discussed in Chapter 17 to include the simultaneous effects of all of these factors. Completion of this effort will provide at least one quantitative prediction of the interactive effects of all of these changes on New England forests.

Population and Land-Use Change:
A Comparison with Environmental Change

As with Hugh Raup's predictions and observations, however, such modeling efforts cannot anticipate the radical or precipitate change that might make many of these predictions irrelevant. As an example, the previous paragraphs describe the response of intact forests that are not disturbed by human use over the period of prediction. For a fifty-year prediction window, the fraction of New England forests to which this applies may be small. Many forests will be harvested for commercial use and others by private landowners for wood or fuel. A major disruption in the global distribution of petroleum (another "oil crisis," as in the 1970s) could result in greater demand for wood-fired power plants and domestic wood burning for heat. It would not take very long, nor a very large increase in the fraction of energy consumed as wood in the region, to completely change the age structure, biomass, species composition, and appearance of the New England forest. In this way, the congruence of a high population density, high energy usage, and large tracts of mature forests could prove a volatile mixture.

Although harvested forests may recover, the ultimate sink for forest-lands is conversion to residential, commercial, or industrial use. Unlike the farmlands and pastures of the 1800s, developed lands are unlikely to be abandoned under anything short of a cataclysmic change in the social structure and are unlikely to revert to forests within any kind of reasonable planning horizon. The U.S. Census Bureau predicts that population in the six New England states will increase from 13.6 million in 2000 to 15.3 million in 2025, an increase of nearly 13 percent. It is certainly risky to convert that to expansion in urban and suburban land use and equate such an expansion into conversion from forests (Dr. Raup's caveat rings in our ears here). Still, assuming that New Englanders maintain their current habits with regard to living in cities or suburbs, it seems that a conservative estimate of a 13 percent increase in urban and suburban land use is reasonable. As most of the agricultural land has already disappeared, through conversion to urban/suburban areas or reversion to forests, much of future conversion must come from currently forested landscapes.

How does this compare with estimated rates of change in environmental parameters over the same period (to 2025)? Carbon dioxide concentrations in the atmosphere are expected to increase by 8 to 22 percent, depending on the degree of control over emissions exerted by the global community, and total nitrogen emissions to the atmosphere in the region might increase by 10 to 15 percent over the same period. The production of ozone is linked to emissions of nitrogen and might be expected to increase by the same percentage, but changes in combustion technologies (and additional Clean Air Amendments) could reduce both of these significantly. The climatic indicators receiving the bulk of popular attention at the turn of the millennium, temperature and precipitation, are expected to change very little according to the latest predictions from the major climate models. From 2000 to 2025, global temperatures should increase by about 1°C, and precipitation should increase by about 1 percent. Regional predictions are more variable among models, but in general temperature increases will be greater at higher latitudes, and the estimate for New England is about 2°C for the period 2021–50, with about a 5 percent increase in precipitation.

Given the mixed and counterbalancing effects of climate, CO_2, nitrogen, and ozone on forests, it seems that a 15 percent expansion of nonforest land over the region, tied directly to human population levels, may represent one of the most important forces for change in the region. Add to this the potential demand for wood as fuel, and the role of human use of the land looms even larger. To researchers at the Harvard Forest, this would signify a continuation of, rather than a break from, the past. History, and the direct human use of the landscape, has dominated the environment in and around the Harvard Forest area for nearly 300 years (and perhaps longer). The indirect effects of human activity through alterations in climate and atmospheric chemistry will continue to alter the biogeochemistry of the forests here in measurable and important ways, ways that in turn contribute to the global response of forests to a changing environment. But, as Hugh Raup's "actors" in the New England landscape, we will determine directly whether forests remain an important part of the "stage."

Bibliographic Essay

To enhance the flow of the text, we have forgone the usual practice in scientific writing of referencing individual statements and facts with numerous citations. However, in almost all cases, the research that is presented, synthesized, and discussed in this volume has been published and is readily available in peer-reviewed scientific journals. Where relevant, the citations supporting the data in figures or tables are supplied in the accompanying captions. In addition, below we point the reader to the major articles from Harvard Forest research and the broader ecological literature that provide important background or specific support for the statements found in each chapter. In this way we seek to provide a general overview to the subject as an entry point to the literature for further consultation. A complete list of all literature cited follows. The complete bibliography of Harvard Forest research, access to all data sets, and further information on our research, educational, and fellowship programs are available on our Web site (http://www.harvardforest.fas.harvard.edu). The Harvard Forest also publishes an annual report that is available as hard copy by request or online at the same Web site.

Chapter 1. *Background and Framework for Long-Term Ecological Research*

For an introduction to the topic of human effects on the environment at local to global scales, there are many places to start that are relevant to New England. *The Earth as Modified by Human Action,* written by Vermont native George Perkins Marsh in 1882, is a classic of environmental literature and one of the first acknowledgments that human land use rivaled geological processes in its scale of impact. More recent contributions on the same subject include B. L. Turner et al.'s *The Earth as Transformed by Human Action,* A. Goudie's *The Human Impact on the Natural Environment,* and G. Woodwell's *The Earth in Transition: Patterns and Processes of Biotic Impoverishment.* The broad elements of the ecological history of the northeastern United States are well-described in G. Whitney's comprehensive *From Coastal Wilderness to Fruited Plain: An Ecological History of Northeastern United States* and W. Cronon's *Changes in the Land: Indians, Colonists and the Ecology of New England.* The Harvard Forest approach to applying land-use history to silvicultural and ecological studies is developed in a long series of papers, including R. Fisher's "New England's Forests: Biological Factors. New England's Prospect" and "Second-Growth White Pine as Related to the Former Uses of the Land"; S. Spurr's "Forest Associations in the Harvard Forest" and (with A. Cline) "Ecological Forestry in Central New England"; H. Raup's "Some Problems in Ecological Theory and Their Relation to Conservation"; J. Goodlett's "The Development of Site Concepts at the Harvard Forest and Their Impact upon Management Policy"; D. Foster and J. O'Keefe's *New England Forests Through Time: Insights From the Harvard Forest Dioramas;* and D. Foster's *Thoreau's Country: Journey Through a Transformed Landscape.*

The Long Term Ecological Research program was established by the National Science Foundation in 1980 to investigate ecological processes over long temporal and broad spatial scales. Background and scientific rationale for the development of this innovative and farsighted program is reviewed by J. T. Callahan, "Long-term Ecological Research"; P. Risser, "Long-Term Ecological Research: An International Perspective"; and J. Franklin, "Importance and Justification of Long-term Studies in Ecology." LTER Network activities and details on the research and educational activities of the individual sites can be located through the LTER Internet home page (http://lternet.edu). A number of publications provide an introduction to the range of approaches that Harvard Forest scientists have brought to their LTER research, including those by D. Foster et al., "Forest Response to Disturbance and Anthropogenic Stress: Rethinking the 1938 Hurricane and the Impact of Physical Disturbance vs. Chemical and Climate Stress on Forest Ecosystems" and "Insights from Historical Geography to Ecology and Conservation: Lessons from the New England Landscape"; J. Aber et al., "Nitrogen Saturation in Temperate Forest Ecosystems: Hypotheses Revisited"; F. Bazzaz, *Plants in Changing Environments: Linking Physiological, Population, and Community Ecology;* J. Melillo, "Carbon and Nitrogen Interactions in the Terrestrial Biosphere: Anthropogenic Effects"; and S. Wofsy et al., "Net Exchange of Carbon Dioxide in a Mid-Latitude Forest."

Chapter 2. *The Physical and Biological Setting for Ecological Studies*
Although the classic description of New England's physiography is provided by N. M. Fenneman in *Physiography of the Eastern United States,* Neil Jorgensen presents a succinct and highly accessible overview in *A Guide to New England's Landscape.* C. Denny's "Geomorphology of New England" includes a fairly detailed description of regional variation in bedrock geology and landscape variation in a more scientific treatment. The development of a simple regression model for New England climate by S. Ollinger et al. in "Modeling Physical and Chemical Climatic Variables Across the Northeastern United States for a Geographic Information System" has been indispensable for our regional LTER work and should be of great interest to all scientists in the area. The standard reference for forest types and variation in New England is M. Westveld's "Natural Forest Vegetation Zones of New England." Westveld chaired a group, which was

working under the auspices of the Society of American Foresters, that took an unusual approach of applying their collective knowledge to describe the natural forest types, the common disturbance processes, and the main patterns of succession or vegetation development after disturbance in each of the major phytogeographic regions in New England. An unusual test and recent refinement of this map is presented in "The Forests of Presettlement New England, USA: Spatial and Compositional Patterns Based on Town Proprietor Surveys," by Charles Cogbill et al.

For the central Massachusetts subregion, which has been the focus of considerable LTER research, the best overviews of landscape dynamics include J. Fuller et al.'s "Impact of Human Activity and Environmental Change on Regional Forest Composition and Dynamics in Central New England" for the history of the past 1,000 years; B. Hall et al.'s "Three Hundred Years of Forest and Land-use Change in Massachusetts, USA"; and D. Foster et al.'s "Land-use History as Long-term Broad-scale Disturbance: Regional Forest Dynamics in Central New England" for the period since European settlement. The Quabbin Reservoir, which serves as the source for metropolitan Boston's drinking water, is a major landmark and environmental resource in central Massachusetts. Its history and major attributes are delightfully described in T. Conuel's *Quabbin: The Accidental Wilderness;* its role in conservation planning in the region is interpreted in A. Golodetz and D. Foster's "History and Importance of Land Use and Protection in the North Quabbin Region of Massachusetts"; and the long-term history of its vegetation is covered in D. Foster et al.'s "Oak, Chestnut and Fire: Climatic and Cultural Controls of Long-Term Forest Dynamics in New England."

In many ways, Petersham is a quintessential New England town, in size, in physical, biological, and cultural attributes, and in general history. This history is well presented in M. Coolidge, *The History of Petersham, Massachusetts;* J. Black and A. Brinser, *Planning One Town; Petersham, a Hill Town in Massachusetts;* H. Raup and R. Carlson, "The History of Land Use in the Harvard Forest"; and D. Foster, "Land-use History (1730–1990) and Vegetation Dynamics in Central New England, USA" and "Land-use History and Forest Transformation in Central New England."

Chapter 3. *Biogeochemistry*

Several text and reference books are available that expand on the simple introduction to carbon, nitrogen, and water interactions presented here. W. Schlesinger's *Biogeochemistry* is the standard introduction to the field in general. J. Aber and J. Melillo's *Terrestrial Ecosystems* concentrates on the biogeochemistry of land-based systems. For an excellent introduction to plant physiology in the context of biogeochemistry and ecosystems, see *Plant Physiological Ecology* by H. Lambers et al. For soils and microbial processes, *Soil Microbiology and Biochemistry,* by E. Paul and F. Clark, is a good source.

F. Golley's *A History of the Ecosystem Concept in Ecology* explores the historical roots of the ecosystem concept, while two early classics in the field, *Biogeochemistry of a Forested Ecosystem* by G. Likens et al. and *Pattern and Pro-*

cess in a Forested Ecosystem by H. Bormann and G. Likens, present pioneering work in measurement and experimentation at the ecosystem scale.

Values for the comparison of cycles in Figure 3.3 come from other sections of this book. The gross and net carbon exchange data are from eddy covariance measurements (Chapters 10 and 17). The nitrogen cycling numbers are from the control plots in the nitrogen saturation study (Chapter 12). The water budget is estimated using the PnET model (Chapter 17).

Chapter 4. *The Environmental and Human History of New England*

The postglacial environmental and biotic history of New England and the northeastern United States provide a fascinating story that has yielded many fundamental ecological insights. M. Davis, who started her career in paleoecology at Harvard University and the Harvard Forest, contributed one of the earliest papers on the subject in "Three Pollen Diagrams from Central Massachusetts," which focused on three sites in the Petersham area. This product of her Ph.D. research with H. Raup was followed by her later works including "Phytogeography and Palynology of Northeastern United States," "Quaternary History and the Stability of Forest Communities," and "Climatic Instability, Timelags, and Community Disequilibrium." R. Davis and G. Jacobson provide a somewhat briefer glimpse into the postglacial history of northern New England in "Late Glacial and Early Holocene Landscapes in Northern New England and Adjacent Areas of Canada," whereas a broader geographical perspective that shows the similarities between many faunal and floristic responses to climate change is presented by A. Graham in "Spatial Response of Mammals to Late Quaternary Environmental Fluctuations."

The ecological and conservation implications of such long-term studies are discussed in papers by M. Hunter et al., "Paleoecology and the Coarse-filter Approach to Maintaining Biological Diversity"; P. Schoonmaker and D. Foster, "Some Implications of Paleoecology for Contemporary Ecology"; and D. Foster et al., "Insights From Paleoecology to Community Ecology" and "Ecological and Conservation Insights From Reconstructive Studies of Temperate Old-growth Forests." Recent studies have shown that shorter-term climate change has exerted a major influence on vegetation and cultural dynamics, most notably the Little Ice Age, which is broadly reviewed by R. Bradley and P. Jones in "'Little Ice Age' Summer Temperature Variations: Their Nature and Relevance to Recent Global Warming Trends." Historical records indicating the variable effect of this climatic event on agricultural yields across New England are interpreted and discussed by W. Baron and D. Smith in "Growing Season Parameter Reconstructions for New England Using Killing Frost Records, 1697–1947," whereas one of the longest-term records of climate in central New England was compiled by R. Bradley et al. in "The Climate of Amherst, Massachusetts, 1836–1985." J. Fuller et al., in "Impact of Human Activity and Environmental Change on Regional Forest Composition and Dynamics in Central New England," advance the suggestion that the vegetation responded strongly to the Little Ice Age and that many of the subsequent changes in vegetation that occur with European settlement were actually initiated by prior climate change.

Although natural disturbance processes have been investigated for many years in New England, many questions remain. W. Patterson's work with K. Sassamann and A. Backman, "Fire and Disease History of Forests" and "Indian Fires in the Prehistory of New England," and a subsequent paper by T. Parshall and D. Foster, "Fire on the New England Landscape: Regional and Temporal Variation, Cultural and Environmental Controls," provide good overviews of fire history, whereas some recent papers by E. Boose et al., "Landscape and Regional Impacts of Hurricanes in New England" and "Hurricane Impacts to Tropical and Temperate Forest Landscapes," bring the discussion and interpretation of hurricane disturbance regimes up-to-date based on a new and innovative approach. In particular, Boose's modeling and reconstructive techniques complement quite nicely the original historical meteorological reconstructions presented by D. M. Smith in his master's thesis at Yale University, "Storm Damage in New England Forests," and the detailed stand-level reconstruction approach that was pioneered by E. Stephens in "The Historical-Developmental Method of Determining Forest Trends" and (with C. Oliver) "Reconstruction of a Mixed-Species Forest in Central New England." D. Henry and M. Swan subsequently extended this approach to Harvard's old-growth Pisgah Forest in "Reconstructing Forest History from Live and Dead Plant Material: An Approach to the Study of Forest Succession in Southwest New Hampshire."

As the only apparent example of a pathogen producing a major decline in a dominant tree species during the pre-European period in North America, the hemlock decline has received considerable attention. M. Davis was the first to attribute the decline to a pathogen rather than climate change in her "Outbreaks of Forest Pathogens in Quaternary History," and she has followed up with collaborators in comparing this event to the more recent decline in chestnut from fungal blight in T. Allison et al., "Pollen in Laminated Sediments Provides Evidence for a Mid-Holocene Forest Pathogen Outbreak." Recent work by N. Bhiry and L. Filion, "Mid-Holocene Hemlock Decline in Eastern North America Linked with Pytophagous Insect Activity," provides more conclusive evidence for the insect pest, and the detailed study by J. Fuller, "Ecological Impact of the mid-Holocene Hemlock Decline in Southern Ontario, Canada," documents the actual pattern and variation in forest response to this major disturbance. The ongoing presence of small populations of hemlock through this event as well as the interaction between fire and hemlock decline on the Prospect Hill tract of the Harvard Forest is described in D. Foster and T. Zebryk, "Long-term Vegetation Dynamics and Disturbance History of a *Tsuga*-dominated Forest in New England."

The effect of the recently introduced hemlock woolly adelgid is reported in many papers by D. Orwig and colleagues, including "Forest Response to the Introduced Hemlock Woolly Adelgid in Southern New England, U.S.A." and "Landscape Patterns of Hemlock Decline in New England Due to the Introduced Hemlock Woolly Adelgid."

Although considerable archaeological and ethnographic work has been conducted on the pre-European history of New England, few good syntheses are both geographically broad and ecologically relevant. Notable exceptions include K. Bragdon's *Native People of Southern New England, 1500–1650,* H. Russell's *Indian New England Before the Mayflower,* and S. Cook's *The Indian Population of New England in the Seventeenth Century.* W. Cronon's *Changes in the Land: Indians, Colonists and the Ecology of New England* explores the transitions in the New England landscape associated with cultural change before and after European arrival, whereas G. Day's "The Indian as an Ecological Factor in the Northeastern Forest" and S. Pyne's *Fire in America: A Cultural History of Wildland and Rural Fire* lay out the argument for Indian burning as an important ecological force. G. Whitney provides a balanced overview of the Indian burning issue as it pertains to the northeastern United States in *From Coastal Wilderness to Fruited Plain: An Ecological History of Northeastern United States.* His work is complemented by the paleoecological studies on fire in W. Patterson's "Indian Fires in the Prehistory of New England" (with K. Sassamann) and "Fire and Disease History of Forests" (with A. Backman).

The ecologically relevant literature dealing with European land-use history and vegetation dynamics is quite large but is accessible through a few overviews. These include G. Whitney's and W. Cronon's books referenced above; M. Williams's *Americans and Their Forests,* which provides particular detail on fuelwood and timber uses; C. Merchant's *Ecological Revolutions: Nature, Gender, and Science in New England;* and H. Russell's *A Long Deep Furrow: Three Centuries of Farming in New England.* H. Barron's *Those Who Stayed Behind: Rural Society in Nineteenth-century New England* and W. Robinson's *Abandoned New England: Its Hidden Ruins and Where to Find Them* and *Mountain New England: Life Past and Present* provide a nice descriptive and pictorial account of this history, with an emphasis on the north and more recent history. For the Harvard Forest the story is well told through H. Raup's "The View from John Sanderson's Farm: A Perspective for the Use of the Land," Raup and R. Carlson's "The History of Land Use in the Harvard Forest," D. Foster and J. O'Keefe's *New England Forests Through Time: Insights From the Harvard Forest Dioramas,* and D. Foster's "Land-use History (1730–1990) and Vegetation Dynamics in Central New England, USA."

The conservation consequences of past and ongoing human activity on the Massachusetts landscape are presented in recent papers by the Massachusetts Audubon Society, The Trustees of Reservations (TTOR), and the Natural Heritage and Endangered Species Program, including C. Leahy et al., *The Nature of Massachusetts;* H. Barbour et al., *Our Irreplaceable Heritage: Protecting Biodiversity in Massachusetts;* TTOR, "Conserving Our Common Wealth"; J. Steel, "Losing Ground: An Analysis of Recent Rates and Patterns of Development and Their Effects on Open Space in Massachusetts"; and P. Weatherbee et al., "A Guide to Invasive Plants in Massachusetts." The implications of New England's land-use history for conservation are addressed in D. Foster and G. Motzkin's "Ecology and Conservation in the Cultural Landscape of New England: Lessons from Nature's History" and D. Foster's *Thoreau's Country: Journey Through a Transformed Landscape* and "Thoreau's Country: A Historical-Ecological Perspective on Conservation in the New England Landscape."

Chapter 5. *Broadscale Forest Response to Land Use and Climate Change*

Details on the long-term changes in vegetation of central New England in relationship to climate change and land-use history come primarily from recent studies by E. Russell et al., "Recent Centuries of Vegetation Change in the Glaciated Northeastern United States"; J. Fuller et al., "Impact of Human Activity and Environmental Change on Regional Forest Composition and Dynamics in Central New England"; and D. Foster et al., "Land-use History as Long-term Broad-scale Disturbance: Regional Forest Dynamics in Central New England." Archaeological data for our central Massachusetts region have been compiled by M. Mulholland at the University of Massachusetts Archaeological Services (see also the Massachusetts Historical Commission's *Historic and Archaeological Resources in the Connecticut Valley*) and are stored in the archives of the Harvard Forest and the Massachusetts Historical Commission. Digital overlays of physical, environmental, biotic, and cultural features are obtained from or available through MassGIS or through Harvard Forest research into regional and national archives. Considerable information on the cultural history, physical environment, and dynamics of rare species and communities in the Connecticut River Valley have been described by G. Motzkin and colleagues through studies at the Harvard Forest and in collaboration with the Massachusetts Natural Heritage and Endangered Species Program. Relevant papers include "Calcareous Fens of Western New England and Adjacent New York State," "Inventory of Uncommon Plant Communities of Western Massachusetts, 1993–1994," "Controlling Site to Evaluate History: Vegetation Patterns of a New England Sand Plain," and "A Historical Perspective on Pitch Pine–Scrub Oak Communities in the Connecticut Valley of Massachusetts."

Chapter 6. *Long-Term Forest and Landscape Dynamics*

The applications of our developing understanding of stand dynamics are varied. Managers and conservationists have an immediate interest in projecting the future of forested landscapes (see K. Kohm and J. Franklin, *Creating a Forestry for the 21st Century: The Science of Ecosystem Management*). Because forests such as those in the Northeast are growing and sequestering carbon, the long-term dynamics of forests also have implications for global change biology; see G. Hurtt et al., "Projecting the Future of the US Carbon Sink." In more basic terms, community dynamics at the stand level provide an intersection between physiology, population biology, and community ecology that has been investigated by forest modelers. S. Pacala et al., in "Forest Models Defined by Field Measurements: Estimation, Error Analysis and Dynamics," provide an overview of this literature and introduce one of the most advanced stand simulators, SORTIE.

In general, stand dynamics are a specific application of the study of disturbance and successional response, which has been the subject of basic ecological research since its inception (see J. Brown and L. Real's *Foundations of Ecology* for the classic works). These topics continue to provide a basis for some of the most compelling questions in ecology (M. Rees et al., "Long-term Studies of Vegetation Dynamics"; S. Pickett and P. White, *The Ecology of Natural Disturbance and Patch Dynamics*). In an applied context, the field of silviculture is fundamentally concerned with managing competition and succession in forests. Works such as *Forest Stand Dynamics* by C. Oliver and B. Larson synthesize our understanding of the structural and compositional development of forests following disturbance.

Many studies of succession and stand development use space for time substitutions to assess the scale and pattern of forest change. A smaller body of work uses reconstructive techniques to estimate the history of individual forests empirically (E. Russell, *People and Land through Time: Linking Ecology and History*). In the Northeast, D. Orwig et al., "Variations in Old-growth Structure and Definitions: Forest Dynamics on Wachusett Mountain, Massachusetts"; C. Ruffner and M. Abrams, "Relating Land-use History and Climate to the Dendroecology of a 326-year-old *Quercus prinus* Talus Slope Forest"; and T. Parshall, "Canopy Mortality and Stand-scale Change in a Northern Hemlock-Hardwood Forest," have used tree rings to understand the trajectory of stand development following disturbance. Longer term compositional histories are available from stand-level pollen records such as those developed by T. Parshall, "Documenting Forest Stand Invasion: Fossil Stomata and Pollen in Forest Hollows," and M. Davis et al., "Patchy Invasion and the Origin of a Hemlock-Hardwoods Forest Mosaic." In long-settled regions, such as the Northeast, archival records also document long-term forest compositional change (see G. Whitney, *From Coastal Wilderness to Fruited Plain: An Ecological History of Northeastern United States,* and D. Foster, "Land-use History [1730–1990] and Vegetation Dynamics in Central New England, USA").

Each approach to stand reconstruction has its own limitations and biases, so applying several approaches to the reconstruction of a single stand's history can provide a more complete picture of stand development (D. Foster et al., "Ecological and Conservation Insights from Reconstructive Studies of Temperate Old-growth Forests"). The Harvard Forest has a long history of such work dating to Earl Stephens's work, subsequently published by C. D. Oliver and Stephens. J. D. Henry and J. M. A. Swan applied his approach to reconstructing the history of an old-growth stand in New Hampshire. Subsequent stand reconstructions at the Harvard Forest include the three studies described in this chapter: D. Foster et al.'s "Post-settlement History of Human Land-use and Vegetation Dynamics of a Hemlock Woodlot in Central New England" and "Oak, Chestnut and Fire: Climatic and Cultural Controls of Long-Term Forest Dynamics in New England, USA" and J. McLachlan et al.'s "Anthropogenic Origins of Late-Successional Structure and Composition in Four New England Hemlock Stands."

Chapter 7. *Wildlife Dynamics in the Changing New England Landscape*
Wildlife changes associated with land-use and land-cover dynamics in New England are surprisingly poorly described, especially given the major changes in the recent past. There are a few notable exceptions, including early

compilations such as Z. Thompson's (1864) *History of the Animals of Vermont* and C. Foss's "Wildlife in a Changing Landscape." Our data come from a major study in the Harvard Forest LTER that has been greatly aided by the cooperation of J. Cardoza and others at the Massachusetts Division of Fisheries and Wildlife. Data from this study are available in the Harvard Forest archives. J. Cardoza has compiled considerable information, much of which he has included in short articles in past issues of *Massachusetts Wildlife* and technical reports to federal agencies and all of which are available through the Massachusetts Division of Fisheries and Wildlife. Other relevant general references include R. DeGraaf and R. Miller, *Conservation of Faunal Diversity in Forested Landscapes;* P. Vickery and P. Dunwiddie, *Grasslands of Northeastern North America: Ecology and Conservation of Native and Agricultural Landscapes;* R. DeGraaf and M. Yamasaki, *New England Wildlife: Habitat, Natural History, and Distribution;* and various articles in *The Northern Forest Forum* and *Restore: The North Woods.* Interesting early papers on the subject of habitat and wildlife dynamics from the Harvard Forest include R. Fisher's "Our Wildlife and the Changing Forest" and N. Hosley's "Some Interrelations of Wildlife Management and Forest Management" and "The Essentials of a Management Plan for Forest Wildlife in New England." Almost weekly, many regional newspapers have an article on the growing conflicts that are developing between the burgeoning, and increasingly wilder, wildlife and the suburban population of New England.

Chapter 8. *Forest Landscape Patterns, Structure, and Composition*
  The influence of historical land use on modern vegetation has been the focus of several important studies in Britain, especially those of G. Peterken and M. Game, "Historical Factors Affecting the Distribution of *Mercurialis perennis* in Central Lincolnshire" and "Historical Factors Affecting the Number and Distribution of Vascular Plant Species in the Woodlands of Central Lincolnshire." M. Hermy's "Effects of Former Land Use on Plant Species Diversity and Pattern in European Deciduous Woodlands" provides a review of relevant studies throughout Europe. G. Matlack's "Plant Species Migration in a Mixed History Forest Landscape in Eastern North America" and J. Brunet and G. Von Oheimb's "Migration of Vascular Plants to Secondary Woodlands in Southern Sweden" discuss migration rates of woodland species.
  Harvard Forest has a long history of incorporating observations of land-use history into interpretations of modern vegetation, including H. Raup and R. Carlson, "The History of Land Use in the Harvard Forest"; S. Spurr, "Forest Associations in the Harvard Forest"; G. Whitney and D. Foster, "Overstorey Composition and Age as Determinants of the Understorey Flora of Central New England's Woods"; D. Foster, "Land-use History (1730–1990) and Vegetation Dynamics in Central New England, USA" and G. Motzkin et al., "Vegetation Patterns in Heterogeneous Landscapes: The Importance of History and Environment." Historical influences on sand plain communities are discussed in G. Motzkin et al., "Controlling Site to Evaluate History: Vegetation Patterns of a New England Sand Plain" and "A Historical Perspective on Pitch Pine–Scrub Oak Communities in

the Connecticut Valley of Massachusetts"; D. Foster and G. Motzkin, "Historical Influences on the Landscape of Martha's Vineyard: Perspectives on the Management of the Manuel F. Correllus State Forest"; G. Motzkin et al., "Vegetation Variations Across Cape Cod, Massachusetts: Environmental and Historical Determinants"; and D. Foster et al., "Cultural, Environmental and Historical Controls of Vegetation Patterns and the Modern Conservation Setting on the Island of Martha's Vineyard." K. Donohue et al., "Effects of the Past and Present on Species Distributions: The Influence of Land-use History on the Demography of Wintergreen," evaluate the biological mechanisms that limit species response to disturbance.

Chapter 9. *Land-Use Legacies in Soil Properties and Nutrients*

A broad discussion of farming practices before the twentieth century is provided by P. Bidwell and J. Falconer's "History of Agriculture in the Northern United States, 1620–1860." The history of New England's forests soon after European settlement, focusing on agricultural history, is provided by W. Cronon's *Changes in the Land: Indians, Colonists and the Ecology of New England.* The long-term changes in soil properties associated with agricultural practices were well-documented in the Rothamsted experiments in England (for example, "The Rothamsted Classical Experiments," by A. Johnston).

The influence of historical land use on present soil conditions has been examined in other areas of the eastern United States and in Europe. The influence of small-scale farming was examined by S. Hamburg, "Effects of Forest Growth on Soil Nitrogen and Organic Matter Pools Following Release from Subsistence Agriculture," and S. Kalisz, "Soil Properties of Steep Appalachian Old Fields." French land surveys of 1830 were used by W. Koerner et al. in "Influence of Past Land Use on the Vegetation and Soils of Present Day Forest in the Vosges Mountains, France." The legacies of logging on soil nitrogen cycling and ecosystem nitrogen retention in the northeastern United States were recently described by C. Goodale and J. Aber in "The Long-term Effects of Land-use History on Nitrogen Cycling in Northern Hardwood Forests" and C. Goodale et al. in "The Long-term Effects of Disturbance on Organic and Inorganic Nitrogen Export in the White Mountains, New Hampshire."

Our recent work at the Harvard Forest has examined the influence of land-use history and modern vegetation on soil properties and processes, including J. Compton et al.'s "Soil Carbon and Nitrogen in a Pine-Oak Sand Plain in Central Massachusetts: Role of Vegetation and Land-use History," and J. Compton and R. Boone's "Long-term Impacts of Agriculture on Soil Carbon and Nitrogen in New England Forests."

Chapter 10. *Exchanges between the Forest and the Atmosphere*

The 2001 International Panel on Climate Change report *Climate Change 2001: The Scientific Basis* by R. Houghton et al. provides a thorough introduction to the global carbon cycle, climate forcing, and reactive trace-gas issues that motivate the work described in Chapter 10. The AmeriFlux Web site at http://

public.ornl.gov/ameriflux/Participants/Sites/Map/index.cfm provides good access to current research on terrestrial carbon budgets in North America. D. Baldocchi et al., in "FLUXNET: A New Tool to Study the Temporal and Spatial Variability of Ecosystem-Scale Carbon Dioxide, Water Vapor and Energy Flux Densities," introduce that global network of carbon flux measurement sites. The basics of atmospheric chemistry are presented in *An Introduction to Atmospheric Chemistry* by Daniel Jacob.

Details of the carbon flux measurements at the Harvard Forest are published in several articles: S. Wofsy et al., "Net Exchange of CO_2 in a Mid-latitude Forest," and M. Goulden et al., "Exchange of Carbon Dioxide by a Deciduous Forest: Response to Interannual Climate Variability" and "Measurements of Carbon Sequestration by Long-term Eddy Covariance: Methods and a Critical Evaluation of Accuracy." This latter paper also discusses the accuracy of long-term carbon flux measurements. Ozone and nitrogen deposition measurements at the Harvard Forest are published in two papers by W. Munger et al., "Atmospheric Deposition of Reactive Nitrogen Oxides and Ozone in a Temperate Deciduous Forest and a Sub-arctic Woodland. I. Measurements and Mechanisms" and "Regional Budgets for Nitrogen Oxides from Continental Sources: Variations of Rates for Oxidation and Deposition with Season and Distance from Source Regions." Measurements of biogenic hydrocarbon fluxes are presented by A. Goldstein et al. in "Seasonal Course of Isoprene Emissions from a Midlatitude Deciduous Forest."

Part IV. *Introduction*

In designing the Harvard Forest LTER we made an early decision to match intensive site-based experimental treatments and observations with broadscale assessments of the history and importance of specific ecological processes, notably hurricane impacts, carbon dynamics, nitrogen deposition, and global warming. The rationale for this approach and the comparative design of the actual experiments are presented in D. Foster et al., "Forest Response to Disturbance and Anthropogenic Stress: Rethinking the 1938 Hurricane and the Impact of Physical Disturbance vs. Chemical and Climate Stress on Forest Ecosystems." Such large experiments run into problems of pseudo-replication or lack of replication, issues that are confronted by many ecosystem scientists who deal with single events or large systems and that are well reviewed in articles by S. Carpenter: "Replication and Treatment Strength in Whole-Lake Experiments" and "Large-scale Perturbations: Opportunities for Innovation."

Chapter 11. *Simulating a Catastrophic Hurricane*

General backgrounds on hurricane meteorology are presented in R. Simpson and H. Riehl, *The Hurricane and Its Impact.* Many compilations of severe storms have been made for North America, with the most informative including D. Ludlum's *Early American Hurricanes, 1492–1870,* S. Perley's *Historic Storms of New England,* C. Neumann et al.'s *Tropical Cyclones of the North Atlantic Ocean, 1871–1986,* and T. Grazulis's "Significant Tornadoes, 1680–

1991: A Chronology and Analysis of Events." The intriguing detective work that lies behind the detailed reconstruction of storm history and meteorology is particularly well-described in H. Lamb's *Historic Storms of the North Sea, British Isles and Northwestern Europe*. Remarkably, considerable detailed information is available on individual major storms, like the hurricanes of 1635 (see Ludlum or D. Smith, "Storm Damage in New England Forests"), 1815 (N. Darling, "Notice of a Hurricane That Passed over New England in September, 1815"), 1938 (C. Brooks, "Hurricanes into New England: Meteorology of the Storm of September 21, 1938," and C. Pierce, "Meteorological History of the New England Hurricane of Sept. 21, 1938"), and 1944 (W. Minsinger and C. Orloff, "The Great Atlantic Hurricane, September 8–16, 1944: An Historical and Pictorial Summary").

The hurricane of 1938 lives on in the memories of many New Englanders, and it changed many perspectives in ecology and forest management. Excellent general considerations of the physical, scientific, and social effects of that storm include NETSA, "Report of the U.S. Forest Service Programs Resulting from the Salvage of Timber After the 1938 Hurricane"; A. Cline, "The Restoration of Watershed Forests in the Hurricane Area"; and D. Smith, "Storm Damage in New England Forests."

Harvard Forest studies that documented the immediate and long-term consequences of the storm include R. Brake and H. Post, "Natural Restocking of Hurricane Damaged 'Old-Field' White Pine Areas in North Central Massachusetts"; S. Spurr, "Natural Restocking of Forests Following the 1938 Hurricane in Central New England"; A. Cline and S. Spurr, "The Virgin Upland Forest of Central New England: A Study of Old Growth Stands in the Pisgah Mountain Section of Southwestern New Hampshire"; D. Hibbs, "Forty Years of Forest Succession in Central New England"; D. Henry and M. Swan, "Reconstructing Forest History from Live and Dead Plant Material: An Approach to the Study of Forest Succession in Southwest New Hampshire"; D. Foster, "Disturbance History, Community Organization and Vegetation Dynamics of the Old-growth Pisgah Forest, Southwestern New Hampshire, U.S.A."; and D. Foster and E. Boose, "Hurricane Disturbance Regimes in Temperate and Tropical Forest Ecosystems." One quite remarkable analysis of the impact of the storm and subsequent salvage logging on regional hydrology is J. Patric's "River Flow Increases in Central New England After the Hurricane of 1938."

The hurricane experiment at the Harvard Forest was developed to mimic the direct effects of the 1938 hurricane as described by W. Rowlands in "Damage to Even-aged Stands in Petersham, Massachusetts, by the 1938 Hurricane as Influenced by Stand Condition"; D. Foster in "Species and Stand Response to Catastrophic Wind in Central New England, U.S.A."; and D. Foster and E. Boose in "Patterns of Forest Damage Resulting from Catastrophic Wind in Central New England, USA"; as well as unpublished data and descriptions in the Harvard Forest Archives. Experiment results that we have drawn from include R. Bowden et al.'s "Fluxes of Greenhouse Gases Between Soils and the Atmosphere in a Temperate Forest Following a Simulated Hurricane Blowdown" and D. Foster et al.'s "Forest Response to Natural Disturbance and Anthropogenic Stress: Rethinking the 1938 Hurricane and the Impact of Physical Disturbance vs. Chemical and Climate Stress on Forest Ecosystems" for ecosystem responses; G. Carlton's "Effects of Microsite Environment on Tree Regeneration Following Disturbance," "Regeneration of Three Sympatric Birch Species on Experimental Hurricane Blowdown Microsites," and "Resource Congruence and Forest Regeneration Following an Experimental Hurricane Blowdown" for resource and seedling dynamics; and S. Cooper-Ellis et al.'s "Vegetation Response to Catastrophic Wind: Results from an Experimental Hurricane" for vegetation responses.

Chapter 12. *Exploring the Process of Nitrogen Saturation*

Much of what is found in the previous chapters sets the stage for the nitrogen saturation experiment. Chapters 4 and 5 discuss the role of human land use in altering species-site relationships present in the presettlement forest. Chapter 9 shows how agricultural usage has left a long-term legacy in soil chemistry and structure. Regional atmospheric chemistry is covered in Chapter 10.

General reviews of the nitrogen cycle can be found in several ecological texts, including J. Aber and J. Melillo's *Terrestrial Ecosystems*. We have reviewed nitrogen cycling in forests in the context of long-term nitrogen additions in several papers, including J. Aber et al.'s "Nitrogen Saturation in Northern Forest Ecosystems: Hypotheses and Implications" and "Nitrogen Saturation in Temperate Forest Ecosystems: Hypotheses Revisited." The first of these papers summarizes the hypotheses presented in Figure 12.2.

The results from Mount Ascutney, Vermont, can be found in S. McNulty and J. Aber, "Effect of Chronic Nitrogen Additions on Nitrogen Cycling in a High Elevation Spruce-Fir Stand," and S. McNulty et al., "Nitrogen Saturation in a High Elevation Spruce-Fir Stand."

We used adsorption of CO_2 into soda lime to measure CO_2 flux from soils in experiment 1 in order to obtain many measurements at the same time. This is an older technique that has important limitations. However, one study conducted in a similar forest setting (J. Raich et al., "Comparison of Two Static Chamber Techniques for Determination of CO_2 Efflux from Forest Soils") suggests that this method and the more generally accepted infrared gas analysis (IRGA) method give comparable results. This near-equality is substantiated by the more recent IRGA analysis presented. Many older studies (reviewed in J. Aber et al., "Nitrogen Saturation in Temperate Forest Ecosystems") support the idea of reduced CO_2 efflux from nitrogen-enriched soils.

Results on DOC and DON flux are presented by S. Currie et al. in "Vertical Transport of Dissolved Organic C and N Under Long-term N Amendments in Pine and Hardwood Forests."

Now-classic papers on the use of short-term (twenty-four-hour) dynamics of the stable isotope of nitrogen (^{15}N) to assess gross rates of nitrogen cycling between microbes and extractable soil nitrogen pools (NH_4 and NO_3) include E. Davidson et al.'s "Microbial Production and Consumption of Nitrate in an Annual Grassland," "Measuring Gross Nitrogen Mineralization, Immobilization, and Nitrification by [15]N Isotope Dilution in Intact Soil Cores," and "Internal Cycling of Nitrate in Soils of a Mature Coniferous Forest"; and J. Stark and S. Hart's "High Rates of Nitrification and Nitrate Turnover in Undisturbed Coniferous Forests." Several papers question some of the assumptions of this technique and emphasize the potential importance of chemical or abiotic reactions in the interpretation of results (for example, D. Schimel and M. Firestone, "Inorganic N Incorporation by Coniferous Forest Floor Material," and G. Berntson and J. Aber, "Fast Nitrate Immobilization in N-saturated Temperate Forest Soils").

Other studies on litter decomposition at the Harvard Forest include J. Melillo et al., "Nitrogen and Lignin Control of Hardwood Leaf Litter Decomposition Dynamics," and J. Aber et al., "Predicting Long-term Patterns of Mass-loss, Nitrogen Dynamics and Soil Organic Matter Formation from Initial Litter Chemistry in Forest Ecosystems."

Mycorrhizae are the absolutely ubiquitous symbiosis between fungi and plant roots. Arguably the most important symbiosis in nature, the physiological dynamics of this critical component of forest ecosystems remains largely unexplored.

Recent papers on the potential importance of abiotic immobilization include D. Johnson et al., "Biotic and Abiotic Nitrogen Retention in a Variety of Forest Soils"; G. Berntson and J. Aber, "Fast Nitrate Immobilization in N-saturated Temperate Forest Soils"; and B. Dail et al., "Rapid Abiotic Transformation of Nitrate in an Acid Forest Soil." Those addressing nitrogen isotope fractionation as a probe into physiological function include E. Hobbie et al., "Correlations Between Foliar ^{15}N and Nitrogen Concentrations May Indicate Plant-Mycorrhizal Interactions," "Foliar Nitrogen Concentration and Natural Abundance of [15]N Suggest Nitrogen Allocation Patterns of Douglas-fir and Mycorrhizal Fungi During Development in Elevated Carbon Dioxide Concentration and Temperature," and "Mycorrhizal vs. Saprotrophic Status of Fungi: The Isotopic Evidence."

Chapter 13. *Soil Warming*

The soil-warming study at the Harvard Forest is but one of a number of ecosystem-level manipulations at the Forest designed to give us insight into how biogeochemical processes are affected by natural and human-induced disturbances of the environment. Ours is not the first soil-warming experiment conducted. In the 1950s, D. R. Redmond studied the effects of soil warming on roots in birch trees in "Studies in Soil Pathology. XV: Rootlets, Mycorrhizae, and Soil Temperatures in Relation to Birch Dieback." More recently, the soil-warming studies have been conducted in an attempt to determine how global warming would affect ecosystem processes. D. Billings and his colleagues observed reduced ecosystem carbon storage in cores of arctic tundra: "Arctic Tundra: A Source or Sink for Atmospheric Carbon Dioxide in a Changing Climate?" E. Chapin and A. Bloom reported that soil warming increased soil inorganic nitrogen concentrations, phosphorus absorption, and plant growth in a wet sedge tundra: "Phosphate Absorption: Adaptation of Tundra Graminoids to a Low

Temperature, Low Phosphorus Environment." Moving south from the tundra, the work of K. Van Cleve et al. in an Alaskan black spruce forest showed that soil warming resulted in increased decomposition, nutrient availability, and foliar nutrient concentration: "Response of Black Spruce *(Picea mariana)* Ecosystems to Soil Temperature Modification in Interior Alaska." Much of the recent work on soil warming has been summarized in "Global Warming and Terrestrial Ecosystems: A Conceptual Framework for Analysis," by G. Shaver et al.

Our work on soil warming at the Harvard Forest is summarized in four publications: W. Peterjohn et al., "Soil Warming and Trace Gas Fluxes: Experimental Design and Preliminary Flux Results" and "Responses of Trace Gas Fluxes and N Availability to Experimentally Elevated Soil Temperatures"; and J. Melillo et al., "Global Change and its Effects on Soil Organic Carbon Stocks" and "Soil Warming and Carbon-Cycle Feedbacks to the Climate System."

Chapter 14. *Comparisons between Physical Disturbances and Novel Stresses*

The discussion of the 1938 hurricane and salvage logging and comparison of novel stresses with historical and physical disturbances draws substantially on D. Foster et al., "Forest Response to Disturbance and Anthropogenic Stress: Rethinking the 1938 Hurricane and the Impact of Physical Disturbances vs. Chemical and Climate Stress on Forest Ecosystems." Relevant discussions emerge in M. Kizlinski et al., "Direct and Indirect Ecosystem Consequences of an Invasive Pest on Forests Dominated by Hemlock."

Chapter 15. *The DIRT Experiment*

The amount and distribution of organic matter in soil profiles influence physical properties such as water-holding capacity and porosity. However, the chemical nature or "quality" of soil organic matter exerts much control over nutrient cycling rates. See, for example, papers by K. Nadelhoffer et al., "Fine Root Production in Relation to Net Primary Production Along a Nitrogen Availability Gradient in Temperate Forests: A New Hypothesis," and J. Aber et al., "Factors Controlling Nitrogen Cycling and Nitrogen Saturation in Northern Temperate Forest Ecosystems." Details of Francis Hole's original experimental designs in oak forest and restored prairies at the University of Wisconsin Arboretum in Madison are described in G. A. Nielsen and F. D. Hole, "A Study of the Natural Processes of Incorporation of Organic Matter Into Soil in the University of Wisconsin Arboretum." Reports of treatment effects on soil carbon and nitrogen stocks and other properties are described in subsequent papers by G. Nielsen, F. Hole, and D. Van Rooyen. K. Nadelhoffer and B. Fry describe how Hole's litter manipulations in forests influenced ^{15}N and ^{13}C distributions in profiles in "Controls on Natural Nitrogen-15 and Carbon-13 Abundances in Forest Soil Organic Matter."

R. Boone et al., in "Roots Determine the Temperature Sensitivity of Soil Respiration," used soil respiration measurements on plots with roots (CONTROL, DOUBLE LITTER, NO LITTER) and with no root growth (NO ROOTS, NO INPUTS) to show that the presence of roots greatly increases the sensitivity of soils to variations in temperature. This finding has important implications for modeling of global carbon cycles and land-atmosphere interactions.

Laboratory incubations of forest floor and soil samples are useful for characterizing the potential releases of respired carbon dioxide and mineralized ammonium and nitrate from soils, as shown by K. Nadelhoffer et al. in "Effects of Temperature and Organic Matter Quality on C, N and P Mineralization in Soils from Arctic Ecosystems." Such incubations provide information on net nitrogen mineralization and nitrification. We also use ^{15}N pool dilution (E. Davidson et al., "Measuring Gross Nitrogen Mineralization, Immobilization and Nitrification by ^{15}N Isotope Dilution in Intact Soil Cores") to characterize gross nitrogen fluxes at shorter timescales between microbial and inorganic pools.

In our Year 5 sampling, we characterized soil communities, particularly numbers and biomass of bacteria, fungi, protozoa, and nematodes, using methods described by R. Ingham et al. in "Trophic Interactions and Nitrogen Cycling in a Semiarid Grassland Soil. Part 1. Seasonal Dynamics of the Natural Populations, Their Interactions and Effects on Nitrogen Cycling" and J. Lodge and R. Ingham in "A Comparison of Agar Film Techniques for Estimating Fungal Biovolumes in Litter and Soil."

Estimating the magnitude of fine root production is problematic. Unfortunately, the most commonly used methods, which involve estimating production by comparing changes in fine root biomass through time, are statistically flawed. Therefore, new methods are needed. The most promising include mass balance techniques (see J. Hendricks et al., "Assessing the Role of Fine Roots in Carbon and Nutrient Cycling") and direct observations of root turnover using fiber optics in "minirhizotrons" (clear plastic tubes that allow camera access). R. Bowden et al., in "Fluxes of Greenhouse Gases Between Soils and the Atmosphere in a Temperate Forest Following a Simulated Hurricane Blowdown" and "Contributions of Aboveground Litter, Belowground Litter, and Root Respiration to Total Soil Respiration in a Temperate Mixed Hardwood Forest," describe how we used measurements of soil respiration on the DIRT plots and mass balance techniques to estimate the annual allocation of carbon to fine root production.

Chance encounters in the early 1980s between Francis Hole and the senior author on bus rides between our east Madison neighborhood and the University of Wisconsin campus motivated the DIRT project. Soils were often the topic of discussion on these rides, but any subject was game. Fellow passengers doubtless learned much from Dr. Hole, and our project was inspired by his work.

Chapter 16. *Experimental Approaches to Understanding Forest Regeneration*

Regeneration processes have been studied intensively by foresters and ecologists in New England for nearly a century and were a primary focus for Harvard Forest professors and students in the first several decades of the Forest's history. Efforts to modify the successional development of cut-over lands and to ensure successful reproduction of white pine after harvest of old-field stands dominated early work on regeneration. The 1938 hurricane led to a broader appreciation for natural regeneration dynamics and the need to base silvicultural practices on a deeper understanding of factors controlling juvenile tree establishment, survival, and growth.

In the four decades following the hurricane, foresters expanded their knowledge of practices that promote the regeneration of desired overstory species. Numerous papers and reports from the U.S. Forest Service's Northeast Forest Experiment Station contributed to this understanding. During the same period, ecologists developed theories about community organization and dynamics based on species-specific responses to multifactored environments (niches) and on species interactions, such as competition and predation, that were regarded, especially by animal ecologists, as the primary regulators of community composition. Physiological ecology began to emerge during these decades as a powerful new discipline with the potential to explain many ecological patterns through a quantitative understanding of integrated organismal responses to field environments. Some ecologists realized that physiological approaches could be used not only to explain adaptation to more extreme environments (for example alpine and desert) but also to establish mechanistic bases for such key ecological theories as niche-based community organization and succession (for example, F. Bazzaz's "The Physiological Ecology of Plant Succession").

Extensive testing during the 1960s and 1970s of the hypotheses underlying community theories revealed that the assumption that communities naturally tend toward, and frequently achieve, some sort of equilibrium state was too simplistic. Ecologists began to realize that natural disturbance regimes were common, were influential, could result in nonequilibrium behavior of communities, and could contribute to the maintenance and possibly the origin of higher species diversity. Consequently, a new emphasis on disturbance regimes and their effects appeared in the late 1970s and 1980s (see papers and references in S. Pickett and P. White's *The Ecology of Natural Disturbance and Patch Dynamics*). Ecologists had studied disturbance before then, particularly in the context of succession, but had not incorporated disturbance processes into the equilibrium-based models of community organization. Plant ecologists were especially aware of the pervasive effects of disturbance and led the way in developing new theory about disturbance ecology. Forest ecologists noted the powerful role of canopy gap disturbance in forest regeneration and began to conceive of gap disturbances as the primary regulator of community composition in mesic to wet closed-canopy forests.

Along with the renewed interest in disturbance regimes came a stronger appreciation for the role of spatial heterogeneity and patch dynamics in ecosystems. Most community theory had ignored heterogeneity because of the complications it introduced into mathematical models. But we now understand that spatial-temporal heterogeneity operates at all scales in nature and must be dealt with in order to make accurate predictions of ecosystem behavior. Useful, broad treatments of heterogeneity include edited volumes by J. Kolasa and S. Pickett *(Ecological Heterogeneity)* and M. Caldwell and R. Pearcy *(Exploitation of En-*

vironmental Heterogeneity by Plants: Ecophysiological Processes Above- and Belowground).

Concern for environmental damage and its effects on human cultures accelerated after 1970. Among the many issues of concern, pollution, biodiversity losses, and global climate change gradually became the most dominant issues. Thus, ecologists began to focus more attention on the origins and implications of biodiversity patterns across biomes and on potential effects of pollution and global change on environments, species, communities, and ecosystems.

All the historical trends just described converge in our research on regeneration processes at the Harvard Forest, which began in the mid 1980s just before the Forest became an LTER site. We brought to this research a resource-based, mechanistic approach toward understanding plant ecology at all hierarchical levels, from leaf to community. The concepts underlying this perspective, and several decades of research on multiple systems, are described in F. Bazzaz's *Plants in Changing Environments: Linking Physiological, Population, and Community Ecology.* This approach stresses spatial-temporal variation of multivariable environments, regulated in part by the community itself, as the dynamic context within which individual plants must integrate their responses over multiple timescales. All the research described in Chapter 16, plus additional studies referred to below, has been aimed at establishing mechanistic bases for higher-level understanding of forest regeneration in response to natural and anthropogenic disturbance in the Harvard Forest ecosystem.

We have used varying combinations of environmental measurement, field experiments, and controlled environment experiments (growth chamber, glasshouse, common garden), a strategy described in F. Bazzaz and T. Sipe's "Physiological Ecology, Disturbance, and Ecosystem Recovery." We have used maples *(Acer pensylvanicum, A. rubrum, A. saccharum)* and birches *(Betula alleghaniensis, B. lenta, B. papyrifera, B. populifolia)* in most of the experiments because these species represent the only two-genus system of co-occurring North American tree species that spans the entire shade-tolerance spectrum, they offer both interspecific and intraspecific contrasts, and the center of their geographic range overlap is in central New England. For other experiments, we have focused on the most dominant overstory species in the Harvard Forest, both coniferous and deciduous.

The gap-partitioning hypothesis was articulated first by Robert Ricklefs ("Environmental Heterogeneity and Plant Species Diversity: A Hypothesis") and then developed by Julie Denslow ("Gap Partitioning Among Tropical Rainforest Trees"). Our work in Prospect Hill was the first experimental test of the hypothesis, and the results are reported in two papers by T. Sipe and F. Bazzaz, "Gap Partitioning Among Maples *(Acer)* in Central New England: Shoot Architecture and Photosynthesis" and "Gap Partitioning Among Maples *(Acer)* in Central New England: Survival and Growth." Seedlings of shade-tolerant tree species may experience damage upon exposure to gap environments, and the effects of leader loss and recovery of damaged stems in our gap study were reported by T. Sipe and F. Bazzaz in "Shoot Damage Effects on Regeneration of Maples *(Acer)* across an Understorey-Gap Microenvironmental Gradient."

The temporally dynamic microclimates in our experimental canopy gaps motivated several experiments on the timing of resource availability and on congruency among multiple resources and environmental factors. The artificial gap experiments on the birches are described in two papers by P. Wayne and F. Bazzaz, "Birch Seedling Responses to Daily Time Courses of Light in Experimental Forest Gaps and Shadehouses" and "Morning vs. Afternoon Sun Patches in Experimental Forest Gaps: Consequences of Temporal Incongruency of Resources to Birch Regeneration." A finely controlled chamber experiment on diurnal timing effects is described by J. Cavender-Bares et al. in "Consequences of Incongruency in Diurnally Varying Resources for Seedlings of *Rumex crispus* (Polygonaceae)." A synthetic view of the work on gap environments and congruency can be found in F. Bazzaz and P. Wayne's "Coping with Environmental Heterogeneity: The Physiological Ecology of Tree Seedling Regeneration Across the Gap-Understory Continuum."

While smaller canopy gaps play continuous roles in forest regeneration, the episodic impacts of major hurricanes are more influential at the Harvard Forest. G. Carlton and F. Bazzaz examined birch responses to multiple resource congruency associated with physically distinct microsites in the experimental hurricane blowdown sites. Two papers describe the work summarized here, "Regeneration of Three Sympatric Birch Species on Experimental Hurricane Blowdown Microsites" and "Resource Congruence and Forest Regeneration Following an Experimental Hurricane Blowdown."

Seedlings of overstory species that establish in the understory in advance of gap formation or hurricane damage contend with not only reduced resource levels, especially light, but also competition with overstory trees, shrubs, and herbs. The effects of ferns with different architectures on tree "seedling banks" were examined by L. George and F. Bazzaz and reported in two papers, "The Fern Understory as an Ecological Filter: Emergence and Establishment of Canopy-Tree Seedlings" and "The Fern Understory as an Ecological Filter: Growth and Survival of Canopy-Tree Seedlings."

Anthropogenic influences such as air pollution and global climate change can be understood as shifts in the spatial-temporal availability of plant resources and environmental modifiers to which co-occurring plant species may respond differently. F. Bazzaz and W. Williams ("Atmospheric Concentrations of CO_2 within a Mixed Forest: Implications for Seedling Growth") measured diurnal and seasonal variation in vertical profiles of ambient CO_2 at the Harvard Forest and stressed the potential impact of naturally enriched CO_2 levels in the herbaceous zone on forest regeneration. We have conducted numerous experiments on juvenile trees, using species from the Harvard Forest system, in which the effects of elevated nitrogen or CO_2 were studied alone or in combination with other controlled variables. The studies most relevant to forest regeneration include the following, in chronological order: F. Bazzaz et al., "Growth Responses of Seven Major Co-Occurring Tree Species of the Northeastern United States to Elevated CO_2"; S. Miao et al., "Elevated CO_2 Differentially Alters the Responses of Co-Occurring Birch and Maple Seedlings to Moisture Gradients"; L. Rochefort and F. Bazzaz, "Growth Response to Elevated CO_2 in Seedlings of Four Co-Occurring Birch Species"; F. Bazzaz and S. Miao, "Successional Status, Seed Size, and Responses of Tree Seedlings to CO_2, Light, and Nutrients"; F. Bazzaz et al., "CO_2-Induced Growth Enhancements of Co-Occurring Tree Species Decline at Different Rates"; R. Crabtree and F. Bazzaz, "Seedling Response of Four Birch Species to Simulated Nitrogen Deposition: Ammonium vs. Nitrate"; S. Bassow et al., "The Response of Temperate Tree Seedlings Grown in Elevated CO_2 to Extreme Temperature Events"; P. Wayne and F. Bazzaz, "Light Acquisition and Growth by Competing Individuals in CO_2-Enriched Atmospheres: Consequences for Size Structure in Regenerating Birch Stands"; G. Berntson and F. Bazzaz, "Regenerating Temperate Forest Microcosms in Elevated CO_2: Species Composition, Belowground Growth, and Nitrogen Cycling"; S. Catovsky and F. Bazzaz, "Elevated CO_2 Influences the Responses of Two Birch Species to Soil Moisture: Implications for Forest Community Structure"; and S. Catovsky and F. Bazzaz, "The Role of Resource Interactions and Seedling Regeneration in Maintaining a Positive Feedback in Hemlock Stands."

Our efforts to connect seedling performance to whole-forest behavior through physiological scaling and simulation modeling reflect the increasing attention to scaling approaches in ecology (for example, see J. Ehleringer and C. Field, eds., *Scaling Physiological Processes: Leaf to Globe*) and the development of individual-based simulation models for forested systems that incorporate resource availability and response functions. Two papers by S. Bassow and F. Bazzaz ("Intra- and Inter-Specific Variation in Canopy Photosynthesis in a Mixed Deciduous Forest" and "How Environmental Conditions Influence Canopy-Level Photosynthesis in Four Deciduous Tree Species") and a third by J. Cavender-Bares and F. Bazzaz ("Changes in Drought Response Strategies with Ontogeny in *Quercus rubra:* Implications for Scaling from Seedlings to Mature Trees") compare seedling performance with overstory tree and whole-canopy behavior so that knowledge of juvenile tree ecophysiology can be coupled to measurements and models of forest-atmosphere exchange. The use of the SORTIE model to explore effects of elevated CO_2 and other resources (particularly light) on long-term forest dynamics was reported by B. Bolker et al. in "Species Diversity and Ecosystem Response to Carbon Dioxide Fertilization: Conclusions from a Temperate Forest Model."

Chapter 17. *Synthesis and Extrapolation*

As a synthesis exercise, much of this chapter builds on information and results already presented. Supporting papers on each topic can be found in the primary references cited in the figure legends throughout and in the earlier chapters. Carbon dioxide and methane fluxes are also covered in Chapter 10 and are part of all of the experiments except those in Chapter 16. Whole-canopy carbon balances are presented in Chapter 10. Litter decay is discussed in Chapter 12 and nitrogen cycling in Chapters 3 and 12. The TRACE model builds on the results of the ^{15}N additions in the chronic nitrogen plots (Chapter 12). Papers presenting the regional PnET applications (see figure legends) contain references to the databases and concepts on which they are constructed.

We have used near infrared reflectance (NIR) spectrometry to determine leaf and litter chemistry in the laboratory as a method for increasing the speed and consistency of analyses as well as a proof-of-concept test bed for remote sensing applications. Demonstrations of the increased precision and decreased analytical error of NIR spectrometry versus wet chemical methods are presented in T. McLellan et al., "Comparison of Wet Chemical and Near Infrared Reflectance Measurements of Carbon Fraction Chemistry and Nitrogen Concentration of Forest Foliage," "Determination of Nitrogen, Lignin and Cellulose Content of Decomposing Leaf Material by Near Infrared Reflectance," and B. Bolster et al., "Interactions Between Precision and Generality in the Development of Calibrations for the Determination of Carbon Fraction and Nitrogen Concentration in Foliage by Near Infrared Reflectance." Extension of this method to fresh foliage was demonstrated by M. Martin and J. Aber in "Analyses of Forest Foliage III: Determining Nitrogen Lignin and Cellulose in Fresh Leaves Using Near Infrared Reflectance Data." The same analytical method was used by M. Martin and J. Aber ("Estimation of Forest Canopy Lignin and Nitrogen Concentration and Ecosystem Processes by High Spectral Resolution Remote Sensing") to generate the images of predicted total foliar nitrogen concentration for stands across the Prospect Hill tract at the Harvard Forest.

Chapter 18. *Insights for Ecology and Conservation*
Many studies have used paleoecological approaches to gain insight into the nature of community assemblages over time, including G. Jacobson et al.'s "Patterns and Rates of Vegetation Change During Deglaciation of Eastern North America," M. Davis's "Climatic Instability, Time Lags, and Community Disequilibrium," and P. Schoonmaker and D. Foster's "Some Implications of Paleoecology for Contemporary Ecology." A. Graham et al., in "Spatial Response of Mammals to Late Quaternary Environmental Fluctuations," extend this approach to faunal assemblages.

For a perspective on British conservation and management, see G. Peterken's *Natural Woodland: Ecology and Conservation in Northern Temperate Regions* and *Woodland Conservation and Management.* The history of American views of wilderness is explored in R. Nash's *Wilderness and the American Mind.* F. Samson and F. Knopf's *Ecosystem Management* provides a compilation of articles that address ecological aspects of land management. The importance of retrospective perspectives for conservation and management is demonstrated in G. Jacobson et al.'s "Conservation of Rare Plant Habitat: Insights from the Recent History of Vegetation and Fire at Crystal Fen, Northern Maine, USA"; M. Hunter et al.'s "Paleoecology and the Coarse-Filter Approach to Maintaining Biological Diversity"; G. Motzkin et al.'s "Fire History and Vegetation Dynamics of a *Chamaecyparis thyoides* Wetland on Cape Cod, Massachusetts" and "A Historical Perspective on Pitch Pine–Scrub Oak Communities in the Connecticut Valley of

Massachusetts"; and D. Foster and G. Motzkin's "Historical Influences on the Landscape of Martha's Vineyard: Perspectives on the Management of the Manuel F. Correllus State Forest." A series of papers by D. Foster and G. Motzkin— "Ecology and Conservation in the Cultural Landscape of New England: Lessons from Nature's History," "Grasslands, Heathlands, and Shrublands in Coastal New England: Historical Interpretations and Approaches to Conservation," "Thoreau's Country: A Historical-Ecological Approach to Conservation of the New England Landscape," and "Interpreting and Conserving the Openland Habitats of Coastal New England: Insights from Landscape History"—outline the conservation challenges faced in highly dynamic and human-influenced landscapes.

Chapter 19. *The Harvard Forest and Understanding the Global Carbon Budget*

An excellent account of the role played by midlatitude forests in taking up CO_2 emitted from combustion of fossil fuel is provided by a series of recent papers. M. Battle et al., in "Global Carbon Sinks and Their Variability Inferred from Atmospheric O_2 and Delta ^{13}C," summarize the evidence showing that CO_2 increases in the atmosphere are smaller than expected from current use of fossil fuels, and they develop the case that forests account for at least half of the "missing" CO_2. S. Pacala et al., in "Consistent Land- and Atmosphere-based US Carbon Sink Estimates," show that forests in North America provide a significant portion of the uptake by analyzing data from the U.S. Forest Service Forest Inventory Analysis (FIA, see L. Heath et al., "Methodology for Estimating Soil Carbon for the Forest Carbon Budget Model of the United States, 2001," and J. Jenkins et al., "Biomass and NPP Estimation for the Mid-Atlantic Region [USA] Using Plot-Level Forest Inventory Data"). S. Fan et al., in "A Large Terrestrial Carbon Sink in North America Implied by Atmospheric and Oceanic Carbon Dioxide Data and Models," highlight the probable role of North American forests.

A key question is why forests are currently taking up CO_2. J. Caspersen et al., in "Contributions of Land-use History to Carbon Accumulation in US Forests," analyze FIA data for several states in the United States. They examine the biomass on thousands of survey plots as a function of the age of each plot and compare the increment in stored carbon with stand age with the increments observed in successive surveys. They conclude that the current growth rates of these forests are virtually the same as growth rates over the past century. They argue that *age structure* of the forests account for most of the current net uptake. Other possible reasons for net uptake, such as stimulation of growth by rising CO_2, by deposition of pollution-derived nutrients especially nitrogen, or by climate warming, would have caused current growth rates in a given stand age to be larger than in the past.

A detailed analysis of biomass changes and eddy fluxes by C. Barford et al., "Factors Controlling Long- and Short-term Sequestration of Atmospheric CO_2 in a Mid-latitude Forest," summarizes the long-term studies at the EMS, establishing that consistent results are obtained for the carbon budget of the forest from traditional biometric methods and eddy fluxes. She and her colleagues show that the pattern of growth at the Harvard Forest is generally consistent with the interpretations of S. Pacala et al. and J. Caspersen et al. mentioned above. Modest increase in growth rates could be attributed to longer growing seasons in recent years. Most of the uptake was due to the expected changes in tree size and stand composition in a sixty-year stand of oak, maple, and white pine, as discussed, for example, by C. Oliver and B. Larson in *Forest Stand Dynamics*. There was also an important component of carbon storage associated with gradual accumulation of coarse woody debris, as expected for a stand of that age that previously had been subjected to agricultural use and salvage of trees downed by the hurricane of 1938 (C. Barford et al., "Factors Controlling Long- and Short-term Sequestration of Atmospheric CO_2 in a Mid-latitude Forest"). The role of species composition of the vegetation assemblage is discussed by S. Bassow and F. Bazzaz in "How Environmental Conditions Affect Canopy Leaf-level Photosynthesis in Four Deciduous Tree Species." Relationships between instantaneous responses to environmental forcing and emergent properties of ecosystems are the focus of many modeling papers, for example, M. Williams et al., "Modelling the Soil-Plant-Atmosphere Continuum in a *Quercus-Acer* Stand at Harvard Forest: The Regulation of Stomatal Conductance by Light, Nitrogen, and Soil/Plant Hydraulic Properties." The approaches developed at the Harvard Forest to understand the carbon cycle of large-stature forests, combining eddy fluxes with basic ecological and forest mensuration methods, are described in S. Wofsy et al.'s "Net Exchange of CO_2 in a Midlatitude Forest" and M. Goulden et al.'s "Measurements of Carbon Sequestration by Long-term Eddy Covariance: Methods and a Critical Evaluation of Accuracy" and "Exchange of Carbon Dioxide by a Deciduous Forest: Response to Interannual Climate Variability." These techniques have been adopted in U.S. and global networks of flux sites as described by D. Baldocchi et al. in "FLUXNET: A New Tool to Study the Temporal and Spatial Variability of Ecosystem-scale Carbon Dioxide, Water Vapor, and Energy Flux Densities."

Chapter 20. *The Long Lens of History*

Hugh Raup's writings remain relevant to modern ecology, conservation, and natural resource management, and many of these are contained in the volume *Forests in the Here and Now* edited by Ben Stout and available on the Harvard Forest Web page. "The View from John Sanderson's Farm: A Perspective for the Use of the Land" captures the historical changes in the people and the land of New England. William Cronon's essay "Foreword: The Turn toward History" appeared in the excellent volume *Humans as Components of Ecosystems: The Ecology of Subtle Human Effects and Populated Areas* that emerged from the Fourth Cary Conference hosted by the Institute of Ecosystem Studies in Millbrook, New York, edited by M. McDonnell and S. Pickett. Contributions in that volume document the influence of human land use in the farthest corners of the globe. *Flatland: A Romance of Many Dimensions,* by E. A. Abbott, first published in 1880, is as interesting a read now as it was then. Both the insights into the limitations of our own three-dimensional view of the world and the satires on social stratification in Victorian England remain fresh. The most recent (2001) authority on projected changes in atmospheric chemistry and climate is the Third Assessment Report of the Intergovernmental Panel on Climate Change (IPCC), *Climate Change 2001: The Scientific Basis,* edited by J. T. Houghton et al. This work represents the result of considerable consensus building within the scientific community. A valuable text on physiological responses to environmental change is H. Lambers et al., *Plant Physiological Ecology.* Projections for human population levels in New England are calculated from data at the Bureau of the Census Web site (http://www.census.gov/population/www/projections/stproj.html). Carbon dioxide and climate change projections are from Houghton et al., *Climate Change 2001: The Scientific Basis.*

Bibliography

Abbott, E. A. 1998. *Flatland: A Romance of Many Dimensions*. Penguin, New York.

Aber, J. D. 1992. Nitrogen cycling and nitrogen saturation in temperate forest ecosystems. *Trends in Ecology and Evolution* 7: 220–223.

———. 1993. Modification of nitrogen cycling at the regional scale: The subtle effects of atmospheric deposition. Pages 163–174 *in* M. J. McDonnell and S. T. A. Pickett (eds.), *Humans as Components of Ecosystems*. Springer-Verlag, New York.

———. 1997. Why don't we believe the models? *Bulletin of the Ecological Society of America* 78: 232–233.

———. 1998. Mostly a misunderstanding, I believe. *Bulletin of the Ecological Society of America* 79: 256–257.

Aber, J. D., and C. T. Driscoll. 1997. Effects of land use, climate variation and N deposition on N cycling and C storage in northern hardwood forests. *Global Biogeochemical Cycles* 11: 639–648.

Aber, J. D., C. Driscoll, C. A. Federer, R. Lathrop, G. Lovett, J. M. Melillo, P. Steudler and J. Vogelmann. 1993. A strategy for the regional analysis of the effects of physical and chemical climate change on biogeographical cycles in northeastern (U.S.) forests. *Ecological Modelling* 67: 37–47.

Aber, J. D., and C. A. Federer. 1992. A generalized, lumped-parameter model of photosynthesis, evapotranspiration and net primary production in temperate and boreal forest ecosystems. *Oecologia* 92: 463–474.

Aber, J. D., and R. Freuder. 2000. Variation among solar radiation data sets for the Eastern US and its effects on predictions of forest production and water yield. *Climate Research* 15: 33–43.

Aber, J. D., A. Magill, R. Boone, J. M. Melillo, P. A. Steudler and R. Bowden. 1993. Plant and soil responses to three years of chronic nitrogen additions at the Harvard Forest, Petersham, MA. *Ecological Applications* 3: 156–166.

Aber, J. D., A. Magill, S. G. McNulty, R. Boone, K. J. Nadelhoffer, M. Downs and R. A. Hallett. 1995. Forest biogeochemistry and primary production altered by nitrogen saturation. *Water, Air and Soil Pollution* 85: 1665–1670.

Aber, J. D., W. H. McDowell, K. J. Nadelhoffer, A. H. Magill, G. Berntson, M. Kamekea, S. G. McNulty, W. Currie, L. Rustad and I. Fernandez. 1998. Nitrogen saturation in temperate forest ecosystems: Hypotheses revisited. *BioScience* 48: 921–934.

Aber, J., and J. Melillo. 2001. *Terrestrial Ecosystems.* International Thomson Publishing, New York.

Aber, J. D., J. M. Melillo and C. A. McClaugherty. 1990. Predicting long-term patterns of mass-loss, nitrogen dynamics and soil organic matter formation from initial litter chemistry in forest ecosystems. *Canadian Journal of Botany* 68: 2201–2208.

Aber, J. D., J. M. Melillo, K. J. Nadelhoffer, J. Pastor and R. Boone. 1991. Factors controlling nitrogen cycling and nitrogen saturation in northern temperate forest ecosystems. *Ecological Applications* 1: 303–315.

Aber, J. D., K. J. Nadelhoffer, P. Steudler and J. M. Melillo. 1989. Nitrogen saturation in northern forest ecosystems—Hypotheses and implications. *BioScience* 39: 378–386.

Aber, J. D., S. V. Ollinger, C. A. Federer and C. Driscoll. 1997. Modeling nitrogen saturation in forest ecosystems in response to land use and atmospheric deposition. *Ecological Modelling* 101: 61–78.

Aber, J. D., S. V. Ollinger, C. A. Federer, P. B. Reich, M. L. Goulden, D. W. Kicklighter, J. M. Melillo and R. G. Lathrop, Jr. 1995. Predicting the effects of climate change on water yield and forest production in the northeastern U.S. *Climate Research* 5: 207–222.

Aber, J. D., P. B. Reich and M. I. Goulden. 1996. Extrapolating leaf CO_2 exchange to the canopy: A generalized model of forest photosynthesis validated by eddy correlation. *Oecologia* 106: 257–265.

Aber, J. D., C. A. Wessman, D. L. Peterson, J. M. Melillo and J. H. Fownes. 1989. Remote sensing of litter and soil organic matter decomposition in forest ecosystems. Pages 87–103 *in* H. Mooney and R. Hobbs (eds.), *Remote Sensing of Biosphere Functioning. Ecological Studies* 79, Springer-Verlag, New York.

Abrams, M. D. 1992. Fire and the development of oak forests. *BioScience* 42: 346–353.

Abrams, M. D., D. A. Orwig and M. J. Dockry. 1997. Dendroecology and successional status of two contrasting old-growth oak forests in the Blue Ridge Mountains, U.S.A. *Canadian Journal of Forest Research* 27: 994–1002.

Akachuku, A. E. 1991. Wood growth determined from growth ring analysis in red pine *(Pinus resinosa)* trees forced to lean by a hurricane. *International Association of Wood Anatomists Bulletin* n.s. 12(3): 263–274.

———. 1993. Recovery and morphology of *Pinus resinosa* Ait. trees 50 years after they were displaced by a hurricane. *Forest Ecology and Management* 56: 113–129.

Allen, A. 1995. Soil science and survey at Harvard Forest. *Soil Survey Horizons* 36: 133–142.

Allison, T. D., R. E. Moeller and M. B. Davis. 1986. Pollen in laminated sediments provides evidence for a mid-Holocene forest pathogen outbreak. *Ecology* 67: 1101–1105.

Altpeter, L. S. 1937. A history of the forests of Cape Cod. M.S. Thesis, Harvard University, Cambridge, Massachusetts.

American Chestnut Foundation. 1986. The Journal of the American Chestnut Foundation. A national foundation dedicated to the restoration of the American chestnut tree. *Journal of the American Chestnut Foundation* 1: 1–28.

Amthor, J. S., M. L. Goulden, J. W. Munger and S. C. Wofsy. 1994. Testing a mechanistic model of forest-canopy mass and energy exchange using eddy correlation: Carbon dioxide and ozone uptake by a mixed oak-maple stand. *Australian Journal of Plant Physiology* 21: 623–651.

Anagnostakis, S. L. 1978. The American chestnut: New hope for a fallen giant. *The Connecticut Agricultural Experiment Station Bulletin* 777.

Andersen, S. T. 1986. Palaeoecological studies of terrestrial soils. Pages 165–180 *in* B. E. Berglund (ed.), *Handbook of Holocene Palaeoecology and Palaeohydrology.* Wiley, New York.

Anderson, J. A., S. Cooper-Ellis and B. C. Tan. 1998. New distribution notes on the mosses of Massachusetts. *Rhodora* 99: 352–367.

Anonymous. 1956. The timber resource in Massachusetts. USDA Forest Service Northeastern Forest Experiment Station, Hamden, Connecticut.

Askins, R. A. 1993. Population trends in grassland, shrubland, and forest birds in eastern North America. *Current Ornithology* 11: 1–34.

Askins, R. A., M. J. Philbrick and D. S. Sugeno. 1987. Relationship between the regional abundance of forest and the composition of forest bird communities. *Biological Conservation* 39: 129–152.

Backman, A. 1984. 1,000-year-record of fire-vegetation interactions in the northeastern United States: A comparison between coastal and inland regions. M.S. Thesis. University of Massachusetts, Amherst.

Baldocchi, D. D., E. Falge, L. H. Gu, R. Olson, D. Hollinger, S. Running, P. Anthoni, Ch. Bernhofer, K. Davis, J. Fuentes, A. Goldstein, G. Katul, B. Law, X. Lee, Y. Malhi, T. Meyers, J. W. Munger, W. Oechel, K. Pilegaard, H. P. Schmid, R. Valentini, S. Verma, T. Vesala, K. Wilson and S. Wofsy. 2001. FLUXNET: A new tool to study the temporal and spatial variability of ecosystem-scale carbon dioxide, water vapor and energy flux densities. *Bulletin of the American Meteorological Society* 82: 2415–2434.

Baldocchi, D. D., B. B. Hicks and T. P. Meyers. 1988. Measuring biosphere-atmosphere exchanges of biologically related gases with micrometeorological methods. *Ecology* 69: 1331–1340.

Baldwin, H. I. 1942. *Forestry in New England.* National Resources Planning Board Publication 70, Boston, Massachusetts.

———. 1949. *Wooden Dollars: A Report on the Forest Resources of New England, Their Condition, Economic Significance and Potentialities.* Federal Reserve Bank of Boston.

Balter, H., and R. E. Loeb. 1983. Arboreal relationships on limestone and gneiss in northern New Jersey and southeastern New York. *Bulletin of the Torrey Botanical Club* 110: 370–379.

Barbour, H., T. Simmons, P. Swain and H. Woolsey. 1998. *Our Irreplaceable Heritage—Protecting Biodiversity in Massachusetts.* Massachusetts Natural Heritage and Endangered Species Program and the Massachusetts Chapter of The Nature Conservancy, Boston.

Barford, C. C., S. C. Wofsy, M. L. Goulden, J. W. Munger, E. H. Pyle, S. P. Urbanski, L. Hutyra, S. R. Saleska, D. Fitzjarrald and K. Moore. 2001. Factors controlling long- and short-term sequestration of atmospheric CO_2 in a mid-latitude forest. *Science* 294: 1688–1691.

Baron, W. R. 1995. Historical climate records from the northeastern United States, 1640 to 1900. Pages 74–91 *in* R. S. Bradley and P. D. Jones (eds.), *Climate Since A.D. 1500.* Routledge, London.

Baron, W. R., and D. C. Smith. 1996. Growing season parameter reconstructions for New England using killing frost records, 1697–1947. *Bulletin,* Maine Agricultural and Forest Experiment Station 846.

Barron, H. S. 1984. *Those Who Stayed Behind: Rural Society in Nineteenth-Century New England.* Cambridge University Press, England.

Barten, P. K., T. Kyker-Snowman, P. J. Lyons, T. Mahlstedt, R. O'Connor and B. A. Spencer. 1998. Massachusetts: Managing a watershed protection forest. *Journal of Forestry* 96: 10–15.

Barten, P. K., and B. Watson. 1997. The Harvard Forest models: Quantitative reconstruction of hydrologic conditions. Manuscript, F&ES 541b: Ecosystem Hydrodynamics. University of Massachusetts, Amherst.

Bassow, S. L. 1995. Canopy photosynthesis and carbon cycling in a deciduous forest: Implications of species composition and rising concentrations of CO_2. Ph.D. Dissertation, Harvard University, Cambridge, Massachusetts.

Bassow, S. L., and F. A. Bazzaz. 1997. Intra- and inter-specific variation in canopy photosynthesis in a mixed deciduous forest. *Oecologia* 109: 507–515.

———. 1998. How environmental conditions affect leaf-level photosynthesis in four deciduous tree species. *Ecology* 79: 2660–2675.

Bassow, S. L., K. D. M. McConnaughay and F. A. Bazzaz. 1994. The response of temperate tree seedlings grown in elevated CO_2 to extreme temperature events. *Ecological Applications* 4: 593–603.

Battle, M., M. L. Bender, P. P. Tans, J. W. C. White, J. T. Ellis, T. Conway and R. J. Francey. 2000. Global carbon sinks and their variability inferred from atmospheric O_2 and delta ^{13}C. *Science* 287: 2467–2470.

Bazzaz, F. A. 1990a. Plant-plant interaction in successional environments. Pages 239–263 *in* J. B. Grace and G. D. Tilman (eds.), *Perspectives on Plant Competition.* Academic Press, San Diego, California.

———. 1990b. The response of natural ecosystems to the rising global CO_2 levels. *Annual Review of Ecology and Systematics* 21: 167–196.

———. 1991. Habitat selection in plants. *American Naturalist* 137: S116–S130.

———. 1993. Scaling in biological systems: Population and community perspectives. Pages 233–254 *in* J. Ehleringer and C. B. Field (eds.), *Scaling Physiological Processes: Leaf to Globe.* Academic Press, San Diego, California.

———. 1996a. *Plants in Changing Environments: Linking Physiological, Population, and Community Ecology.* Cambridge University Press, England.

———. 1996b. The physiological ecology of plant succession. *Annual Review of Ecology and Systematics* 10: 351–371.

———. 1997. Allocation of resources in plants: State-of-the-science and critical questions. Pages 1–37 *in* F. A. Bazzaz and J. Grace (eds.), *Plant Resource Allocation.* Physiological Ecology Series, Academic Press, San Diego, California.

Bazzaz, F. A., S. L. Bassow, G. M. Berntson and S. C. Thomas. 1996. Elevated CO_2 and terrestrial vegetation: Implications for and beyond the global carbon budget. Pages 43–76 *in* B. Walker and W. Steffen (eds.), *Global Change and Terrestrial Ecosystems.* Cambridge University Press, England.

Bazzaz, F. A., C. Ceballos, M. Davis, R. Dirzo, P. R. Ehrlich, T. Eisner, S. Levin, J. H. Lawton, J. Lubchenco, P. A. Matson, H. A. Mooney, P. H. Raven, J. E. Roughgarden, J. Sarukhan, G. D. Toam, P. Vitousek, B. Walker, D. H. Wall, E. O. Wilson and G. M. Woodwell. 1998. Editorial: Ecological science and the human predicament. *Science* 282: 879.

Bazzaz, F. A., J. S. Coleman and S. R. Morse. 1990. Growth response of seven major co-occurring tree species of the northeastern United States to elevated CO_2. *Canadian Journal of Forest Research* 20: 1479–1484.

Bazzaz, F. A., and E. D. Fajer. 1992. Plant life in a CO_2-rich world. *Scientific American* 266: 68–77.

Bazzaz, F. A., and J. Grace (eds.). 1997. *Plant Resource Allocation.* Physiological Ecology Series, Academic Press, San Diego, California.

Bazzaz, F. A., and S. L. Miao. 1993. Successional status, seed size, and responses of tree seedlings to CO_2, light, and nutrients. *Ecology* 74: 104–112.

Bazzaz, F. A., S. L. Miao and P. M. Wayne. 1993. CO_2-induced growth enhancements of co-occurring tree species decline at different rates. *Oecologia* 96: 478–482.

Bazzaz, F. A., and T. W. Sipe. 1987. Physiological ecology, disturbance, and ecosystem recovery. Pages 203–227 in E.-D. Schulze and H. Zwölfer (eds.), *Potentials and Limitations of Ecosystem Analysis.* Springer-Verlag, Berlin, Germany.

Bazzaz, F. A., and W. G. Sombroek (eds.). 1996. *Global Climate Change and Agricultural Production.* John Wiley & Sons, Chichester.

Bazzaz, F. A., and K. A. Stinson. 2001. Genetic versus environmental control of ecophysiological processes: Some challenges for predicting community responses to global change. Pages 283–295 in M. C. Pres, J. D. Scholes and M. G. Barker (eds.), *Physiological Plant Ecology.* British Ecological Society and Blackwell Science, Oxford, England.

Bazzaz, F. A., and P. M. Wayne. 1994. Coping with environmental heterogeneity: The physiological ecology of tree seeding regeneration across the gap-understory continuum. Pages 349–390 in M. Caldwell and R. Pearcy (eds.), *Exploitation of Environmental Heterogeneity by Plants: Ecophysiological Processes Above- and Belowground.* Academic Press, New York.

Bazzaz, F. A., and W. E. Williams. 1991. Atmospheric concentrations of CO_2 within a mixed forest: Implications for seedling growth. *Ecology* 72: 12–16.

Behre, K.-E. 1988. The role of man in European vegetation history. *Handbook of Vegetation Science,* Volume 7. *Vegetation History,* Section IV.2: 633–672.

Belknap, J. 1792. *The History of New Hampshire,* Volume III. Pages 89–107. Dover, Boston, Massachusetts.

Bellemare, J., G. Motzkin and D. R. Foster. 2002. Legacies of the agricultural past in the forested present: An assessment of historical land-use effects on rich mesic forests of western Massachusetts, U.S.A. *Journal of Biogeography* 29: 1401–1420.

Benzinger, J. 1994. Hemlock decline and breeding birds. II. Effects of habitat change. *Records of New Jersey Birds* 20 (2): 34–51.

Bergeron, Y., and S. Archambault. 1993. Decreasing frequency of forest fires in the southern boreal zone of Quebec and its relation to global warming since the end of the 'Little Ice Age.' *The Holocene* 3: 255–259.

Bergh, J., S. Linder, T. Lundmark, and B. Elfving. 1999. The effect of water and nutrient availability on the productivity of Norway spruce in northern and southern Sweden. *Forest Ecology and Management* 119: 51–62.

Berglund, B. E. (ed.). 1991. *The Cultural Landscape During 6000 Years in Southern Sweden—The Ystad Project. Ecological Bulletins* 41.

Berlik, M. M., D. B. Kittredge and D. R. Foster. 2002a. The illusion of preservation: A global environmental argument for the local production of natural resources. Harvard Forest Paper No. 26. Petersham, Massachusetts.

———. 2002b. The illusion of preservation: A global environmental argument for the local production of natural resources. *Journal of Biogeography* 29: 1557–1568.

Berliner, R., and J. G. Torrey. 1989a. On tripartite *Frankia:* Mycorrhizal associations in the Myricaceae. *Canadian Journal of Botany* 67: 1708–1712.

———. 1989b. Studies on mycorrhizal associations in Harvard Forest, Massachusetts. *Canadian Journal of Botany* 67: 2245–2251.

Berntson, G. M. 1997. Topological scaling and plant root system architecture: Developmental and functional hierarchies. *New Phytologist* 135: 621–634.

Berntson, G. M., and J. D. Aber. 2000. Fast nitrate immobilization in N-saturated temperate forest soils. *Soil Biology and Biochemistry* 32: 151–156.

Berntson, G. M., and F. A. Bazzaz. 1996a. The allometry of root production and loss in seedlings of *Acer rubrum* (Aceraceae) and *Betula papyrifera* (Betulaceae): Implications for root dynamics in elevated CO_2. *American Journal of Botany* 83: 608–616.

———. 1996b. Belowground positive and negative feedbacks on CO_2 growth enhancement. *Plant and Soil* 187: 119–131.

———. 1997a. Nitrogen cycling in microcosms of yellow birch exposed to elevated CO_2: Simultaneous positive and negative belowground feedbacks. *Global Change Biology* 3: 247–258.

———. 1997b. Elevated CO_2 and the magnitude and seasonal dynamics of root production and loss in *Betula papyrifera. Plant and Soil* 190: 211–216.

———. 1998. Regenerating temperate forest microcosms in elevated CO_2: Species composition, belowground growth and nitrogen cycling. *Oecologia* 113: 115–125.

Berntson, G. M., N. Rajakaruna and F. A. Bazzaz. 1998. Species- and community-level growth and nitrogen acquisition in elevated CO_2 atmospheres in an experimental annual community. *Global Change Biology* 4: 101–120.

Berntson, G. M., and P. Stoll. 1997. Correcting for finite spatial scales of self-similarity when calculating the fractal dimensions of real-world structures. *Proceedings of the Royal Society, Biological Sciences* 264: 1531–1537.

Berntson, G. M., P. M. Wayne and F. A. Bazzaz. 1997. Belowground architectural and mycorrhizal responses to elevated CO_2 in *Betula alleghaniensis* populations. *Functional Ecology* 11: 684–695.

Bhiry, N., and L. Filion. 1996. Mid-Holocene hemlock decline in eastern North America linked with phytophagous insect activity. *Quaternary Research* 45: 312–320.

Bickford, W. E., and U. J. Dymon (eds.). 1990. *An Atlas of Massachusetts River Systems, Environmental Designs for the Future.* University of Massachusetts Press, Amherst.

Bidwell, P. W., and J. L. Falconer. 1941. *History of Agriculture in the Northern United States 1620–1860.* Peter Smith, New York.

Billings, W. D., J. O. Luken, D. A. Mortensen, and K. M. Peterson. 1982. Arctic tundra: A source or sink for atmospheric carbon dioxide in a changing climate? *Oecologia* 53: 7–11.

Binford, M. W., M. Brenner, T. J. Whitmore, A. Higuera-Gundy, E. S. Deevey and B. Leyden. 1987. Ecosystems, paleoecology and human disturbance in subtropical and tropical America. *Quaternary Science Reviews* 6: 115–128.

Birdsey, R. A., and L. S. Heath. 1995. Carbon changes in U.S. forests. Pages 56–70 *in* L. A. Joyce (ed.), *Productivity of America's Forests and Climate Change,* No. 271 Edition. USDA Forest Service General Technical Report RM-271. Rocky Mountain Forest and Range Experiment Station, Fort Collins, Colorado.

Bishop, G. D., M. R. Church, J. D. Aber, R. P. Neilson, S. V. Ollinger and C. Daley. 1998. A comparison of mapped estimates of long-term runoff in the northeastern United States. *Journal of Hydrology* 206: 176–190.

Black, J. D., and A. Brinser. 1952. *Planning One Town—Petersham: A Hill Town in Massachusetts.* Harvard University Press, Cambridge, Massachusetts.

Bogart, E. L. 1948. *Peacham: The Story of a Vermont Hill Town.* Vermont Historical Society, Montpelier.

Bolker, B. M., S. W. Pacala, F. A. Bazzaz, C. D. Canham and S. A. Levin. 1995. Species diversity and ecosystem response to carbon dioxide fertilization: Conclusions from a temperate forest model. *Global Change Biology* 1: 373–381.

Bolster, K. L., M. E. Martin and J. D. Aber. 1996. Interactions between precision and generality in the development of calibrations for the determination of carbon fraction and nitrogen concentration in foliage by near infrared reflectance. *Canadian Journal of Forest Research* 26: 590–600.

Bond, R. S. 1998. Professional forestry, forestry education and research. Pages 220–255 *in* C. H. W. Foster (ed.), *Stepping Back to Look Forward: A History of the Massachusetts Forest.* Harvard Forest, Petersham, Massachusetts.

Boone, R. D., K. J. Nadelhoffer, J. D. Canary and J. P. Kaye. 1998. Roots determine the temperature sensitivity of soil respiration. *Nature* 396: 571–572.

Boose, E. R., E. F. Boose and A. L. Lezberg. 1998. A practical method for mapping trees using distance measurements. *Ecology* 79: 819–827.

Boose, E. R., K. E. Chamberlin and D. R. Foster. 1997. Reconstructing historical hurricanes in New England. Pages 388–389 *in Preprints of the 22nd Conference on Hurricanes and Tropical Meteorology,* American Meteorological Society, Boston, Massachusetts.

———. 2001. Landscape and regional impacts of hurricanes in New England. *Ecological Monographs* 71: 27–48.

Boose, E. R., D. R. Foster and M. Fluet. 1994. Hurricane impacts to tropical and temperate forest landscapes. *Ecological Monographs* 64: 369–400.

Bormann, F. H., and G. E. Likens. 1979a. Catastrophic disturbance and the steady state in northern hardwood forests. *American Scientist* 67: 660–669.

———. 1979b. *Pattern and Process in a Forested Ecosystem.* Springer-Verlag, New York.

Bormann, F. H., G. E. Likens, T. G. Siccama, R. S. Pierce and J. S. Eaton. 1974. The export of nutrients and recovery of stable conditions following deforestation at Hubbard Brook. *Ecological Monographs* 44: 255–277.

Bowden, R., M. C. Castro, J. M. Melillo, P. A. Steudler and J. D. Aber. 1993. Fluxes of greenhouse gases between soils and the atmosphere in a temperate forest following a simulated hurricane blowdown. *Biogeochemistry* 21: 61–71.

Bowden, R. D., J. M. Melillo, P. A. Steudler and J. D. Aber. 1991. Effects of nitrogen additions on annual nitrous oxide fluxes from temperate forest soils in the northeastern United States. *Journal of Geophysical Research* 96: 9321–9328.

Bowden, R. D., K. J. Nadelhoffer, R. D. Boone, J. M. Melillo and J. B. Garrison. 1993. Contributions of aboveground litter, belowground litter, and root respiration to total soil respiration in a temperate mixed hardwood forest. *Canadian Journal of Forest Research* 123: 1402–1407.

Bowden, R. D., K. M. Newkirk and G. M. Rullo. 1998. Carbon dioxide and methane fluxes by a forest soil under laboratory-controlled moisture and temperature conditions. *Soil Biology Biochemistry* 30: 1591–1597.

Bowden, R. D., P. A. Steudler, J. M. Melillo and J. D. Aber. 1990. Annual nitrous oxide fluxes from temperate forest soils in the northeastern United States. *Journal of Geophysical Research* 95: 13997–14005.

Bradford, W. 1856 (1981). *Of Plymouth Plantation, 1620–1647.* Reprint. The Modern Library, New York.

Bradley, R. S., J. K. Eisheid and P. T. Ives. 1987. The climate of Amherst, Massachusetts, 1836–1985. Contribution No. 50; Department of Geology and Geophysics, University of Massachusetts, Amherst.

Bradley, R. S., and P. D. Jones. 1993. 'Little Ice Age' summer temperature variations: Their nature and relevance to recent global warming trends. *The Holocene* 3: 367–376.

Bradshaw, R. H. W. 1988. Spatially-precise studies of forest dynamics. Pages 725–751 in B. Huntley and T. Webb III (eds.), *Handbook of Vegetation Science,* Volume 7. *Vegetation History.* Kluwer Academic Publishers, Dordrecht.

Bradshaw, R. H. W., and N. G. Miller. 1988. Recent successional processes investigated by pollen analysis of closed canopy forested sites. *Vegetatio* 76: 45–54.

Bragdon, K. J. 1996. *Native People of Southern New England 1500–1650.* University of Oklahoma Press, Norman.

Brake, R., and H. Post. 1941. Natural restocking of hurricane damaged "old-field" white pine areas in north central Massachusetts. M.F.S. Thesis. Harvard University, Cambridge, Massachusetts.

Braun, E. K., and D. P. Braun. 1994. *The First Peoples of the Northeast.* Lincoln Historical Society, Lincoln, Massachusetts.

Bromley, S. W. 1935. The original forest types of southern New England. *Ecological Monographs* 5: 61–89.

Brooks, C. F. 1939. Hurricanes into New England: Meteorology of the Storm of September 21, 1938. *The Geographical Review* 29: 119–127.

Brown, J. H., and L. Real. 1991. *Foundations of Ecology.* University of Chicago Press, Chicago.

Brunet, J. 1993. Environmental and historical factors limiting the distribution of rare forest grasses in south Sweden. *Forest Ecology and Management* 61: 263–275.

Brunet, J., and G. Von Oheimb. 1998. Migration of vascular plants to secondary woodlands in southern Sweden. *Journal of Ecology* 86: 429–438.

Burden, E. T., J. H. McAndrews and G. Norris. 1985. Palynology of Indian and European forest clearance and farming in lake sediment cores from Awenda Provincial Park, Ontario. *Canadian Journal of Earth Science* 23: 43–54.

Butcher, G. S., W. A. Niering, W. J. Barry and R. H. Goodwin. 1981. Equilibrium biogeography and the size of nature preserves: An avian case study. *Oecologia* 49: 29–37.

Butzer, K. W. 1993. No Eden in the New World. *Nature* 362: 15–16.

Caldwell, M. M., and R. W. Pearcy (eds.). 1994. *Exploitation of Environmental Heterogeneity by Plants: Ecophysiological Processes Above- and Belowground.* Academic Press, San Diego, California.

Callahan, J. T. 1984. Long-term ecological research. *BioScience* 34: 363–367.

Campbell, C. A., and W. Souster. 1982. Loss of organic matter and potentially mineralizable nitrogen from Saskatchewan soils due to cropping. *Canadian Journal of Soil Science* 62: 651–656.

Campbell, I. D. 1993. Forest disequilibrium caused by rapid Little Ice Age cooling. *Nature* 366: 336–338.

Carlton, G. C. 1993. Effects of microsite environment on tree regeneration following disturbance. Ph.D. Dissertation, Harvard University, Cambridge, Massachusetts.

Carlton, G. C., and F. A. Bazzaz. 1998a. Regeneration of three sympatric birch species on experimental hurricane blowdown microsites. *Ecological Monographs* 68: 99–120.

———. 1998b. Resource congruence and forest regeneration following an experimental hurricane blowdown. *Ecology* 79: 1305–1319.

Carpenter, S. R. 1989. Replication and treatment strength in whole-lake experiments. *Ecology* 70: 453–463.

———. 1990. Large-scale perturbations: Opportunities for innovation. *Ecology* 71: 2038–2043.

Caspersen, J. P., S. W. Pacala, J. C. Jenkins, G. C. Hurtt, P. R. Moorcroft and R. A. Birdsey. 2000. Contributions of land-use history to carbon accumulation in US forests. *Science* 290: 1148–1151.

Castro, M. S., J. M. Melillo, J. D. Aber and R. D. Bowden. 1995. Soil moisture as a predictor of methane uptake by temperate forest soils. *Canadian Journal of Forest Research* 24: 1805–1810.

Castro, M. S., P. A. Steudler, J. M. Melillo, J. D. Aber and R. D. Bowden. 1995. Factors controlling atmospheric methane consumption by temperate forest soils. *Global Biogeochemical Cycles* 9: 1–10.

Castro, M., P. A. Steudler, J. M. Melillo, J. D. Aber and S. Millham. 1993. Exchange of N_2O and CH_4 between the atmosphere and soils in spruce-fir forests in the northeastern U.S. *Biogeochemistry* 18: 119–135.

Catovsky, S., and F. A. Bazzaz. 1999a. Elevated CO_2 influences the responses of two birch species to soil moisture: Implications for forest community structure. *Global Change Biology* 5: 507–518.

———. 1999b. The role of resource interactions and seedling regeneration in maintaining a positive feedback in hemlock stands. *Journal of Ecology* 88: 100–112.

———. 1999c. Contributions of coniferous and broad-leaved species to temperate forest carbon uptake: A bottom-up approach. *Canadian Journal of Forest Research* 30: 100–111.

Cavender-Bares, J., and F. A. Bazzaz. 2000. Changes in drought response strategies with ontogeny in *Quercus rubra:* Implications for scaling from seedlings to mature trees. *Oecologia* 124: 8–18.

Cavender-Bares, J., M. Potts, E. Zacharias and F. A. Bazzaz. 1999. Consequences of CO_2 and light interactions for leaf phenology, growth, and senescence in *Quercus rubra*. *Global Change Biology* 6: 877–887.

Cavender-Bares, J., P. B. Voss and F. A. Bazzaz. 1998. Consequences of incongruency in diurnally varying resources for seedlings of *Rumex crispus* (Polygonaceae). *American Journal of Botany* 85: 1216–1223.

Ceci, L. 1975. Fish fertilizer: A native North American practice? *Science* 188: 26–30.

———. 1979–80. Maize cultivation in coastal New York: The archaeological, agronomical, and documentary evidence. *North American Archaeologist* 1: 45–74.

Champlain, S. de. 1922. Les voyages du Sieur de Champlain. *In* H. P. Biggar (ed.), *The Works of Samuel de Champlain*. Reprinted from 1613 edition. Volume 1. Champlain Society, Toronto.

Chapin F. S. III, and A. J. Bloom. 1976. Phosphate absorption: Adaptation of tundra graminoids to a low temperature, low phosphorus environment. *Oikos* 26: 111–121.

Ciais, P., P. P. Tans, M. Trolier, J. W. C. White and R. J. Francey. 1995. A large northern-hemisphere terrestrial CO_2 sink indicated by the C-13/C-12 ratio of atmospheric CO_2. *Science* 269(5227): 1098–1102.

Clark, J. S., and P. D. Royall. 1995. Transformation of a northern hardwood forest by aboriginal (Iroquois) fire: Charcoal evidence from Crawford Lake, Ontario, Canada. *The Holocene* 5: 1–9.

Clark, W. C., R. W. Kates, B. L. Turner, J. F. Richards and W. B. Meyer. 1993. *The Earth as Transformed by Human Action*. Cambridge University Press, England.

Cline, A. C. 1939. The restoration of watershed forests in the hurricane area. *Journal of the New England Water Works Association* 53(2): 223–237.

Cline, A. C., and S. H. Spurr. 1942. The virgin upland forest of central New England: A study of old growth stands in the Pisgah mountain section of southwestern New Hampshire. *Harvard Forest Bulletin,* No. 21: 1–45.

Cogbill, C., J. Burk and G. Motzkin. 2002. The forests of pre-settlement New England, USA: Spatial and compositional patterns based on town proprietor surveys. *Journal of Biogeography* 29: 1279–1304.

Cole, D. W., J. E. Compton, P. S. Homann, R. L. Edmonds and H. Van Miegroet. 1995. Comparison of carbon accumulation in Douglas fir and red alder forests. *In* J. M. Kelly and W. W. McFee (eds.), *Carbon Forms and Functions in Forest Soils*. Soil Science Society of America, Madison, Wisconsin.

Compton, J. E., and R. D. Boone. 2000. Long-term impacts of agriculture on soil carbon and nitrogen in New England forests. *Ecology* 81: 2314–2330.

Compton, J. E., R. D. Boone, G. Motzkin and D. R. Foster. 1998. Soil carbon and nitrogen in a pine-oak sand plain in central Massachusetts: Role of vegetation and land-use history. *Oecologia* 116: 536–542.

Conuel, T. 1990. *Quabbin: The Accidental Wilderness*. University of Massachusetts Press, Amherst.

Cook, H. O. 1917. The forests of Worcester County. The results of a forest survey of the fifty-nine towns in the county and a study of their lumber industry. State Printing Office, Boston, Massachusetts.

Cook, S. F. 1976. The Indian population of New England in the seventeenth century. *Publications in Anthropology*, no. 12: 1–91. University of California, Berkeley.

Coolidge, M. C. 1948. *The History of Petersham, Massachusetts*. The Petersham Historical Society.

Cooper-Ellis, S. 1998. Bryophytes in old-growth forests of western Massachusetts. *Journal of the Torrey Botanical Society* 125: 117–132.

Cooper-Ellis, S., D. R. Foster, G. Carlton and A. Lezberg. 1999. Vegetation response to catastrophic wind: Results from an experimental hurricane. *Ecology* 80: 2683–2696.

Coutts, M. P., and J. Grace (eds.). 1995. *Wind and Trees*. Cambridge University Press, England.

Cowell, C. M. 1993. Environmental gradients in secondary forests of the Georgia Piedmont, U.S.A. *Journal of Biogeography* 20: 199–207.

Crabtree, R. C. 1992. Birch regeneration in a changing nitrogen environment. Ph.D. Dissertation, Harvard University, Cambridge, Massachusetts.

Crabtree, R. C., and F. A. Bazzaz. 1993a. Seedling response of four birch species to simulated nitrogen deposition: Ammonium versus nitrate. *Ecological Applications* 3: 315–321.

———. 1993b. Black birch (*Betula lenta* L.) seedlings as foragers for nitrogen. *New Phytologist* 122: 617–625.

Crampton, A. B., O. Stutter, K. J. Kirby and R. C. Welch. 1998. Changes in the composition of Monks Wood National Nature Reserve, Cambridgeshire, UK, 1964–1996. *Arboricultural Journal* 22: 229–245.

Cronon, W. 1983. *Changes in the Land: Indians, Colonists and the Ecology of New England*. Hill and Wang, New York.

Crosby, A. W. 1976. Virgin soil epidemics as a factor in the aboriginal depopulation in America. *William and Mary Quarterly* 33: 289–299.

———. 1986. *Ecological Imperialism. The Biological Expansion of Europe 900–1900*. Cambridge University Press, England.

Crowley, T. J. 2000. Causes of climate change over the past 1000 years. *Science* 289 (5477): 270–277.

Currie, W. S., and J. D. Aber. 1997. Modeling leaching as a decomposition process in humid, montane forests. *Ecology* 78: 1844–1860.

Currie, W. S., J. D. Aber and C. T. Driscoll. 1999. Leaching of nutrient cations from the forest floor: Effects of nitrogen saturation in two long-term manipulations. *Canadian Journal of Forest Research* 29: 609–620.

Currie, W. S., J. D. Aber, W. H. McDowell, R. D. Boone and A. H. Magill. 1996. Vertical transport of dissolved organic C and N under long-term N amendments in pine and hardwood forests. *Biochemistry* 35: 471–505.

Currie, W. S., and K. J. Nadelhoffer. 1999. Dynamic redistribution of isotopically labeled cohorts of nitrogen inputs in two temperate forests. *Ecosystems* 2: 4–18.

Currie, W. S., K. J. Nadelhoffer and J. D. Aber. 1999. Soil detrital processes controlling the movement of ^{15}N tracers to forest vegetation. *Ecological Applications* 9: 87–102.

Dail, D. B., E. A. Davidson and J. Chorover. 2001. Rapid abiotic transformation of nitrate in an acid forest soil. *Biogeochemistry* 54: 131–146.

Darling, N. 1842. Notice of a hurricane that passed over New England in September 1815. *American Journal of Science and Arts* 42: 243–252.

Davidson, E. A., and I. Ackerman. 1993. Changes in soil carbon inventories following cultivation of previously untilled soils. *Biogeochemistry* 20: 161–193.

Davidson, E. A., E. Belk and R. D. Boone. 1998. Soil and water content and temperature as independent or confounded factors controlling soil respiration in a temperate mixed hardwood forest. *Global Change Biology* 4: 217–228.

Davidson, E. A., S. C. Hart and M. K. Firestone. 1992. Internal cycling of nitrate in soils of a mature coniferous forest. *Ecology* 73: 1148–1156.

Davidson, E. A., S. C. Hart, C. A. Shanks and M. K. Firestone. 1991. Measuring gross nitrogen mineralization, immobilization, and nitrification by ^{15}N isotope dilution in intact soil cores. *Journal of Soil Science* 42: 335–349.

Davidson, E. A., J. M. Stark and M. K. Firestone. 1990. Microbial production and consumption of nitrate in an annual grassland. *Ecology* 71: 1968–1975.

Davis, M. B. 1958. Three pollen diagrams from central Massachusetts. *American Journal of Science* 256: 540–570.

———. 1965. Phytogeography and palynology of northeastern United States. Pages 377–401 *in* H. E. Wright and D. G. Frey (eds.), *The Quaternary of the United States*. Princeton University Press, Princeton, New Jersey.

———. 1969. Climatic changes in southern Connecticut recorded by pollen deposition at Roger Lake. *Ecology* 50: 400–422.

———. 1976a. Pleistocene biogeography of temperate deciduous forests. *Geoscience and Man* 13: 13–26.

———. 1976b. Outbreaks of forest pathogens in quaternary history. *Proceedings of the IV International Conference on Palynology*. Lucknow, India.

———. 1981. Quaternary history and the stability of forest communities. Pages 132–153 *in* D. C. West, H. H. Shugart and D. B. Botkin (eds.), *Forest Succession: Concepts and Application*. Springer-Verlag, New York.

———. 1986. Climatic instability, time lags, and community disequilibrium. Pages 269–284 *in* J. Diamond and T. J. Case (eds.), *Community Ecology*. Harper and Row, New York.

———. 1989. Lags in vegetation response to greenhouse warming. *Climatic Change* 15: 75–82.

Davis, M. B., and D. B. Botkin. 1985. Sensitivity of cool-temperature forests and their fossil pollen record to rapid temperature change. *Quaternary Research* 23: 327–340.

Davis, M. B., R. E. Calcotte, S. Sugita and H. Takahara. 1998. Patchy invasion and the origin of a hemlock-hardwoods forest mosaic. *Ecology* 79: 2641–2659.

Davis, M. B., R. W. Spear and L. C. K. Shane. 1980. Holocene climate in New England. *Quaternary Research* 14: 240–250.

Davis, R. B., and G. L. Jacobsen, Jr. 1985. Late glacial and early Holocene landscapes in northern New England and adjacent areas of Canada. *Quaternary Research* 23: 341–368.

Day, G. M. 1953. The Indian as an ecological factor in the northeastern forest. *Ecology* 34: 329–343.

DeGraaf, R., and R. Miller. 1996. *Conservation of Faunal Diversity in Forested Landscapes*. Chapman and Hall, New York.

DeGraaf, R., and M. Yamasaki. 2001. *New England Wildlife: Habitat, Natural History, and Distribution*. University Press of New England, Hanover, New Hampshire.

Denevan, W. M. 1992. The pristine myth: The landscape of the Americas in 1492. *Annals of the Association of American Geographers* 82: 369–385.

Denny, C. S. 1982. Geomorphology of New England. U.S. Geological Survey Professional Paper 1208.

Denslow, J. S. 1980. Gap partitioning among tropical rainforest trees. *Biotropica* (Supplement) 12: 47–55.

Diaz, H. F., and R. S. Pulwarty (eds.). 1997. *Hurricanes: Climate and Socioeconomic Impacts.* Springer-Verlag, New York.

Dickson, D. R., and C. L. McAfee. 1988. Forest statistics for Massachusetts—1972 and 1985. USDA Forest Service Northeastern Forest Experiment Station Resource Bulletin NE-106.

Dise, N. B., and R. F. Wright. 1995. Nitrogen leaching in European forests in relation to nitrogen deposition. *Forest Ecology and Management* 71: 153–162.

Donohue, K., D. R. Foster and G. Motzkin. 2000. Effects of the past and the present on species distributions: The influence of land-use history on the demography of wintergreen. *Journal of Ecology* 88: 303–316.

Doolittle, W. E. 1992. Agriculture in North America on the eve of contact: A reassessment. *Annals of the Association of American Geographers* 83: 386–401.

Downs, M. R., R. H. Michener, B. Fry and K. J. Nadelhoffer. 1999. Routine measurement of dissolved inorganic ^{15}N in streamwater. *Environmental Monitoring and Assessment* 55: 211–220.

Downs, M. R., K. J. Nadelhoffer, J. M. Melillo and J. D. Aber. 1993. Foliar and fine root nitrate reductase activity in seedlings of four forest tree species in relation to nitrogen availability. *Trees, Structure and Function* 7: 233–236.

———. 1996. Immobilization of a ^{15}N labelled nitrate addition by decomposing forest litter. *Oecologia* 105: 141–150.

Dunwiddie, P. 1989. Forest and heath: The shaping of the vegetation on Nantucket Island. *Journal of Forest History* 33: 126–133.

Dunwiddie, P., D. Foster, D. Leopold and R. Leverett. 1996. Old-growth forests of southern New England, New York and Pennsylvania. Pages 126–143 *in* M. B. Davis (ed.), *Eastern Old Growth Forests.* Island Press, Covelo, California.

Dwight, T. 1969. *Travels in New England and New York.* 4 vols. 1823. Reprint; B. M. Solomon and P. M. King (eds.). Harvard University Press, Cambridge.

Ehleringer, J. E., and C. B. Field (eds.). 1993. *Scaling Physiological Processes: Leaf to Globe.* Academic Press, San Diego, California.

Emerson, G. B. 1846. *A Report on the Trees and Shrubs Growing Naturally in the Forests of Massachusetts.* Little, Brown and Company, Boston.

Engstrom, D. R., E. B. Swain, and J. C. Kingston. 1985. A palaeolimnological record of human disturbance from Harvey's Lake, Vermont: Geochemistry, pigments and diatoms. *Freshwater Biology* 15: 261–288.

Fan, S.-M., M. Gloor, J. Mahlman, S. Pacala, J. Sarmiento, T. Takahashi and P. Tans. 1998. A large terrestrial carbon sink in North America implied by atmospheric and oceanic carbon dioxide data and models. *Science* 282: 442–446.

Fan, S.-M., M. L. Goulden, J. W. Munger, B. C. Daube, P. S. Bakwin, S. C. Wofsy, J. S. Amthor, D. R. Fitzjarrald, K. E. Moore and T. R. Moore. 1995. Environmental controls on the photosynthesis and respiration of a boreal lichen woodland: A growing season of whole-ecosystem exchange measurements by eddy correlation. *Oecologia* 102: 443–452.

Fenn, M. E., M. Poth, J. D. Aber, J. S. Baron, B. T. Bormann, D. W. Johnson, A. D. Lemly, S. G. McNulty, D. F. Ryan and R. Stottlemeyer. 1998. Nitrogen excess in North American ecosystems: A review of predisposing factors, geographic extent, ecosystem responses and management strategies. *Ecological Applications* 8: 706–733.

Fenneman, N. M. 1938. *Physiography of the Eastern United States.* McGraw-Hill, New York.

Filion, L., and F. Quinty. 1993. Macrofossil and tree-ring evidence for a long-term forest succession and mid-Holocene hemlock decline. *Quaternary Research* 40: 89–97.

Finnie, T. C., C. D. Tomlin, J. Bossler, D. Cowen, B. Petchenik, H. Thomas, T. Wilbanks and J. Estes. 1990. *Spatial Data Needs: The Future of the National Mapping Program.* National Academy Press, Washington, D.C.

Fisher, R. T. 1918. Second-growth white pine as related to the former uses of the land. *Journal of Forestry* 16: 253–254.

———. 1929. Our wildlife and the changing forest. *The Sportsman,* March, pages 2–11.

———. 1933. New England's forests: Biological factors. New England's prospect. *American Geographical Society Special Publication* 16: 213–223.

Forbush, E. H. 1905. *Useful Birds and Their Protection.* Wright and Potter Printing Co., State Printers, Boston.

———. 1925. *Birds of Massachusetts and Other New England States.* Part 1. Massachusetts Department of Agriculture, Boston.

Forman, R. T. T., and E. W. B. Russell. 1983. Commentary: Evaluation of historical data in ecology. *Bulletin of the Ecological Society of America* 64: 5–7.

Foss, C. R. 1992. Wildlife in a changing landscape. Pages 14–22 *in* R. Ober (ed.), *At What Cost? Shaping the Land We Call New Hampshire.* Historical Society and Society for the Protection of New Hampshire Forests, Concord, New Hampshire.

Foster, C. H. W. (ed.). 1998. *Stepping Back to Look Forward: A History of the Massachusetts Forest.* Harvard Forest, Petersham, Massachusetts.

Foster, C. H. W., and D. R. Foster. 1994. *Managing the Greenwealth: The Forests of Quabbin.* John F. Kennedy School of Government, Center for Science and International Affairs, Environment and Natural Resources Program, Harvard University, Cambridge, Massachusetts.

———. 1999. Thinking in forest time. A vision for the Massachusetts forest. Harvard Forest Paper No. 24, Petersham, Massachusetts.

Foster, D. R. 1983. The history and pattern of fire in the boreal forest of southeastern Labrador. *Canadian Journal of Botany* 61: 2459–2471.

———. 1985. Vegetation development following fire in *Picea mariana* (black spruce): *Pleurozium* forests of south-eastern Labrador, Canada. *Journal of Ecology* 73: 517–534.

———. 1988a. Species and stand response to catastrophic wind in central New England, U.S.A. *Journal of Ecology* 76: 135–151.

———. 1988b. Disturbance history, community organization and vegetation dynamics of the old-growth Pisgah Forest, southwestern New Hampshire, U.S.A. *Journal of Ecology* 76: 105–134.

———. 1992. Land-use history (1730–1990) and vegetation dynamics in central New England, USA. *Journal of Ecology* 80: 753–772.

————. 1993. Land-use history and forest transformations in central New England. Pages 91–110 *in* S. Pickett and M. McDonald (eds.), *Humans as Components of Ecosystems*. Springer-Verlag, New York.

————. 1995. Land-use history and four hundred years of vegetation change in New England. Pages 253–319 *in* B. L. Turner, A. G. Sal, F. G. Bernaldez and F. DiCastri (eds.), *Global Land Use Change: A Perspective from the Colombian Encounter*. SCOPE Publication. Consejo Superior de Investigaciones Cientificas, Madrid.

————. 1999a. Forests the way they used to be. *New York Times,* June 26, C-4.

————. 1999b. *Thoreau's Country: Journey Through a Transformed Landscape.* Harvard University Press, Cambridge, Massachusetts.

————. 2002a. Conservation issues and approaches for dynamic cultural landscapes. *Journal of Biogeography* 29: 1533–1535.

————. 2002b. Historical-geographical approaches to ecology and conservation. Application to the New England landscape. *Journal of Biogeography* 29: 1269–1275.

————. 2002c. Thoreau's country: A historical-ecological approach to conservation of the New England landscape. *Journal of Biogeography* 29: 1537–1555.

Foster, D. R., J. D. Aber, J. M. Melillo, R. Bowden and F. Bazzaz. 1997. Forest response to disturbance and anthropogenic stress: Rethinking the 1938 Hurricane and the impact of physical disturbance vs. chemical and climate stress on forest ecosystems. *BioScience* 47: 437–445.

————. 1998. Forest response to natural disturbance versus human-induced stresses. *Arnoldia* 58: 35–40.

Foster, D. R., and E. Boose. 1992. Patterns of forest damage resulting from catastrophic wind in central New England, USA. *Journal of Ecology* 80: 79–99.

————. 1995. Hurricane disturbance regimes in temperate and tropical forest ecosystems. Pages 305–339 *in* M. Coutts and J. Grace (eds.), *Wind and Trees.* Cambridge University Press, New York.

Foster, D. R., S. Clayden, D. A. Orwig, B. Hall, and S. Barry. 2002. Oak, chestnut and fire: Climatic and cultural controls of long-term forest dynamics in New England, USA. *Journal of Biogeography* 29: 1359–1379.

Foster, D. R., M. Fluet and E. Boose. 1999. Human or natural disturbance: Landscape-scale dynamics of the tropical forests of Puerto Rico. *Ecological Applications* 9: 555–572.

Foster, D. R., B. Hall, S. Barry, S. Clayden, and T. Parshall. 2002. Cultural, environmental, and historical controls of vegetation patterns and the modern conservation setting on the island of Martha's Vineyard, U.S.A. *Journal of Biogeography* 29: 1381–1400.

Foster, D. R., and G. A. King. 1986. Vegetation pattern and diversity in S.E. Labrador, Canada: *Betula papyrifera* (birch) forest development in relation to fire history and physiography. *Journal of Ecology* 74: 465–483.

Foster, D. R., D. Knight and J. Franklin. 1998. Landscape patterns and legacies of large infrequent disturbances. *Ecosystems* 1: 497–510.

Foster, D. R., and G. Motzkin. 1998. Ecology and conservation in the cultural landscape of New England: Lessons from nature's history. *Northeastern Naturalist* 5: 111–126.

————. 1999. Historical influences on the landscape of Martha's Vineyard: Perspectives on the management of the Manuel F. Correllus State Forest. Harvard Forest Paper No. 23, Petersham, Massachusetts.

———. 2003. Interpreting and conserving the openland habitats of coastal New England: Insights from landscape history. *Forest Ecology and Management.* In press.

Foster, D., G. Motzkin, D. Bernardos and J. Cardoza. 2002. Wildlife dynamics in the changing New England landscape. *Journal of Biogeography* 29: 1337–1357.

Foster, D., G. Motzkin and B. Slater. 1998. Land-use history as long-term broad-scale disturbance: Regional forest dynamics in central New England. *Ecosystems* 1: 96–119.

Foster, D. R., and J. O'Keefe. 2000. *New England Forests Through Time. Insights from the Harvard Forest Dioramas.* Harvard Forest and Harvard University Press, Petersham and Cambridge, Massachusetts.

Foster, D. R., D. A. Orwig and J. S. McLachlan. 1996. Ecological and conservation insights from reconstructive studies of temperate old-growth forests. *Trends in Ecology and Evolution* 11: 419–424.

Foster, D. R., P. K. Schoonmaker and S. T. A. Pickett. 1990. Insights from paleoecology to community ecology. *Trends in Ecology and Evolution* 5: 119–122.

Foster, D. R., and T. M. Zebryk. 1993. Long-term vegetation dynamics and disturbance history of a *Tsuga*-dominated forest in New England. *Ecology* 74: 982–998.

Foster, D. R., T. Zebryk, P. Schoonmaker and A. Lezberg. 1992. Post-settlement history of human land-use and vegetation dynamics of a hemlock woodlot in central New England. *Journal of Ecology* 80: 773–786.

Fowells, H. A. 1965. *Silvics of Forest Trees of the United States.* USDA Forest Service Agricultural Handbook No. 271.

Franklin, J. F. 1989. Importance and justification of long-term studies in ecology. Pages 3–19 *in* G. E. Likens (ed.), *Long-Term Studies in Ecology.* Springer-Verlag, New York.

Franklin, J. F., C. S. Bledsoe and J. T. Callahan. 1990. Contributions to the long-term ecological research program. *BioScience* 40(7): 509–523.

Fujita, T. T. 1971. Proposed characterization of tornadoes and hurricanes by area and intensity. SMRP Research Paper 91, University of Chicago.

Fuller, J. L. 1997. Holocene forest dynamics in southern Ontario, Canada: Fine-resolution pollen data. *Canadian Journal of Botany* 75: 1714–1727.

———. 1998. Ecological impact of the mid-Holocene hemlock decline in southern Ontario, Canada. *Ecology* 79: 2337–2351.

Fuller, J., D. R. Foster, J. McLachlan and N. Drake. 1998. Impact of human activity and environmental change on regional forest composition and dynamics in central New England. *Ecosystems* 1: 76–95.

George, L. O., and F. A. Bazzaz. 1999a. The fern understory as an ecological filter: Emergence and establishment of canopy tree seedings. *Ecology* 80: 833–845.

———. 1999b. The fern understory as an ecological filter: Growth and survival of canopy tree seedlings. *Ecology* 80: 846–856.

Gerhardt, F. 1993. Physiographic and historical influences of forest composition in central New England, U.S.A. M.F.S. Thesis, Harvard University, Cambridge, Massachusetts.

Gerhardt, F., and D. R. Foster. 2002. Physiographic and historical effects on forest vegetation in central New England, U.S.A. *Journal of Biogeography* 29: 1421–1437.

Glitzenstein, J. S., C. D. Canham, M. J. McDonnell and D. R. Streng. 1990. Effects of environment and land-use history on upland forests of the Cary Arboretum, Hudson Valley, New York. *Bulletin of the Torrey Botanical Club* 117: 106–122.

Godbold, D. L., and G. M. Berntson. 1997. Elevated atmospheric CO_2 concentrations lead to changes in ectomycorrhizal morphotype assemblages in *Betula papyrifera*. *Tree Physiology* 17: 347–350.

Godbold, D. L., G. M. Berntson and F. A. Bazzaz. 1997. Growth and mycorrhizal colonization of three North-American tree species under elevated atmospheric CO_2. *New Phytologist* 137: 433–440.

Goldstein, A. H., B. C. Daube, J. W. Munger and S. C. Wofsy. 1995. Automated *in situ* monitoring of atmospheric non-methane hydrocarbon concentrations and gradients. *Journal of Atmospheric Chemistry* 21: 43–59.

Goldstein, A. H., S.-M. Fan, M. L. Goulden, J. W. Munger and S. C. Wofsy. 1996. Emissions of ethane, propene and 1–butene by a midlatitude forest. *Journal of Geophysical Research* 101: 9149–9157.

Goldstein, A. H., M. L. Goulden, J. W. Munger, S. C. Wofsy and C. D. Geron. 1998. Seasonal course of isoprene emissions from a midlatitude deciduous forest. *Journal of Geophysical Research* 103: 31045–31051.

Goldstein, A. H., M. Spivokovsky and S. C. Wofsy. 1995. Seasonal variations of non-methane hydrocarbons in rural New England: Constraints on OH concentrations in northern midlatitudes. *Journal of Geophysical Research* 100: 21023–21033.

Golley, F. B. 1993. *A History of the Ecosystem Concept in Ecology.* Yale University Press, New Haven, Connecticut.

Golodetz, A. 1993. Historical patterns of land protection in north-central Massachusetts: The emergence of a greenway. Senior Thesis, Hampshire College, Amherst, Massachusetts.

Golodetz, A., and D. R. Foster. 1997. History and importance of land use and protection in the North Quabbin Region of Massachusetts. *Conservation Biology* 11: 227–235.

Gomez-Pompa, A., and A. Kaus. 1992. Taming the wilderness myth. *BioScience* 42: 271–278.

Goodale, C. L., and J. D. Aber. 2000. The long-term effects of land-use history on nitrogen cycling in northern hardwood forests. *Ecological Applications* 11: 253–267.

Goodale, C. L., J. D. Aber and E. P. Farrell. 1998. Applying an uncalibrated, physiologically based model of forest productivity to Ireland. *Climate Research* 10: 51–67.

Goodale, C. L., J. D. Aber and W. H. McDowell. 2000. The long-term effects of disturbance on organic and inorganic nitrogen export in the White Mountains, New Hampshire. *Ecosystems* 3: 433–450.

Goodale, C. L., J. D. Aber and S. V. Ollinger. 1998. Mapping monthly precipitation, temperature and solar radiation for Ireland with polynomial regression and a digital elevation model. *Climate Research* 10: 35–49.

Goodlett, J. C. 1960. The development of site concepts at the Harvard Forest and their impact upon management policy. *Harvard Forest Bulletin,* No. 28: 1–56.

Gookin, D. 1972. *An Historical Account of the Doings and Sufferings of the Christian Indians in New England in the Years 1675, 1676, 1677.* Reprinted from 1836 edition. Arno Press, New York.

Goudie, A. 2000. *The Human Impact on the Natural Environment.* MIT Press, Cambridge, Massachusetts.

Gould, E. M. 1960. Fifty years of management at the Harvard Forest. *Harvard Forest Bulletin,* No. 29: 1–48.

Goulden, M. L., J. W. Munger, S.-M. Fan, B. C. Daube and S. C. Wofsy. 1996a. Measurements of carbon sequestration by long-term eddy covariance: Methods and a critical evaluation of accuracy. *Global Change Biology* 2: 169–182.

———. 1996b. Exchange of carbon dioxide by a deciduous forest: Response to interannual climate variability. *Science* 271: 1576–1578.

Goulden, M. L., S. C. Wofsy, J. W. Harden, S. E. Trumbore, P. M. Crill, S. T. Gower, T. Fires, B. Daube, S.-M. Fan, D. J. Sutoon, F. A. Bazzaz and J. W. Munger. 1998. Sensitivity of boreal forest carbon balance to soil thaw. *Science* 279: 214–217.

Graham, A. 1996. Spatial response of mammals to Late Quaternary environmental fluctuations. *Science* 272: 1601–1606.

Grazulis, T. P. 1991. *Significant Tornadoes 1680–1991.* The Tornado Project, St. Johnsbury, Vermont.

Hall, B., G. Motzkin, D. Foster, M. Syfert and J. Burk. 2002. Three hundred years of forest and land-use change in Massachusetts, U.S.A. *Journal of Biogeography* 29: 1319–1335.

Hamburg, S. P. 1984. Effects of forest growth on soil nitrogen and organic matter pools following release from subsistence agriculture. Pages 145–158 *in* E. L. Stone (ed.), *Forest Soils and Treatment Impacts.* Proceedings of the Sixth North American Forest Soils Conference, University of Tennessee, Knoxville.

Harmon, M. E., K. J. Nadelhoffer and J. M. Blair. 1999. Measuring decomposition, nutrient turnover and stores in plant litter. Pages 202–240 *in* G. P. Robertson, C. S. Bledsoe, D. C. Coleman and P. Sollins (eds.), *Standard Soil Methods for Long Term Ecological Research.* Oxford University Press, New York.

Hawes, A. F. 1933. *The Present Condition of Connecticut Forests, A Neglected Resource.* State of Connecticut, Hartford.

Heath, L. S., R. A. Birdsey and D. W. Williams. 2002. Methodology for estimating soil carbon for the forest carbon budget model of the United States, 2001. *Environmental Pollution* 116: 373–380.

Heeley, R. W., and W. S. Motts. 1973. Surficial geologic map of Massachusetts. Pages 25–28 *in* J. S. Larson (ed.), *A Guide to Important Characteristics and Values of Freshwater Wetlands.* Pub. 31, Water Resources Research Center, University of Massachusetts, Amherst.

Hendricks, J. J., J. D. Aber, K. J. Nadelhoffer and R. D. Hallett. 2000. Nitrogen controls on fine root substrate quality in temperate forest ecosystems. *Ecosystems* 3: 57–69.

Hendricks, J. J., K. J. Nadelhoffer and J. D. Aber. 1993. Assessing the role of fine roots in carbon and nutrient cycling. *Trends in Ecology and Evolution* 8(5): 174–178.

———. 1997. A ^{15}N tracer technique for assessing fine root production and turnover. *Oecologia* 112: 300–304.

Henry, J. D., and J. M. A. Swan. 1974. Reconstructing forest history from live and dead plant material: An approach to the study of forest succession in southwest New Hampshire. *Ecology* 55: 772–783.

Hermy, M. P. 1994. Effects of former land use on plant species diversity and pattern in European deciduous woodlands. Pages 123–144 *in* T. J. B. Boyle and C. E. B. Boyle (eds.), *Biodiversity, Temperate Ecosystems, and Global Change.* Springer-Verlag, Berlin, Germany.

Hermy, M. P., P. van dem Bremt and G. Tack. 1993. Effects of site history on woodland vegetation. Pages 219–232 *in* M. E. A. Broekmeyer, W. Vos and H. Koop (eds.), *European Forest Reserves.* Pudoc Scientific, Wageningen, The Netherlands.

Hibbs, D. E. 1983. Forty years of forest succession in central New England. *Ecology* 64(6): 1394–1401.

Hirsch, A. I., J. W. Munger, D. J. Jacob, L. W. Horowitz and A. H. Goldstein. 1996. Seasonal variation of the ozone production efficiency per unit NO_x at Harvard Forest, Massachusetts. *Journal of Geophysical Research* 101: 12659–12666.

Hobbie, E. A., S. A. Macko and M. Williams. 2000. Correlations between foliar ^{15}N and nitrogen concentrations may indicate plant-mycorrhizal interactions. *Oecologia* 122: 273–283.

Hobbie, E. A., D. M. Olszyk, P. T. Rygiewicz, D. T. Tingey and M. G. Johnson. 2001. Foliar nitrogen concentration and natural abundance of ^{15}N suggest nitrogen allocation patterns of Douglas-fir and mycorrhizal fungi during development in elevated carbon dioxide concentration and temperature. *Tree Physiology* 21: 1113–1122.

Hobbie, E. A., N. S. Weber and J. M. Trappe. 2001. Mycorrhizal vs saprotrophic status of fungi: The isotopic evidence. *New Phytologist* 150: 601–610.

Hornbeck, J., C. W. Martin, R. S. Pierce, F. H. Bormann, G. E. Likens and J. S. Eaton. 1986. Clearcutting northern hardwoods: Effects on hydrologic and nutrient ion budgets. *Forest Science* 32: 667–686.

Hosley, N. W. 1931. Wildlife damage to New England forests. *Journal of Forestry* 29: 700–708.

———. 1935. The essentials of a management plan for forest wildlife in New England. *Journal of Forestry* 33: 985–989.

———. 1937. Some interrelations of wildlife management and forest management. *Journal of Forestry* 35: 674–678.

Houghton, J. T., Y. Ding, D. J. Griggs, M. Noguer, P. J. van der Linden and D. Xiaosu (eds.). 2001. *Climate Change 2001: The Scientific Basis: Contribution of Working Group I to the Third Assessment Report of the Intergovernmental Panel on Climate Change.* Cambridge University Press, England.

Hunter, M. L., Jr., G. L. Jacobson, Jr., and T. Webb III. 1988. Paleoecology and the coarse-filter approach to maintaining biological diversity. *Conservation Biology* 2: 375–385.

Hurlbert, S. H. 1984. Pseudoreplication and the design of ecological field experiments. *Ecological Monographs* 54: 187–211.

Hurtt, G. C., S. W. Pacala, P. R. Moorcroft, J. Caspersen, E. Shevliakova, R. Houghton and B. Moore. 2002. Projecting the future of the US carbon sink. *Proceedings of the National Academy of Sciences of the United States (PNAS)* 99: 1389–1394.

Jackson, S. T., and M. E. Lyford. 1999. Pollen dispersal models in quaternary plant ecology: Assumptions, parameters, and prescriptions. *Palaeogeography, Palaeoclimatology, Palaeoecology* 153: 179–201.

Jacob, D. J. 1999. *Introduction to Atmospheric Chemistry.* Princeton University Press, Princeton, New Jersey.

Jacobson, G. L., H. Almquist-Jacobson and J. C. Winne. 1991. Conservation of rare plant habitat: Insights from the recent history of vegetation and fire at Crystal Fen, northern Maine, USA. *Biological Conservation* 57: 287–314.

Jacobson, G. L., Jr., and R. H. W. Bradshaw. 1981. The selection of sites for paleo-environmental studies. *Quaternary Research* 16: 80–96.

Jacobson, G. L., T. Webb III and E. C. Grimm. 1987. Patterns and rates of vegetation change during deglaciation of eastern North America. Pages 277–288 *in* W. F. Ruddiman and H. E. Wright (eds.), *The Geology of North America.* Vol. K-3. Geological Society of America, Boulder, Colorado.

Jasienski, M., S. C. Thomas and F. A. Bazzaz. 1998. Blaming the trees: A critique of research on forest responses to high CO_2. *Tree* 13: 427.

Jenkins, J. C., R. A. Birdsey and Y. Pan. 2001. Biomass and NPP estimation for the mid-Atlantic region (USA) using plot-level forest inventory data. *Ecological Applications* 11: 1174–1193.

Johnson, D. W. 1992a. Effects of forest management on soil carbon storage. *Water Air Soil Pollution* 64: 83–120.

———. 1992b. Nitrogen retention in forest soils. *Journal of Environmental Quality* 21: 1–12.

Johnson, D. W., W. Cheng and I. C. Burke. 2000. Biotic and abiotic nitrogen retention in a variety of forest soils. *Soil Science Society of America Journal* 64: 1503–1514.

Johnson, W. C., and T. Webb III. 1989. The role of blue jays (*Cyanocitta cristata* L.) in the postglacial dispersal of fagaceous trees in eastern North America. *Journal of Biogeography* 16: 561–572.

Johnston, A. E. 1994. The Rothamsted classical experiments. Pages 9–37 *in* R. A. Leigh and A. E. Johnston (eds.), *Long-term Experiments in Agricultural and Ecological Sciences.* CABI Publishing, Oxford.

Jones, L. R., and F. V. Rand. 1979. *The Handbook of Vermont Shrubs and Woody Vines.* Charles E. Tuttle, Rutland, Vermont.

Jones, N. 1991. An ecological history of agricultural land use in two Massachusetts towns. Undergraduate Thesis, Hampshire College, Amherst, Massachusetts.

Jorgensen, N. 1977. *A Guide to New England's Landscape.* Barre Publishers, Barre, Vermont.

Josselyn, J. [1672] 1860. *New England's Rarities Discovered.* Edited by E. Tuckerman. Reprinted *in Transactions and Collections of the American Antiquarian Society* 4: 105–238.

Kalisz, P. 1986. Soil properties of steep Appalachian old fields. *Ecology* 67: 1011–1023.

Keeling, C. D., T. P. Whorf, M. Whalen and J. van der Plicht. 1995. Interannual extremes in the rate of rise of atmospheric carbon dioxide since 1980. *Nature* 375: 666–670.

Kicklighter, D. W., J. M. Melillo, W. T. Peterjohn, E. B. Rastetter, A. D. McGuire, P. A. Steudler and J. D. Aber. 1994. Scaling field measurements to estimate regional carbon dioxide fluxes from temperate forest soils. *Journal of Geophysical Research* 99: 1303–1315.

Kirby, K. J., C. M. Reid, R. C. Thomas and F. B. Goldsmith. 1998. Preliminary estimates of falled dead wood and standing dead trees in managed and unmanaged forests in Britain. *Journal of Applied Ecology* 35: 148–155.

Kittredge, D. B., A. O. Finley and D. R. Foster. 2003. Timber harvesting as ongoing disturbance in a landscape of diverse ownership. *Forest Ecology and Management.* 180: 425–442.

Kittredge, J. 1913. Notes on the chestnut bark disease (*Diaportha parasitica* Murvill) in Petersham, Massachusetts. *Harvard Forestry Club Bulletin* 2: 13–22.

Kizlinski, M., D. A. Orwig, R. A. Cobb and D. R. Foster. 2002. Direct and indirect ecosystem consequences of an invasive pest on forests dominated by eastern hemlock. *Journal of Biogeography* 29: 1489–1503.

Koerner, W., J. L. Dupouey, E. Dambrine and M. Benoit. 1997. Influence of past land use on the vegetation and soils of present day forest in the Vosges Mountains, France. *Journal of Ecology* 85: 351–358.

Kohm, K. A., and J. F. Franklin. 1996. *Creating a Forestry for the 21st Century: The Science of Ecosystem Management.* Island Press, Washington, D.C.

Kolasa, J., and S. T. A. Pickett (eds.). 1991. *Ecological Heterogeneity.* Springer-Verlag, New York.

Körner, C., and F. A. Bazzaz (eds.). 1996. *Carbon Dioxide, Population, and Communities.* Physiological Ecology Series, Academic Press, London.

Kuikman, P. J., A. G. Jansen and J. A. vanVeena. 1991. ^{15}N-nitrogen mineralization from bacteria by protozoan grazing at different soil moisture regimes. *Soil Biology and Biochemistry* 23: 193–200.

Kuikman, P. J., A. G. Jansen, J. A. vanVeena and A. J. B. Zehnder. 1990. Protozoan predation and the turnover of soil organic carbon and nitrogen in the presence of plants. *Biology and Fertility of Soils* 10: 22–28.

Lajtha, K., B. Seely, and I. Valiela. 1995. Retention and leaching losses of atmospherically-derived nitrogen in the aggrading coastal watershed of Waquoit Bay, MA. *Biogeochemistry* 28: 33–54.

Lamb, H. 1991. *Historic Storms of the North Sea, British Isles and Northwest Europe.* Cambridge University Press, England.

Lambers, H., T. L. Pons and F. S. Chapman (eds.). 1998. *Plant Physiological Ecology.* Springer-Verlag, New York.

Leahy, C., J. H. Mitchell and T. Conuel. 1996. *The Nature of Massachusetts.* Addison-Wesley, New York.

Lefer, B. L., R. W. Talbot and J. W. Munger. 1999. Nitric acid and ammonia at a rural northeastern U.S. site. *Journal of Geophysical Research* 104: 1645–1661.

Likens, G., and F. H. Bormann. 1995. *Biogeochemistry of a Forested Ecosystem.* Springer-Verlag, New York.

Likens, G. E., F. H. Bormann, N. M. Johnson, D. W. Fischer and R. S. Pierce. 1970. The effect of forest cutting and herbicide treatment on nutrient budgets in the Hubbard Brook watershed-ecosystem. *Ecological Monographs* 40: 23–47.

Likens, G. E., F. H. Bormann, R. S. Pierce and W. A. Reiners. 1978. Recovery of a deforested ecosystem. *Science* 199: 492–496.

Linkins, A. E., R. L. Sinsabaugh, C. M. McClaugherty and J. M. Melillo. 1990. Comparison of cellulase activity on decomposing leaves in a hardwood forest and woodland stream. *Soil Biology and Biochemistry* 22: 423–425.

Litvaitis, J. A. 1993. Response of early successional vertebrates to historic changes in land use. *Conservation Biology* 7: 866–873.

Lodge, D. J., and E. R. Ingham. 1991. A comparison of agar film techniques for estimating fungal biovolumes in litter and soil. *Agriculture Ecosystems and the Environment* 5: 31–37.

Louis, I., S. Racette and J. G. Torrey. 1990. Occurrence of cluster roots on *Myrica cerifera* L. (Myricaceae) in water culture in relation to phosphorus nutrition. *New Phytologist* 115: 311–317.

Ludlum, D. M. 1963. *Early American Hurricanes 1492–1870.* American Meteorological Society, Boston.

Lyford, W. H., J. C. Goodlett, and W. H. Coates. 1963. Landforms, soils with fragipans, and forest on a slope in the Harvard Forest. *Harvard Forest Bulletin,* No. 30: 1–29.

Mabry, C., and T. Korsgren. 1998. A permanent plot study of vegetation and vegetation-site factors fifty-three years following disturbance in central New England, U.S.A. *Ecoscience* 5: 232–240.

MacCleery, D. W. 1992. American forests: A history of resiliency and recovery. USDA Forest Service, FS-540.

MacConnell, W. P. 1973. *Massachusetts Map Down: Land-Use and Vegetation Cover Mapping Classification Manual.* Cooperative Extension Service, University of Massachusetts, Amherst.

———. 1975. Remote sensing 20 years of change in Massachusetts. *Massachusetts Agricultural Experiment Station Bulletin,* No. 630: 1–110.

MacConnell, W. P., and W. Niedzwiedz. 1974. Remote sensing 20 years of change in Worcester County, Massachusetts, 1951–1971. Massachusetts Agricultural Experiment Station, University of Massachusetts, Amherst.

Magill, A. H., and J. D. Aber. 1998. Long-term effects of chronic nitrogen additions on foliar litter decay and humus formation in forest ecosystems. *Plant and Soil* 203: 301–311.

———. 2000a. Dissolved organic carbon and nitrogen relationships in forest litter as affected by nitrogen deposition. *Soil Biology and Biochemistry* 32: 603–613.

———. 2000b. Variation in soil net mineralization rates with dissolved organic carbon additions. *Soil Biology and Biochemistry* 32: 597–601.

Magill, A. H., J. D. Aber, G. M. Berntson, W. H. McDowell, K. J. Nadelhoffer, J. M. Melillo and P. A. Steudler. 2000. Long-term nitrogen additions and nitrogen saturation in two temperate forests. *Ecosystems* 3: 238–253.

Magill, A. H., J. D. Aber, J. J. Hendricks, R. D. Bowden, J. M. Melillo and P. A. Steudler. 1997. Biogeochemical response of forest ecosystems to simulated chronic nitrogen deposition. *Ecological Applications* 7: 402–415.

Magill, A. H., M. R. Downs, K. J. Nadelhoffer, R. A. Hallett and J. D. Aber. 1996. Forest ecosystem response to four years of chronic nitrate and sulfate additions at Bear Brooks Watershed, Maine, USA. *Forest Ecology and Management* 84: 29–37.

Magnuson, J. J. 1990. Long-term ecological research and the invisible present: Uncovering the processes hidden because they occur slowly or because effects lag years behind causes. *BioScience* 40: 495–501.

Maherali, H., E. H. DeLucia and T. W. Sipe. 1997. Hydraulic adjustment of maple saplings to canopy gap formation. *Oecologia* 112: 472–480.

Marks, P. L. 1983. On the origin of the field plants of the northeastern United States. *American Naturalist* 122: 210–227.

Marsh, G. P. 1882 (1976). *The Earth as Modified by Human Action*. Reprint. Ayer Publishing, New York.

Martin, M. E., and J. D. Aber. 1994. Analyses of forest foliage III: Determining nitrogen lignin and cellulose in fresh leaves using near infrared reflectance data. *Journal of Near Infrared Spectroscopy* 2: 25–32.

———. 1997. Estimation of forest canopy lignin and nitrogen concentration and ecosystem processes by high spectral resolution remote sensing. *Ecological Applications* 7: 431–443.

———. 1999. Estimating forest canopy characteristics as inputs for models of forest carbon exchange by high spectral resolution remote sensing. Pages 341–349 *in* H. Gholz (ed.), *The Use of Remote Sensing in the Modeling of Forest Productivity at Scales from the Stand to the Globe*. Kluwer Academic Publishers, The Netherlands.

Martin, M. E., S. D. Newman, J. D. Aber and R. Congalton. 1998. Determining forest species composition using high spectral resolution remote sensing data. *Remote Sensing of the Environment* 65: 249–254.

Maser, C., R. F. Tarrant, J. M. Trappe and J. F. Franklin (eds.). 1988. From the forest to the sea: A story of fallen trees. USDA, Forest Service, PNW-GTR-229.

Massachusetts Historic Commission. 1985. Historic and archaeological resources in the Connecticut Valley. Massachusetts Historical Commission State Survey Team Regional Reports. Commonwealth of Massachusetts, Boston.

MassGIS. 2002. *Land Use Datalayer for the Year 1999*. MassGIS, Commonwealth of Massachusetts, Executive Office of Environmental Affairs, Boston. Web page: *www.state.ma.us/mgis/lus.htm*.

Matlack, G. R. 1994. Plant species migration in a mixed-history forest landscape in eastern North America. *Ecology* 75: 1491–1502.

McAndrews, J. H. 1988. Human disturbance of North American forests and grasslands: The fossil pollen record. *Handbook of Vegetation Science,* Volume 7. Vegetation History, Section IV.3, pages 673–697.

McClure, M. S. 1987. Biology and control of hemlock woolly adelgid. *The Connecticut Agricultural Experiment Station Bulletin* 851, December, pages 3–9.

———. 1990. Role of wind, birds, deer, and humans in the dispersal of hemlock woolly adelgid (Homoptera: Adelgidae). *Environmental Entomology* 19: 36–43.

McConnaughay, K. D. M., S. L. Bassow, G. M. Berntson and F. A. Bazzaz. 1996. Leaf senescence and decline of end-of-season gas exchange in five temperate deciduous tree species grown in elevated CO_2 concentrations. *Global Change Biology* 2: 25–34.

McConnaughay, K. D. M., A. B. Nicotra and F. A. Bazzaz. 1996. Rooting volume, nutrient availability and CO_2-induced growth enhancement in temperate forest tree seedlings. *Ecological Applications* 6: 619–627.

McDonnell, M. J., and S. T. A. Pickett (eds.). 1993. *Humans as Components of Ecosystems: The Ecology of Subtle Human Effects and Populated Areas*. Springer-Verlag, New York.

McDowell, W. H., W. S. Currie, J. D. Aber and Y. Yano. 1998. Effects of chronic nitrogen amendment on production of dissolved organic carbon and nitrogen in forest soils. *Water, Air and Soil Pollution* 105: 175–182.

McKibben, B. 1995. An explosion of green. *Atlantic Monthly,* April, pages 61–83.

McLachlan, J., D. R. Foster and F. Menalled. 1999. Anthropogenic origins of late-successional structure and composition in four New England hemlock stands. *Ecology* 81: 717–733.

McLellan, T., J. D. Aber, M. E. Martin, J. M. Melillo and K. J. Nadelhoffer. 1991. Determination of nitrogen, lignin and cellulose content of decomposing leaf material by near infrared reflectance. *Canadian Journal of Forest Research* 21: 1684–1688.

McLellan, T., M. E. Martin, J. D. Aber, J. M. Melillo, K. J. Nadelhoffer and B. Dewey. 1991. Comparison of wet chemical and near infrared reflectance measurements of carbon fraction chemistry and nitrogen concentration of forest foliage. *Canadian Journal of Forest Research* 21: 1689–1693.

McManis, D. R. 1975. *Colonial New England: An Historical Geography.* Oxford University Press, New York.

McNulty, S. G., and J. D. Aber. 1993. Effect of chronic nitrogen additions on nitrogen cycling in a high elevation spruce-fir stand. *Canadian Journal of Forest Research* 23: 1252–1263.

McNulty, S. G., J. D. Aber and R. D. Boone. 1991. Spatial changes in forest floor and foliar chemistry of spruce-fir forests across New England. *Biogeochemistry* 14: 13–29.

McNulty, S. G., J. D. Aber, T. M. McLellan and S. M. Katt. 1990. Nitrogen cycling in high elevation forests of the U.S. in relation to nitrogen deposition. *Ambio* 19: 38–40.

McNulty, S. G., J. D. Aber and S. D. Newman. 1996. Nitrogen saturation in a high elevation spruce-fir stand. *Forest Ecology and Management* 84: 109–121.

Melillo, J. M. 1995. Human influences on the global nitrogen budget and their implications for the global carbon budget. Pages 117–133 *in* S. Murai and M. Kimura (eds.), *Toward Global Planning of Sustainable Use of the Earth: Development of Global Eco-Engineering.* Elsevier, Amsterdam, The Netherlands.

Melillo, J. M. 1996. Carbon and nitrogen interactions in the terrestrial biosphere: Anthropogenic effects. Pages 431–450 *in* B. Walker and W. Steffen (eds.), *Global Change and Terrestrial Ecosystems.* International Geosphere-Biosphere Programme Book Series 2, Cambridge University Press, England.

Melillo, J. M., J. D. Aber, A. E. Linkins, A. Ricca, B. Fry and K. Nadelhoffer. 1989. Carbon and nitrogen dynamics along the decay continuum: Plant litter to soil organic matter. *Plant and Soil* 115: 189–198.

Melillo, J. M., J. D. Aber, and J. M. Muratore. 1982. Nitrogen and lignin control of hardwood leaf litter decomposition dynamics. *Ecology* 63: 621–626.

Melillo, J. M., D. W. Kicklighter, A. D. McGuire, W. T. Peterjohn and K. Newkirk. 1995. Global change and its effects on soil organic carbon stocks. Pages 175–189 *in* R. Zepp et al. (eds.), *Report of the Dahlem Workshop on the Role of Nonliving Organic Matter in the Earth's Carbon Cycle.* John Wiley & Sons, New York.

Melillo, J. M., and P. A. Steudler. 1989. The effect of nitrogen fertilization on the COS and CS$_2$ emissions from temperate forest soils. *Journal of Atmospheric Chemistry* 9: 411–417.

Melillo, J. M., P. A. Steudler, J. D. Aber and R. D. Bowden. 1989. Atmospheric deposition and nutrient cycling. Pages 47–59 *in* M. O. Andreae and D. S. Schimel (eds.), *Exchange of Trace Gases Between Terrestrial Ecosystems and the Atmosphere.* John Wiley & Sons, New York.

Melillo, J. M., P. A. Steudler, J. D. Aber, K. Newkirk, H. Lux, F. P. Bowles, C. Catricala, A. Magill, T. Ahrens and S. Morrisseau. 2002. Soil warming and carbon-cycle feedbacks to the climate system. *Science* 298:2173–2176.

Merchant, C. 1989. *Ecological Revolutions: Nature, Gender, and Science in New England.* The University of North Carolina Press, Chapel Hill.

Miao, S. L., P. M. Wayne and F. A. Bazzaz. 1992. Elevated CO_2 differentially alters the responses of co-occurring birch and maple seedlings to moisture gradients. *Oecologia* 90: 300–304.

Micks, P. 1994. Soil respiration and nitrogen application in two forest stands: Short- and long-term responses. M.S. Thesis. University of New Hampshire, Durham, New Hampshire.

Millikin, C. S., and R. D. Bowden. 1996. Soil respiration in pits and mounds following an experimental hurricane. *Soil Science Society of America Journal* 60: 1951–1953.

Mills, A. V. 1993. Predicting forest growth and composition: A test of the JABOWA model using data from Earl Stephens' study in the Tom Swamp tract. Senior Thesis, Hampshire College, Amherst, Massachusetts.

Minsinger, W., and C. T. Orloff. 1994. The Great Atlantic Hurricane, Sept. 8–16, 1944: An historical and pictorial summary. Blue Hill Meteorological Observatory, Milton, Massachusetts.

Moody, J. L., J. W. Munger, A. H. Goldstein, D. J. Jacob and S. C. Wofsy. 1998. Harvard Forest regional-scale air mass composition by patterns in atmospheric transport history. *Journal of Geophysical Research* 103: 13181–13194.

Moore, G. C., and G. R. Parker. 1992. Colonization by the Eastern Coyote. *In* A. H. Boer (ed.), *Ecology and Management of the Eastern Coyote.* Wildlife Research Unit. University of New Brunswick, Fredericton, New Brunswick.

Morse, S. R., P. Wayne, S. Miao and F. A. Bazzaz. 1993. Elevated CO_2 and drought alter tissue water relations of birch (*Betula populifolia* Marsh.) seedlings. *Oecologia* 95: 599–602.

Mott, J. R., and D. C. Fuller. 1967. Soil survey of Franklin County. USDA Soil Conservation Service.

Motzkin, G. M. 1994. Calcareous fens of western New England and adjacent New York state. *Rhodora* 96: 44–68.

———. 1995. *Inventory of Uncommon Plant Communities of Western Massachusetts: 1993–1994.* Massachusetts Natural Heritage and Endangered Species Program, Boston.

Motzkin, G. M., R. Eberhardt, B. Hall, D. R. Foster. J. Harrod and D. MacDonald. 2002. Vegetation variation across Cape Cod, Massachusetts: Environmental and historical determinants. *Journal of Biogeography* 29: 1439–1454.

Motzkin, G. M., and D. R. Foster. 1998. How land use determines vegetation: Evidence from a New England sandplain. *Arnoldia* 58: 32–34.

———. 2002. Grasslands, heathlands, and shrublands in coastal New England: Historical interpretations and approaches to conservation. *Journal of Biogeography* 29: 1569–1590.

Motzkin, G. M., D. R. Foster, A. Allen, J. Harrod and R. D. Boone. 1996. Controlling site to evaluate history: Vegetation patterns of a New England sand plain. *Ecological Monographs* 66: 345–365.

Motzkin, G. M., D. A. Orwig and D. R. Foster. 2002. Vegetation and disturbance history of a rare dwarf pitch pine community in western New England, U.S.A. *Journal of Biogeography* 29: 1455–1467.

Motzkin, G. M., W. A. Patterson III and N. E. R. Drake. 1993. Fire history and vegetation dynamics of a *Chamaecyparis thyoides* wetland on Cape Cod, Massachusetts. *Journal of Ecology* 81: 391–402.

Motzkin, G. M., W. A. Patterson III and D. R. Foster. 1999. A historical perspective on pitch pine–scrub oak communities in the Connecticut Valley of Massachusetts. *Ecosystems* 3: 255–273.

Motzkin, G. M., P. Wilson, D. R. Foster and A. Allen. 1999. Vegetation patterns in heterogeneous landscapes: The importance of history and environment. *Journal of Vegetation Science* 10: 903–920.

Mulholland, M. T. 1985. Patterns of change in prehistoric southern New England: A regional approach. Ph.D. Dissertation, University of Massachusetts, Amherst.

Munger, J. W., S.-M. Fan, P. S. Bakwin, M. L. Goulden, A. H. Goldstein, A. S. Colman and S. C. Wofsy. 1998. Regional budgets for nitrogen oxides from continental sources: Variations of rates for oxidation and deposition with season and distance from source regions. *Journal of Geophysical Research* 103: 8355–8368.

Munger, J. W., S. C. Wofsy, P. S. Bakwin, S.-M. Fan, M. L. Goulden, B. C. Daube and A. H. Goldstein. 1996. Atmospheric deposition of reactive nitrogen oxides and ozone in a temperate deciduous forest and a sub-arctic woodland. I: Measurements and mechanisms. *Journal of Geophysical Research* 101: 12639–12657.

Nadelhoffer, K. J. 1990. A microlysimeter for measuring N mineralization and microbial respiration in aerobic soil incubations. *Soil Science Society of America Journal* 54: 411–415.

Nadelhoffer, K. J., J. D. Aber, M. R. Downs, B. Fry and J. M. Melillo. 1993. Biological sinks for nitrogen additions to a forested catchment. Pages 322–329 *in CEC Report of the Proceedings of the International Symposium on Experimental Manipulation of Biota and Biogeochemical Cycling in Ecosystems,* 18–20 May 1992, Copenhagen.

Nadelhoffer, K. J., J. D. Aber and J. M. Melillo. 1983. Leaf-litter production and soil organic matter dynamics along a nitrogen availability gradient in southern Wisconsin (USA). *Canadian Journal of Forest Research* 13: 12–21.

———. 1985. Fine root production in relation to net primary production along a nitrogen availability gradient in temperate forests: A new hypothesis. *Ecology* 66: 1377–1390.

Nadelhoffer, K. J., M. R. Downs and B. Fry. 1999. Sinks for N additions to an oak forest and a red pine plantation at the Harvard Forest, Massachusetts, USA. *Ecological Applications* 9: 72–86.

Nadelhoffer, K. J., M. R. Downs, B. Fry, A. Magill and J. D. Aber. 1999. Controls on N retention and exports in a fertilized watershed. *Environmental Monitoring and Assessment* 55: 187–210.

Nadelhoffer, K. J., B. A. Emmett, P. Gundersen, O. J. Kjønaas, C. J. Koopmans, P. Schleppi, A. Tietema and R. F. Wright. 1999. Nitrogen deposition makes a minor contribution to carbon sequestration in temperate forests. *Nature* 398: 145–148.

Nadelhoffer, K. J., and B. Fry. 1988. Controls on natural nitrogen-15 and carbon-13 abundances in forest soil organic matter. *Soil Science Society of America Journal* 52(5): 1633–1640.

Nadelhoffer, K. J., A. E. Giblin, G. R. Shaver and J. Laundre. 1991. Effects of temperature and organic matter quality on C, N and P mineralization in soils from arctic ecosystems. *Ecology* 72: 242–253.

Nadelhoffer, K. J., and J. W. Raich. 1992. Fine root production estimates and belowground carbon allocation in forest ecosystems. *Ecology* 73: 1139–1147.

Nadelhoffer, K. J., J. W. Raich and J. D. Aber. 1998. Ecosystem stoichiometry: Carbon budgets and fine root production in forests. *Ecology* 79: 1822–1825.

Nash, R. 1971. *Wilderness and the American Mind.* Yale University Press, New Haven, Connecticut.

Neelon, S. E. 1996. The response of understory vegetation to simulated hurricane disturbance. Honors Thesis, Smith College, Northampton, Massachusetts.

NETSA (New England Timber Salvage Administration). 1943. Report of the U.S. Forest Service programs resulting from the salvage of timber after the 1938 hurricane. U.S. Forest Service, Washington, D.C.

Neumann, C. J., B. R. Jarvinen and A. C. Pike. 1987. *Tropical Cyclones of the North Atlantic Ocean 1871–1986.* Third revised edition. NOAA–National Climatic Data Center, Asheville, North Carolina.

Nielsen, G. A. 1960. Some quantitative measurements of annual natural organic deposits and their effects on prairie and forest soil properties. M.S. Thesis, University of Wisconsin, Madison.

———. 1963. Incorporation of organic matter into the A horizon of some Wisconsin soils under native vegetation. Ph.D. Dissertation, University of Wisconsin, Madison.

Nielsen, G. A., and F. D. Hole. 1963. A study of the natural processes of incorporation of organic matter into soil in the University of Wisconsin Arboretum. *Wisconsin Academic Review* 52: 213–227.

Niering, W. A. 1981. The role of fire management in altering ecosystems. Pages 489–510 *in* H. A. Mooney, T. M. Bonnicksen and N. L. Christensen (eds.), *Fire Regimes and Ecosystem Properties.* USDA Forest Service General Technical Report WO-26. Department of Agriculture, Washington, D.C.

Noger, P. 1995. Effects of vesicular arbuscular mycorrhizae on plant growth. "Diplomarbeit," Swiss Eidgenössiche Technische Hochschule of Zürich.

O'Keefe, J., and D. R. Foster. 1998a. An ecological history of Massachusetts forests. *Arnoldia* 58: 2–31.

———. 1998b. An ecological and environmental history of Massachusetts forests. Pages 19–66 *in* C. H. W. Foster (ed.), *Stepping Back to Look Forward: A History of the Massachusetts Forest.* Harvard Forest, Petersham, Massachusetts.

Oliver, C. D. 1981. Forest development in North America following major disturbances. *Forest Ecology and Management* 3: 153–168.

Oliver, C. D., and B. C. Larson. 1996. *Forest Stand Dynamics.* John Wiley & Sons, New York.

Oliver, C. D., and E. P. Stephens. 1977. Reconstruction of a mixed-species forest in central New England. *Ecology* 58: 562–572.

Ollinger, S. V., J. D. Aber and C. A. Federer. 1998. Estimating regional forest productivity and water yield using an ecosystem model linked to a GIS. *Landscape Ecology* 13: 323–334.

Ollinger, S. V., J. D. Aber, C. A. Federer, G. M. Lovett and J. M. Ellis. 1995. Modeling physical and chemical climatic variables across the northeastern U.S. for a geographic information system. USDA Forest Service General Technical Report NE-191. Department of Agriculture, Washington, D.C.

Ollinger, S. V., J. D. Aber, G. M. Lovett, S. E. Millham, R. G. Lathrop and J. E. Ellis. 1993. A spatial model of atmospheric deposition for the northeastern U.S. *Ecological Applications* 3: 459–472.

Ollinger, S. V., J. D. Aber and P. B. Reich. 1996. Predicting the effects of tropospheric ozone on forest productivity in the northeastern U.S. Pages 217–225 *in Proceedings, 1995 Meeting of the Northern Global Change Program,* USDA Forest Service General Technical Report NE-214. Department of Agriculture, Washington, D.C.

———. 1997. Simulating ozone effects on forest productivity: Interactions between leaf-, canopy-, and stand-level processes. *Ecological Applications* 7: 1237–1251.

Ollinger, S. V., J. D. Aber, P. B. Reich and R. J. Freuder. 2002. Interactive effects of nitrogen deposition, tropospheric ozone, elevated CO_2 and land use history on the carbon dynamics of northern hardwood forests. *Global Change Biology* 8: 545–562.

Orwig, D. A. Ecosystem to regional impacts of introduced pests and pathogens: Historical context, questions and issues. *Journal of Biogeography* 29: 1471–1474.

Orwig, D. A., and M. D. Abrams. 1997. Variation of radial growth responses to drought among species, site, and canopy strata. *Trees* 11: 474–484.

———. 1999. Impacts of early selective logging on the dendroecology of an old-growth, bottomland hemlock–white pine–northern hardwood forest on the Allegheny Plateau. *Bulletin of the Torrey Botanical Club* 126: 234–244.

Orwig, D. A., C. V. Cogbill, D. R. Foster and J. F. O'Keefe. 2001. Variations in old-growth structure and definitions: Forest dynamics on Wachusett Mountain, Massachusetts. *Ecological Applications* 11: 437–452.

Orwig, D. A., and D. R. Foster. 1998a. Forest response to the introduced hemlock woolly adelgid in southern New England, U.S.A. *Bulletin of the Torrey Botanical Club* 125: 60–73.

———. 1998b. Ecosystem response to an imported pathogen: The hemlock woolly adelgid. *Arnoldia* 58: 41–44.

Orwig, D. A., D. R. Foster and D. L. Mausel. 2002. Landscape patterns of hemlock decline in New England due to the introduced hemlock woolly adelgid. *Journal of Biogeography* 29: 1475–1487.

Overpeck, J. T., D. Rind and R. Goldberg. 1990. Climate-induced changes in forest disturbance and vegetation. *Nature* 343: 51–53.

Pacala, S. W., C. D. Canham, J. Saponara, J. A. Silander, R. K. Kobe and E. Ribbens. 1996. Forest models defined by field measurements: Estimation, error analysis and dynamics. *Ecological Monographs* 66: 1–43.

Pacala, S. W., G. C. Hurtt, D. Baker, P. Peylin, R. A. Houghton, R. A. Birdsey, L. Heath, E. T. Sundquist, R. F. Stallard, P. Ciais, P. Moorcroft, J. P. Caspersen , E. Shevliakova, B. Moore, G. Kohlmaier, E. Holland, M. Gloor, M. E. Harmon, S.-M. Fan, J. L. Sarmiento, C. L. Goodale, D. Schimel and C. B. Field. 2001. Consistent land- and atmosphere-based US carbon sink estimates. *Science* 292: 2316–2320.

Paillet, F. 1982. Ecological significance of American chestnut in the Holocene forests of Connecticut. *Bulletin of the Torrey Botanical Club* 109: 457–473.

———. 1984. Growth-form and ecology of American chestnut sprout clones in northeastern Massachusetts. *Bulletin of the Torrey Botanical Club* 111: 316–328.

———. 2002. American chestnut: The history and ecology of a transformed species. *Journal of Biogeography* 29: 1517–1530.

Parshall, T. 1995. Canopy mortality and stand-scale change in a northern hemlock–hardwood forest. *Canadian Journal of Forest Research* 25: 1466–1478.

———. 1999. Documenting forest stand invasion: Fossil stomata and pollen in forest hollows. *Canadian Journal of Botany* 77: 1529–1538.

———. 2002. Late-Holocene stand-scale invasion by hemlock *(Tsuga canadensis)* at its western range limit. *Ecology* 83:1382–1391.

Parshall, T., and D. R. Foster. 2002. Fire on the New England landscape: Regional and temporal variation, cultural and environmental controls. *Journal of Biogeography* 29: 1305–1317.

Parshall, T., D. R. Foster, E. Faison, D. MacDonald, and B. C. S. Hansen. 2003. Long-term vegetation and fire dynamics of pitch pine–oak forests on Cape Cod, Massachusetts. *Ecology* 84: 736–748.

Pastor, J., J. D. Aber, C. A. McClaugherty and J. M. Melillo. 1984. Aboveground production and N and P cycling along a nitrogen mineralization gradient on Blackhawk Island, Wisconsin. *Ecology* 65: 256–268.

Pastor, J., M. A. Stillwell and D. Tilman. 1987. Nitrogen mineralization and nitrification in four Minnesota old fields. *Oecologia* 71: 481–485.

Patric, J. H. 1974. River flow increases in central New England after the hurricane of 1938. *Journal of Forestry* 72: 21–25.

Patric, J. H., and E. M. Gould. 1976. Shifting land use and the effects on river flow in Massachusetts. *Journal of the American Water Works Association* 68: 41–45.

Patterson, W. A., III, and A. E. Backman. 1988. Fire and disease history of forests. Pages 603–632 *in* B. Huntley and T. Webb III (eds.), *Handbook of Vegetation Science,* Volume 7. *Vegetation History.* Kluwer Academic Publishers, Dordrecht.

Patterson, W. A., III, and D. R. Foster. 1990. "Tabernacle Pines": The rest of the story. *Journal of Forestry* 89: 23–25.

Patterson, W. A., III, and K. E. Sassaman. 1988. Indian Fires in the Prehistory of New England. Pages 107–135 *in* G. Nicholas (ed.), *Holocene Human Ecology in Northeastern North America.* Plenum Press, New York.

Paul, E. A., and F. E. Clark. 1996. *Soil Microbiology and Biochemistry.* Academic Press, New York.

Peet, R. K., and N. L. Christensen. 1980. Hardwood forest vegetation of the North Carolina Piedmont. *Veroffingen Geobotanisches Institut ETH, Stiftung Rubel* 69: 14–39.

Perley, S. 1891. *Historic Storms of New England.* The Salem Press Publishing and Printing Co., Salem, Massachusetts.

Peterjohn, W. T., M. B. Adams and F. S. Gilliam. 1996. Symptoms of nitrogen saturation in two central Appalachian hardwood ecosystems. *Biogeochemistry* 35: 507–522.

Peterjohn, W. T., J. M. Melillo, F. P. Bowles and P. A. Steudler. 1993. Soil warming and trace gas fluxes: Experimental design and preliminary flux results. *Oecologia* 93: 18–24.

Peterjohn, W. T., J. M. Melillo, P. A. Steudler, K. M. Newkirk, F. P. Bowles and J. D. Aber. 1994. Responses of trace gas fluxes and N availability to experimentally elevated soil temperatures. *Ecological Applications* 4: 617–625.

Peterken, G. F. 1993. *Woodland Conservation and Management.* Second Edition. Chapman and Hall, London.

——. 1999. *Natural Woodland.* Cambridge University Press, England.

Peterken, G. F., and M. Game. 1981. Historical factors affecting the distribution of *Mercurialis perennis* in central Lincolnshire. *Journal of Ecology* 69: 781–796.

——. 1984. Historical factors affecting the number and distribution of vascular plant species in the woodlands of central Lincolnshire. *Journal of Ecology* 72: 155–182.

Peterken, G. F., and P. T. Harding. 1975. Woodland conservation in eastern England: Comparing the effects of changes in three study areas since 1946. *Biological Conservation* 8: 279–298.

Peters, J. R., and T. M. Bowers. 1977. Forest statistics for Massachusetts. USDA Forest Service Resource Bulletin NE-48.

Pickett, S. T. A., and P. S. White. 1985. *The Ecology of Natural Disturbance and Patch Dynamics.* Academic Press, New York.

Pierce, C. H. 1939. Meteorological history of the New England hurricane of September 21, 1938. *Monthly Weather Review* 67: 237–285.

Post, W. M., and L. K. Mann. 1990. Changes in soil organic carbon and nitrogen as a result of cultivation. Pages 401–407 *in* A. F. Bowman (ed.), *Soil and the Greenhouse Effect.* John Wiley & Sons, New York.

Potosnak, M. J., S. C. Wofsy, A. S. Denning, T. J. Conway and D. H. Barnes. 1999. Influence of biotic exchange and combustion sources on atmospheric CO_2 concentrations in New England from observations at a forest flux tower. *Journal of Geophysical Research* 104: 9561–9569.

Pruitt, B. H. 1984. Self-sufficiency and the agricultural economy of eighteenth century Massachusetts. *William and Mary Quarterly* 41: 333–364.

Pyne, S. J. 1982. *Fire in America.* Princeton University Press, Princeton, New Jersey.

Rackham, O. 1979. Documentary evidence for the historical ecologist. *Landscape History* 1: 29–33.

——. 1986. *The History of the Countryside.* J. M. Dent, London.

Raich, J. W., R. D. Bowden and P. A. Steudler. 1990. Comparison of two static chamber techniques for determination of CO_2 efflux from forest soils. *Soil Science Society of America Journal* 54: 1754–1757.

Raich, J. W., and W. H. Schlesinger. 1992. The global carbon dioxide flux in soil respiration and its relationship to vegetation and climate. *Tellus* 44B: 81–99.

Rainey, S. M., K. J. Nadelhoffer, S. L. Silver and M. R. Downs. 1999. Effects of chronic nitrogen additions on understory species abundance and nutrient content in a red pine plantation. *Ecological Applications* 9: 949–957.

Rane, F. W. 1908. Fourth Annual Report of the State Forester of Massachusetts for the Year 1907. Wright and Potter, Boston, Massachusetts.

Rasche, H. 1958. Temperature differences in Harvard Forest and their significance. Harvard Forest Paper No. 4, Petersham, Massachusetts.

Raup, H. M. 1964. Some problems in ecological theory and their relation to conservation. *Journal of Ecology* (Suppl.) 52: 19–28.

———. 1966. The view from John Sanderson's farm: A perspective for the use of the land. *Forest History* 10: 2–11.

Raup, H. M., and R. E. Carlson. 1941. The history of land use in the Harvard Forest. *Harvard Forest Bulletin* 20: 1–64.

Redmond, D. R. 1955. Studies in forest pathology. XV: Rootlets, mycorrhiza, and soil temperatures in relation to birch dieback. *Canadian Journal of Botany* 33: 595–627.

Rees, M., R. Condit, M. Crawley, S. Pacala and D. Tilman. 2001. Long-term studies of vegetation dynamics. *Science* 293: 650–655.

Richter, D., D. Markewitz, C. G. Wells, H. L. Allen, R. April, P. R. Heine and B. Urrego. 1994. Soil chemical changes during three decades in an old-field loblolly pine (*Pinus taeda* L.) ecosystem. *Ecology* 75(5): 1463–1473.

Ricklefs, R. E. 1977. Environmental heterogeneity and plant species diversity: A hypothesis. *American Naturalist* 111: 376–381.

Risser, P. (ed.) 1991. Introduction. *Long-term Ecological Research: An International Perspective.* SCOPE 47. John Wiley & Sons, New York.

Robertson, G. P., D. Wedin, P. M. Groffman, J. M. Blair, E. Holland, K. J. Nadelhoffer and D. Harris. 1999. Soil carbon and nitrogen availability: Nitrogen mineralization, nitrification and soil respiration potentials. Pages 258–271 *in* G. P. Robertson, C. S. Bledsoe, D. C. Coleman and P. Sollins (eds.), *Standard Soil Methods for Long Term Ecological Research.* Oxford University Press, New York.

Robinson, W. F. 1976. *Abandoned New England: Its Hidden Ruins and Where to Find Them.* New York Graphic Society. Little, Brown, Boston.

———. 1988. *Mountain New England: Life Past and Present.* New York Graphic Society. Little, Brown, Boston.

Rochefort, L., and F. A. Bazzaz. 1992. Growth response to elevated CO_2 in seedlings of four co-occurring birch species. *Canadian Journal of Forest Research* 22: 1583–1587.

Rowlands, W. 1939. A study of forest damage caused by the 1938 hurricane in the vicinity of Petersham, Massachusetts. Harvard Forest Archives: HF 1939–05.

———. 1941. Damage to even-aged stands in Petersham, Massachusetts, by the 1938 hurricane as influenced by stand condition. M.F.S. Thesis, Harvard University, Cambridge, Massachusetts.

Ruffner, C. M., and M. D. Abrams. 1998. Relating land-use history and climate to the dendroecology of a 326-year-old *Quercus prinus* talus slope forest. *Canadian Journal of Forest Research* 28: 347–358.

Russell, E. W. B. 1983. Indian-set fires in the forests of the northeastern United States. *Ecology* 64: 78–88.

———. 1997a. Review of International Conference on Advances in Forest and Woodland History, University of Nottingham, 2–7 September 1996. *Journal of Historical Geography* 23: 205–208.

———. 1997b. *People and the Land Through Time: Linking Ecology and History.* Yale University Press, New Haven, Connecticut.

Russell, E. W. B., R. B. Davis, R. S. Anderson, T. E. Rhodes and D. S. Anderson. 1993. Recent centuries of vegetational change in the glaciated northeastern United States. *Journal of Ecology* 81: 647–664.

Russell, E. W. B., and R. T. T. Forman. 1984. Indian burning, "The Unlikely Hypothesis." *Bulletin of the Ecological Society of America* 65: 281–282.

Russell, H. 1976. *A Long, Deep Furrow: Three Centuries of Farming in New England.* University Press of New England, Hanover, New Hampshire.

———. 1980. *Indian New England Before the Mayflower.* University Press of New England, Hanover, New Hampshire.

Sala, O. E., M. E. Biondi and W. K. Lauenroth. 1988. Bias in estimates of primary production: An analytical solution. *Ecological Modelling* 44: 43–55.

Samson, F. B., and F. L. Knopf. 1996. *Ecosystem Management.* Springer-Verlag, New York.

Saxon, E. C., T. Parris and C. D. Elvidge. 1997. Satellite surveillance of national CO_2 emissions from fossil fuels. Development Discussion Paper No. 608, Harvard Institute for International Development, Cambridge, Massachusetts.

Schimel, D. S. 1986. Carbon and nitrogen turnover in adjacent grassland and cropland ecosystems. *Biogeochemistry* 2: 345–357.

Schimel, D. S., and C. B. Field. 2001. Consistent land- and atmosphere-based US carbon sink estimates. *Science* 292: 2316–2320.

Schimel, J. P., and M. K. Firestone. 1989. Inorganic N incorporation by coniferous forest floor material. *Soil Biology and Biochemistry* 21: 41–46.

Schlesinger, W. H. 1997. *Biogeochemistry.* Second edition. Academic Press, San Diego.

Schoonmaker, P. K. 1992. Long-term vegetation dynamics in southwestern New Hampshire. Ph.D. Dissertation, Harvard University, Cambridge, Massachusetts.

Schoonmaker, P. K., and D. R. Foster. 1991. Some implications of paleoecology for contemporary ecology. *Botanical Review* 57: 204–245.

Schulze, E.-D., F. A. Bazzaz, K. Nadelhoffer, T. Koike and S. Takatsuki. 1996. Biodiversity and ecosystem function of temperate deciduous broad-leaved forests. Pages 71–98 *in* H. A. Mooney, J. H. Cushman, E. A. Medina, O. E. Sala and E.-D. Schulze (eds.), *Functional Roles of Biodiversity: A Global Perspective.* John Wiley & Sons, New York.

Shaver, G. R., J. Canadell, F. S. Chapin III, J. Gurevitch, J. Harte, G. Henry, P. Ineson, S. Jonasson, J. Melillo, K. Pitelka and L. Rustad. 2000. Global warming and terrestrial ecosystems: A conceptual framework for analysis. *BioScience* 50: 871–882.

Simberloff, D. 1996. Impacts of introduced species in the United States. *Consequences* 2: 13–22.

Simpson, R. H., and H. Riehl. 1981. *The Hurricane and Its Impact.* Louisiana State University Press, Baton Rouge

Singh, J. S., W. K. Lauenroth, H. W. Hunt and D. M. Swift. 1984. Bias and random errors in estimators of net root production: A simulation approach. *Ecology* 65: 1760–1764.

Sipe, T. W. 1990. Gap partitioning among maples *(Acer)* in the forests of central New England. Ph.D. Dissertation, Harvard University, Cambridge, Massachusetts.

Sipe, T. W., and F. A. Bazzaz. 1994. Gap partitioning among maples *(Acer)* in central New England: Shoot architecture and photosynthesis. *Ecology* 75: 2318–2332.

————. 1995. Gap partitioning among maples *(Acer)* in central New England: Survival and growth. *Ecology* 76(5): 1587–1602.

————. 2001. Shoot damage effects on regeneration of maples *(Acer)* across a gap-understorey gradient. *Journal of Ecology* 89: 761–773.

Smith, B. D. 1989. Origins of agriculture in eastern North America. *Science* 246: 1566–1571.

Smith, D. M. 1946. Storm damage in New England forests. M.F.S. Thesis, Yale University, New Haven, Connecticut.

Smith, K. A., G. P. Robertson and J. M. Melillo. 1994. Exchange of trace gases between the terrestrial biosphere and the atmosphere in the midlatitudes. Pages 179–203 *in* R. G. Prinn (ed.), *Global Atmospheric-Biospheric Chemistry.* Plenum Press, New York.

Sprugel, D. G. 1991. Disturbance, equilibrium, and environmental variability: What is 'natural' vegetation in a changing environment? *Biological Conservation* 58: 1–18.

Spurr, S. H. 1950. Stand composition in the Harvard Forest as influenced by site and forest management. Ph.D. Dissertation, Yale University, New Haven, Connecticut.

————. 1956a. Forest associations in the Harvard Forest. *Ecological Monographs* 26: 245–262.

————. 1956b. Natural restocking of forests following the 1938 hurricane in central New England. *Ecology* 37: 443–451.

Spurr, S. H., and A. C. Cline. 1942. Ecological forestry in central New England. *Journal of Forestry* 40: 418–420.

Stafford, R. 1992. Heterogeneity in forest structure, composition and dynamics following catastrophic wind disturbance in southwestern New Hampshire. M.F.S. Thesis, Harvard University, Cambridge, Massachusetts.

Stark, J. M., and S. C. Hart. 1997. High rates of nitrification and nitrate turnover in undisturbed coniferous forests. *Nature* 385: 61–64.

Steel, J. 1999. Losing ground: An analysis of recent rates and patterns of development and their effects on open space in Massachusetts. 2d ed. Massachusetts Audubon Society, Lincoln.

Steer, H. B. 1948. Lumber production in the United States, 1799–1946. USDA Miscellaneous Publications No. 669.

Steinbeck, J. 1962. *Travels with Charley: In Search of America.* Viking Press, New York.

Stephens, E. P. 1955. The historical-development method of determining forest trends. Ph.D. Dissertation, Harvard University, Cambridge, Massachusetts.

————. 1956. The uprooting of trees: A forest process. *Soil Science Society of America Proceedings* 20: 113–116.

Steudler, P. A., R. D. Bowden, J. M. Melillo and J. D. Aber. 1989. Influence of nitrogen fertilization on methane uptake in temperate forest soils. *Nature* 341: 314–316.

Steudler, P. A., R. D. Jones, M. S. Castro, J. M. Melillo and D. L. Lewis. 1996. Microbial controls of methane oxidation in temperate forest and agricultural soils. Pages 69–84 *in* J. C. Murrell and D. P. Kelly (eds.), *Microbiology of Atmospheric Trace Gases.* Springer-Verlag, Berlin/Heidelberg.

Stilgoe, J. R. 1976. Pattern on the land: The making of a colonial landscape 1633–1800. Ph.D. Dissertation, Harvard University, Cambridge, Massachusetts.

————. 1982. *Common Landscape of America, 1580 to 1845.* Yale University Press, New Haven.

Stott, P. 1991. Harvard Forest. *Global Ecology and Biogeography Letters* 1: 99–101.

Stout, B. 1952. Species distribution and soils in the Harvard Forest. *Harvard Forest Bulletin,* No. 24: 1–58.

Stout, B. B. (ed.). 1981. Forests in the Here and Now: A Collection of Writings of Hugh Miller Raup, Bullard Professor of Forestry, Emeritus, Harvard University. Pages 46–58 *in The Montana Forest and Conservation Experiment Station.* School of Forestry, University of Montana, Missoula.

Sumner, H. C. 1944. The North Atlantic hurricane of September 8–16, 1944. *Monthly Weather Review* 72: 187–189.

Thomas, S. C., and F. A. Bazzaz. 1996. Effects of elevated CO_2 on leaf shapes: Are dandelions getting toothier? *American Journal of Botany* 83: 106–111.

————. 1999. Asymptotic height as a predictor of photosynthetic characteristics in Malaysian rain forest trees. *Ecology* 80: 1607–1622.

Thomas, S. C., M. Jasienski and F. A. Bazzaz. 1999. Early vs asymptotic growth responses of herbaceous plants to elevated CO_2. *Ecology* 80: 1552–1567.

Thompson, Z. 1864. *History of the Animals of Vermont.* Chauncey Goodrich, Burlington, Vermont.

Thoreau, H. D. [1842, 1856] 1962. *In* B. Torrey and F. H. Allen (eds.), *The Journal of Henry D. Thoreau.* Dover Publications, New York. (Reissue of 1906 edition published by Houghton Mifflin).

Thorne, J. F., and S. P. Hamburg. 1985. Nitrification potentials of an old-field chronosequence in Campton, New Hampshire. *Ecology* 66: 1333–1338.

Tiessen, H., J. Stewart and J. Moir. 1983. Changes in organic and inorganic phosphorus composition of two grassland soils and their particle size fractions during 60–90 years of cultivation. *Journal of Soil Science* 34: 815–823.

Tingley, M., D. Orwig, R. Field and G. Motzkin. 2002. Avian response to removal of a forest dominant: Consequences of hemlock woolly adelgid infestations. *Journal of Biogeography* 29: 1505–1516.

Torrey, J. G. 1993. Can plant productivity be increased by inoculation of tree roots with soil microorganisms? *Canadian Journal of Forest Research* 22: 1815–1823.

Torrey, J. G., and L. J. Winship (eds.). 1989. *Applications of Continuous and Steady-State Methods to Root Biology.* Kluwer Academic Publishers, Dordrecht, The Netherlands.

Traw, M. B., R. L. Lindroth and F. A. Bazzaz. 1996. Decline in gypsy moth *(Lymantria dispar)* performance in an elevated CO_2 atmosphere depends upon host plant species. *Oecologia* 108: 113–120.

The Trustees of Reservations (TTOR). 1999. *Conserving Our Common Wealth—A Vision for the Massachusetts Landscape.* The Trustees of Reservations, Beverly, Massachusetts.

Turner, B. L., W. Clark, R. Kates, J. Richards, J. Matthews and W. Meyer. 1990. *The Earth as Transformed by Human Action.* Cambridge University Press, England.

Twery, M. J., and W. A. Patterson. 1984. Effect of beech bark disease on northern hardwood stands of central New England. *Canadian Journal of Forest Research* 14: 565–574.

U.S. Department of Agriculture. 2002. Hemlock Woolly Adelgid Map: 2002. USDA Forest Service GIS Group, Durham, New Hampshire.

U.S. Forest Service. 1958. Timber resources for America's future. USDA Forest Service Forest Resource Report No. 14.

———. 1965. Timber trends in the United States. USDA Forest Service Forest Resource Report No. 17.

———. 1973. The outlook for timber in the United States. USDA Forest Service Forest Resource Report No. 20.

———. 1990. New England's forests. USDA Forest Service Northeastern Forest Experiment Station NE-INF-91-90.

U.S. Geological Survey, 1993. *Digital Elevation Models Data Users Guide 5.* U.S. Department of the Interior, U.S. Geological Survey, Reston, Virginia.

Van Cleve, K., W. C. Oechel and J. L. Hom. 1990. Response of black spruce (*Picea mariana* [Mill] B.S.P.) ecosystems to soil temperature modification in interior Alaska. *Canadian Journal of Forest Research* 20: 1530–1535.

Van Rooyen, D. J. 1972. I. Organic carbon and nitrogen status in two hapludalfs under prairie and deciduous forest, as related to moisture regime, some morphological features, and response to manipulation of cover. Ph.D. Dissertation, University of Wisconsin, Madison.

Verrazano, G. da. 1968. Voyages. Pages 1–24 *in* G. P. Winship (ed.). *Sailors' Narratives of Voyages along the New England Coast 1524–1624.* B. Franklin, New York.

Vickery, P., and P. Dunwiddie. 1997. Grasslands of northeastern North America: Ecology and conservation of native and agricultural landscapes. Massachusetts Audubon Society, Lincoln, Massachusetts.

Vitousek, P. M., J. D. Aber, S. E. Bayley, R. W. Howarth, G. E. Likens, P. A. Matson, D. W. Schindler, W. H. Schlessinger and G. D. Tilman. 1997. Human alteration of the global nitrogen cycle: Causes and consequences. *Ecological Issues* 1: 1–15.

Walker, B. H. 1975. Vegetation: Site relationships in the Harvard Forest. *Vegetatio* 29(3): 169–178.

Waring, R. H., B. E. Law, M. L. Goulden, S. L. Bassow, R. W. McCreight, S. C. Wofsy and F. A. Bazzaz. 1995. Scaling gross ecosystem production at Harvard Forest with remote sensing: A comparison of estimates from a constrained quantum-use efficiency model and eddy correlation. *Plant, Cell and Environment* 18: 1201–1213.

Wayne, P. M. 1992. Effects of the daily timecourse of light availability on the sun-shade responses and regeneration of birch seedlings. Ph.D. Dissertation, Harvard University, Cambridge, Massachusetts.

Wayne, P. M., and F. A. Bazzaz. 1991. Assessing diversity in plant communities: The importance of within-species variations. *Trends in Ecology and Evolution* 6: 400–404.

———. 1993a. Morning vs afternoon sun patches in experimental forest gaps: Consequences of temporal incongruency of resources to birch regeneration. *Oecologia* 94: 235–243.

———. 1993b. Birch seedling responses to daily time courses of light in experimental forest gaps and shadehouses. *Ecology* 74: 1500–1515.

———. 1997. Light acquisition and growth by competing individuals in CO_2-enriched atmospheres: Consequences for size structure in regenerating birch stands. *Journal of Ecology* 85: 29–42.

Wayne, P. M., A. L. Carnelli, J. Connolly and F. A. Bazzaz. 1999. The density dependence of plant responses to elevated CO_2. *Journal of Ecology* 87: 1283–1292.

Wayne, P. M., E. G. Reekie and F. A. Bazzaz. 1998. Elevated CO_2 ameliorates birch response to high temperature and frost stress: Implications for modelling climate-induced geographic range shifts. *Oecologia* 114: 335–342.

Webb, S. L. 1986. Potential role of passenger pigeons and other vertebrates in the rapid Holocene migrations of nut trees. *Quaternary Research* 26: 367–375.

Wessman, C. A., J. D. Aber and D. L. Peterson. 1989. An evaluation of imaging spectrometry for estimating forest canopy chemistry. *International Journal of Remote Sensing* 10: 1293–1316.

Wessman, C. A., J. D. Aber, D. L. Peterson and J. M. Melillo. 1988a. Foliar analysis using near infrared spectroscopy. *Canadian Journal of Forest Research* 18: 6–11.

———. 1988b. Remote sensing of canopy chemistry and nitrogen cycling in temperate forest ecosystems. *Nature* 335: 154–156.

Westveld, M. 1956. Natural forest vegetation zones of New England. *Journal of Forestry* 54: 332–338.

Whitney, G. G. 1984. Fifty years of change in the arboreal vegetation of Heart's Content, an old-growth hemlock–white pine–northern hardwood stand. *Ecology* 65: 403–408.

———. 1991. Relation of plant species to substrate, landscape position, and aspect in north central Massachusetts. *Canadian Journal of Forest Research* 21: 1245–1252.

———. 1994. *From Coastal Wilderness to Fruited Plain: An Ecological History of Northeastern United States*. Cambridge University Press, England.

Whitney, G. G. (ed.). 1989. *Harvard Forest Bibliography, 1907–1989*. Harvard Forest, Petersham, Massachusetts.

Whitney, G. G., and D. R. Foster. 1988. Overstorey composition and age as determinants of the understorey flora of central New England's woods. *Journal of Ecology* 76: 867–876.

Wilkie, R. W., and J. Tager. 1991. *Historical Atlas of Massachusetts*. University of Massachusetts Press, Amherst.

Williams, M. 1989. *Americans and Their Forests: A Historical Geography*. Cambridge University Press, New York.

Williams, M., E. B. Rastetter, D. N. Fernandes, M. L. Goulden, S. C. Wofsy, G. R. Shaver, J. M. Melillo, J. W. Munger, S.-M. Fan and K. J. Nadelhoffer. 1996. Modeling the soil-plant atmosphere continuum in a *Quercus-Acer* stand at Harvard Forest: The regulation of stomatal conductance by light, nitrogen and soil/plant hydraulic properties. *Plant, Cell and Environment* 19: 911–927.

Wilson, B. F., and R. R. Archer. 1979. Tree design: Some biological solutions to mechanical problems. *BioScience* 29: 293–298.

Winer, H. I. 1955. History of the Great Mountain Forest, Litchfield County, Connecticut. Ph.D. Dissertation, Yale University, New Haven, Connecticut.

Winkler, M. G. 1985. A 12,000-year history of vegetation and climate for Cape Cod, Massachusetts. *Quaternary Research* 23: 301–312.

Winthrop, J. 1908. Journal: History of New England. Pages 211–229 *in* J. K. Hosmer (ed.) *Original Narratives in Early American History*, 2 vols., reprinted from 1630–1649 journal edition. Charles Scribner's Sons, New York.

Wofsy, S. C., M. L. Goulden, J. W. Munger, S.-M. Fan, P. S. Bakwin, B. C. Daube, S. L. Bassow and F. A. Bazzaz. 1993. Net exchange of CO_2 in a mid-latitude forest. *Science* 260: 1314–1317.

Woodwell, G. 1991. *The Earth in Transition: Patterns and Processes of Biotic Impoverishment.* Cambridge University Press, England.

Wright, H. E. 1974. Landscape development, forest fires, and wilderness management. *Science* 186: 487–495.

———. 1984. Sensitivity and response time of natural systems to climatic change in the late Quaternary. *Quaternary Science Reviews* 3: 91–131.

Wright, J. K. 1933. Regions and landscapes of New England. New England's prospect. American Geographical Society Special Publication 16: 14–49.

Xiao, X., J. M. Melillo, D. W. Kicklighter, A. D. McGuire, R. G. Prinn, C. Wang, P. H. Stone and A. Sokolov. 1998. Transient climate change and net ecosystem production of the terrestrial biosphere. *Global Biogeochemical Cycles* 12: 345–360.

Yanes, C. V., A. Orozco, M. Rojas, J. E. Sánchez and V. Cervantes. 1997. La Reproducción de las Plantas: Semillas y Meristemos. Secretaría de Educación Pública Fondo de Cultura Económica consejo Nacional de Ciencia y Technologial.

Yano, Y., W. H. McDowell and J. D. Aber. 2000. Biodegradable dissolved organic carbon in forest soil solution and effects of chronic nitrogen deposition. *Soil Biology and Biochemistry* 32: 1743–1751.

York, E. C. 1996. Fisher population dynamics in north-central Massachusetts. M.S. Thesis, University of Massachusetts, Amherst.

Zak, D. R., D. F. Grigal, S. Gleeson and D. Tilman. 1990. Carbon and nitrogen cycling during old-field succession: Constraints on plant and microbial biomass. *Biogeochemistry* 11: 111–129.

Zebryk, T. 1991. Holocene development of a forested wetland in central Massachusetts. M.F.S. Thesis, Harvard University, Cambridge, Massachusetts.

Zwieniecki, M. A., and N. M. Holbrook. 1998. Diurnal variation in xylem hydraulic conductivity in white ash (*Fraxinus americana* L.), red maple (*Acer rubrum* L.) and red spruce (*Picea rubens* Sarg.). *Plant Cell and Environment* 21: 1173–1180.

Contributors

John Aber
  Complex Systems Research
    Center
  University of New Hampshire

Toby Ahrens
  Department of Geological and
    Environmental Sciences
  Stanford University

Jacqueline Aitkenhead
  Department of Natural
    Resources
  University of New Hampshire

Arthur Allen
  EcoTec Inc.
  Worcester, Massachusetts

Carol Barford
  Gaylord Institute for
    Environmental Studies
  University of Wisconsin

Audrey Barker Plotkin
  Harvard Forest
  Harvard University

Sylvia Barry
  Harvard Forest
  Harvard University

Susan Bassow
  Littleton, Colorado

Fakhri Bazzaz
  Organismic and Evolutionary
    Biology
  Harvard University

Debra Bernardos
  Missouri Department of
    Conservation

Glenn Berntson
  Boston, Massachusetts

Richard Boone
  Institute of Arctic Biology
  University of Alaska

Emery Boose
  Harvard Forest
  Harvard University

Richard Bowden
  Environmental Sciences
  Allegheny College

Frank Bowles
  Research Designs
  Woods Hole, Massachusetts

Elizabeth Burrows
  The Ecosystems Center
  Marine Biological Laboratory

Jana Canary
  Institute of Arctic Biology
  University of Alaska

James Cardoza
  Massachusetts Division of
    Fisheries and Wildlife

Gary Carlton
  Biological Sciences Department
  California State Polytechnic
    University

Mark Castro
  Appalachian Laboratory
  University of Maryland

Sebastian Catovsky
  Department of Environment,
    Food and Rural Affairs
  London, U.K.

Christina Catricala
  Forest and Rangeland Ecosystem
    Science Center
  Corvallis, Oregon

Susan Clayden
  Univerisity of New Brunswick
  Fredericton, New Brunswick,
    Canada

Jana Compton
  U.S. Environmental Protection
    Agency
  Corvallis, Oregon

Sarah Cooper-Ellis
  Brattleboro, Vermont

Rose Crabtree
  Oxford, U.K.

William Currie
  School of Natural Resources &
    Environment
  University of Michigan

Kathleen Donohue
  Organismic and Evolutionary
    Biology
  Harvard University

David Foster
  Harvard Forest
  Harvard University

Janice Fuller
  Department of Botany
  University College, Galway
  Ireland

Lisa George
  Houston, Texas

Brian Hall
  Harvard Forest
  Harvard University

Joseph Hendricks
  Department of Biology
  State University of West Georgia

Jason Kaye
  Department of Biology
  Colorado State University

Heidi Lux
  The Ecosystems Center
  Marine Biological Laboratory

Alison Magill
  Complex Systems Research
    Center
  University of New Hampshire

Mary Martin
  Complex Systems Research
    Center
  University of New Hampshire

William McDowell
  Department of Natural Resources
  University of New Hampshire

Jason McLachlan
  Department of Botany
  Duke University

Jerry Melillo
The Ecosystems Center
Marine Biological Laboratory

Patricia Micks
The Ecosystems Center
Marine Biological Laboratory

Sarah Morrisseau
The Ecosystems Center
Marine Biological Laboratory

Glenn Motzkin
Harvard Forest
Harvard University

J. William Munger
Department of Earth and
Planetary Sciences
Harvard University

Knute Nadelhoffer
Biological Station
University of Michigan

Kathleen Newkirk
Winslow, Maine

John O'Keefe
Harvard Forest
Harvard University

Scott Ollinger
Complex Systems Research
Center
University of New Hampshire

David Orwig
Harvard Forest
Harvard University

Andrea Ricca
Woods Hole, Massachusetts

Timothy Sipe
Department of Biology
Franklin and Marshall College

Paul Steudler
The Ecosystems Center
Marine Biological Laboratory

Paul Wilson
California State University,
Northridge

Steven Wofsy
Department of Earth and
Planetary Sciences
Harvard University

Index

gray birch *(Betula populifolia),* 85, 108, 114, 184, 322, 324, 334, 335
gray wolf, 154, 160, 161
greenhouse gases, 10–11, 38, 203, 280
Green Mountain Uplands, as physiographic division, 21, 22
gross carbon exchange (GCE), 37
gross ecosystem exchange (GEE), 210, 211, 215, 216–217, 345; measurements of, 385–388
gross photosynthesis, 37, 344. *See also* photosynthesis
gross primary production (GPP), 37
Gulf Stream, 45
gypsy moth, as forest pest, 10, 61–62, 93, 164

Hardwick, Massachusetts, 84
hardwood forest: and landscape patterns, 29, 86; New England vegetation zones, 25; and nitrogen saturation experiment, 263–270; and wind damage, 56, 57
Harlock Pond, Martha's Vineyard, 69, 72
Harvard Forest: and CO_2 flux measurements, 380–393; Environmental Management Site (EMS), 204–208; establishment of, 19; land-use history, 28–31; Long Term Ecological Research (LTER) program, 3, 14–18, 30
hay-scented fern *(Dennstaedtia punctilobula),* 329–332
hemlock *(Tsuga canadensis):* characteristics of, 94–95; distribution, 25, 45, 108, 178, 179, 380; fire sensitivity of, 69, 111–112; landscape patterns, 29, 86, 108–114; late-successional role, 125, 138–139, 140; outlook for, 98–99; pollen record, 44, 117; prehistoric decline, 5, 59–62; seedling regeneration dynamics, 329, 334
Hemlock Hollow, and forest dynamics, 129–131, 132, 134
hemlock woolly adelgid (HWA), 5, 10, 93, 94–98, 138, 299, 369
Henry, David, 51
heron, great blue, 156, 166
heterogeneity of resources, and wind damage, 248–251
hickory *(Carya* spp.), 25, 45, 69
historical perspectives, and conservation management, 372–375

historical studies, and land use, 104–108
Hole, Francis D., 301–302, 315
Holyoke, Massachusetts, 87
house sparrows, 163–164
Hubbard Brook research group, 23; and PnET model, 351–352
Hubbard Brook watershed, 240, 251
huckleberry *(Gaylussacia baccata),* 181, 183, 184
human activity: and atmospheric chemistry, 10–11, 39, 332, 400; and conservation planning, 372–379; and disturbance history, 10–11, 299, 370–372; ecosystem response to, 73, 394–400; and wildlife distribution, 164–168
humus, and soil nutrient dynamics, 36, 37, 39, 301
hunting, and wildlife populations, 155, 166
hurricane experiment: design of, 241–245; ecosystem consequences, 248–251, 296–297, 299; effects of, on soils and biogeochemistry, 250, 251–253; interpretations and insights from, 253–258; seedling response, 325–328; vegetation response, 245–248
Hurricane Hugo, 56, 240, 241, 251–252
Hurricane Mitch, 241
Hurricane of *1938,* 8, 59, 176–177, 180, 235–241, 254–256
hurricanes: ecological role of, 5, 6, 235–241; history and impact of, in New England, 48, 51–59, 370
hurricane surface wind model (HURRECON), 54
hydrocarbons, 38, 223–226, 228–229
hydrology: and hurricane disturbance, 239–241, 298–299; and land conversion, 91–93

ice sheets, melting of, 44, 64
Indians. *See* Native Americans
industrialization: and atmospheric changes, 10, 260; and reforestation, 82–86
industry, local, and land-use patterns, in Massachusetts, 77–82
insect pests, 10, 61. *See also* gypsy moth, as forest pest; hemlock woolly adelgid

interrupted fern *(Osmunda claytoni-ana),* 329–332
introduced pests and pathogens, 93–94, 137–138
introduced species of wildlife, 147 (Table 7.1), 163–164
isoprene, 38, 228, 229

Jones, L. R., 124

lake sediments: paleoecology studies of, 69, 102–104, 108, 111, 112; and pollen data, 102–104, 116–119
land conservation, and forest harvesting, 376–378
landscape dynamics: agriculture and local industry, 77–82; forest clearance, 73–76; Indian use of fire, 67–72; introduced pests and pathogens, 93–99; and reforestation, 82–86, 190; suburbanization and development, 91–93, 99; timber harvesting, 86–91
landscape scale, and integrated measurements, 17
land-use history: and ecosystems, 395, 399; and forest composition, 112–114; of Harvard Forest, 28–31; and hurricane damage, 56; and soil legacies, 196–201; sources for, 104–108; and species distribution, 172–175, 177–188; and vegetation patterns, 181–188, 195
Laurentide ice sheet, 44
Leer, Xuhui, 390
light, seasonal variation in, 318–321
light availability: and gap heterogeneity, 321–323; seedling response to, 318–323, 327, 334
lignin: concentrations of, in foliage, 360–362; and litter decomposition, 274, 346–349
litter decomposition: measurement of, and modeling process, 346–349; and nitrogen dynamics, 35–40, 273–275, 276–278, 311–312; and soil nutrients in DIRT experiment, 300–315
Little Ice Age, 57, 117, 118, 119, 370
liverwort *(Lepidozia reptans),* 178
logging: after *1938* hurricane, 254–256, 299; and forest stand dynamics, 135–137; patterns of, and timber harvest, 86–91; preemptive, of hemlock, 98

Long Term Ecological Research Program (LTER): and DIRT experiment, 302; and integrated measurements, 16; and LIDET study, 349; Luquillo project, 241; and regional analysis, 338–339
Ludlum, David, 53
Lyford, Walter, 19, 192
Lyme disease, 155, 168

Maine: Bear Brooks watershed, 16; disease epidemics, 66; regional physiographic divisions, 21–25
maize, and Native American agriculture, 8, 64, 66, 67, 72, 118
mammals, historic dynamics of, 147–148 (Table 7.1)
manure, as fertilizer: and land-use history, 259–260; soil legacy of, 190–191, 196–197, 198, 200, 201
maple *(Acer* spp.): as early successional species, 111; fire sensitivity of, 69, 72; harvesting of *(1885),* 88; vegetation zones of, 25. *See also* red maple *(A. rubrum);* striped maple *(A. pennsylvanicum);* sugar maple *(A. saccharum)*
maps, historical, and land-use analysis, 104
Martha's Vineyard, 45, 66, 69, 72, 158
Massachusetts: beaver distribution in, 151–154; coyote population, 161–163; deer population, 154–155; development and suburbanization of, 91–93, 99, 149–150; forest area of, 75, 81–82; and LTER program, 17; physiographic divisions of, 25–28; population size and distribution, 80; settlement patterns of, 74; timber harvesting in, 89, 90; wildlife dynamics of, 144–146
Massachusetts Division of Fisheries and Wildlife, 146, 153, 155
meadowlarks, 142, 157, 158
measurements, long-term, of ecosystem patterns, 16–18
Merrimack River, 298
mesic forests, and fire occurrence, 69, 72
meteorological modeling and reconstruction, 51–59
methane (CH_4), 38; and blowdown experiment, 253; as greenhouse gas, 10–11, 280; measurement of, and

modeling process, 344; and nitrogen saturation experiment, 266–267; and soil-warming experiments, 283, 288–289, 295

microbial decomposition, and nitrogen cycle, 35, 36, 259, 261–262, 354–355

mineralization, and nitrogen release, 35, 38–40

mineral soils, 308, 309, 314–315

models (computer models): ecosystem processes, 338–343; regional analysis, 356–360; and remote sensing, 360–362

Montague Plain: land-use legacies, 181–186, 193–196, 199; as study site, 191–192

moose, 142, 160, 163, 165, 167, 168

Morton, Thomas, 67–68, 164

Mount Washington, 22

Muir, John, 49

mycorrhizal processes, 40, 276, 277–278, 312, 355

Nantucket, 45, 66, 77, 158

National Aeronautics and Space Administration (NASA), and AVIRIS, 360

National Oceanic and Atmospheric Administration, 54

Native Americans: and disease epidemics, 65, 66, 149; distribution and population of, 8, 27, 62–64, 66, 112; and European settlement, 65, 73, 370–371; and fire, 48, 62, 67–72; impact of, on landscape, 62–67; maize agriculture of, 8, 64, 66, 67, 72, 118; and wildlife populations, 148, 154, 156

"natural experiments," 15

nematodes, in forest soils, 314–315

net carbon exchange: eddy flux studies of, 383–384, 385–388; spatial patterns of, and AVIRIS, 360

net ecosystem exchange (NEE), 37–38, 210; measurements of, 210, 211, 215–219, 385–388

net ecosystem production (NEP), 38; and CO_2 exchange, 211, 215–216

net primary production (NPP), 37

New England region: climate and vegetative change in, 43–48; and conservation approaches, 375–379; fire in, 67–72; forest composition of, 23–25, 29, 45; forest ecosystem dynamics,

5–12, 91–93; forest resources of, and logging, 86–91; hemlock mortality in, 59–62, 94–99; land-use history of, 72–82, 373–374; and LTER program, 14–18; Native American impact on, 8, 62–72; physiographic divisions of, 21–25; population of, and development, 399; and reforestation, 82–86; species distribution in, 367–370; wildlife dynamics of, 142–168; wind damage and hurricanes in, 49–59

New Hampshire: Pisgah Forest tract, 19, 51, 59, 89–90, 238; pulp and paper industry, 87; and regional physiographic divisions, 21–25

Nichols, George, 49

nitrate (NO_3^-): deposition of, and global atmosphere, 332–333; experimental nitrification, 261–262, 265, 269–270, 289, 292; nitrogen cycle, 38–40, 253; production of, and land-use history, 190, 199, 201

nitric acid (HNO_3), 223–226, 227–228, 260

nitric oxide (NO), 223–226

nitrification: and DIRT experiment, 302–312; and nitrogen saturation experiment, 262, 265–270

nitrogen: additions of, and ecosystem changes, 32, 33, 297–298; atmospheric conversion of, and ozone, 397; interactions of, in photosynthesis and transpiration, 34–35; and land-use legacies, 189–191, 194–196, 197, 201; modeling structure for, and litter decay, 348; and nitrogen-carbon dynamics, 35, 36, 273–275; and plant growth, 259; and seedling growth, 327–328; and soil-warming experiment, 289–295

nitrogen cycle: and forest ecosystems, 33–36, 38–40, 260–262, 278–279; and land-use history, 196–198, 199, 200; and soil-warming experiment, 298

nitrogen deposition, and environmental change, 279, 332–337, 356, 398

nitrogen flux: and soil-warming experiments, 291–295; and TRACE model, 354–355

nitrogen mineralization: and blow-down consequences, 253; and DIRT experiment, 305, 311–312

population dynamics, in New England, 73–74, 207
Post, Howard, 239, 241
potash, as fertilizer, 75, 87, 190
primary forests, 51, 78–79, 95, 126
Pring, Martin, 66, 67
proprietors' survey records, 107
Prospect Hill tract, 30, 31; disturbance history of, and species composition, 174, 175–181; experimental blowdown on, 325; land-use legacies of, 31, 82, 185–186, 192–193, 196–198; nitrogen saturation experiment, 263–264; vegetation change over time, 71, 75
protozoa, in forest soils, 314–315

Quabbin Reservoir, 26–27, 28; and conservation, 91, 93, 378; deer population of, 155; and silviculture, 257, 378

Rand, F. V., 124
range extension of wildlife, 160–163
Raup, Hugh, 19, 49, 50, 128, 192, 400; "The View from John Sanderson's Farm," 31, 394, 396–397
red maple *(Acer rubrum)*, 25, 45, 108; carbon storage by, 216; disturbance experiments, 246, 316–331 passim; and forest dynamics, 239, 246, 247, 380, 389, 391; and land-use history, 29, 85, 114, 117, 134, 178, 183; seedling regeneration experiments, 329–337 passim; successional pattern of, 85, 111, 140, 183
red oak *(Quercus rubra)*, 25, 90, 108, 125, 140, 216; disturbance experiments, 239, 246, 247, 328, 331; and forest dynamics, 380, 389, 391, 392; and land-use history, 29, 192; seedling regeneration experiments, 329–337 passim
red pine *(Pinus resinosa)*, and nitrogen saturation experiments, 263–269 *passim*
red spruce *(Picea rubens)*, 25, 122
reforestation: and atmospheric CO_2, 201, 203; in New England, 82–86, 190; and soil properties, 190, 199; and wildlife dynamics, 151–156
remote sensing: and ecosystem models, 360–362; and ecosystem patterns, 16; and regional analysis, 338–362

A Report on the Ornithology of Massachusetts, 157
resource congruence, plant response to, 323–324
respiration: and climate variation, 216–219, 291–295; DIRT experiment, 305–312; rates of, and ecosystem properties, 385–386
rhizosphere, 305, 307–308
Rhode Island, 66, 67
river flow, after *1938* hurricane, 240, 241, 298
root inputs, and soil respiration, 305, 310–312, 315
Rowlands, Willett, 56, 236, 238–239, 241, 242, 244

salvage logging: after *1938* hurricane, 236, 254, 299; ecosystem response to, 256–257
Sanderson, John, 395
Sanderson Farm, 82
sassafras *(Sassafras albidum),* 25, 67, 121–122
scarlet oak *(Quercus coccinea),* 192
scrub oak *(Quercus ilicifolia; Quercus prinoides),* 25, 69, 72, 184, 192
seedling regeneration: and environmental change, 317, 332–337; and forest structure, 328–332
seedling response: to hurricane disturbance, 317, 325–328; to light availability, 317, 321–324
Shaler, Nathaniel, 19, 20
shrub species, distribution patterns of, 120–124
Slab City tract, 31, 128, 135, 136, 140
Smith, David (D. M.), 236, 238, 257
Smith, John, 66, 151
"smog," 397
soil dynamics, 36, 37–40, 300–302, 305–308; DIRT experiment, 304, 305–313; experimental blowdown, 250, 251–253; nineteenth-century agriculture, 189–191; nitrogen retention experiments, 270–278; species distribution and, 177–178, 180
soil temperature: and ecosystem respiration, 216–219, 291–295, 385–388; and global warming, 280–281, 291–295; and methane consumption, 344; soil-warming experiments, 16, 281–291, 297–298,

waterpower, and industrialization, 83, 84

watersheds, and forest modification, 91, 93, 378

Westfield classification of forest vegetation, 23, 25

Westveld, Marinius, 23

White Mountains, 22

white oak *(Quercus alba)*, 25, 29, 108

white pine *(Pinus strobus):* fire tolerance of, 111; and *1938* hurricane, 246, 254, 256, 299; seedling regeneration experiments, 332, 334; as successional species, 29, 56, 85–86, 178, 196; timber harvest of, 90, 256, 299; and wind damage, 56, 57, 238–239, 246, 254; zones and distribution of, 25, 45, 108

white pine blister rust, 62, 93

wildlife dynamics, in New England: cultural and environmental setting for, 146–150; data sources for, 145–146; ecological and social consequences of, 164–168; general trends in, 150–164

Williams, Roger, 67

wind damage: and hurricane experiment, 235–258; as natural disturbance, 49–59

wintergreen *(Gaultheria procumbens)*, 175, 181, 183–185

witch hazel *(Hamamelis virginiania)*, 178

"witness" tree data, 23, 106, 108–109, 111

wolves, 143, 154, 160, 161, 163, 167

Wood, William, 65, 67–68, 164

wood ash, 87

wood consumption and fuel, 87

woodlots, 79, 82, 178, 196, 260

wood products, 87

yellow birch *(Betula alleghaniensis)*, 25, 250–251, 380; drought sensitivity of, 324, 335; seedling regeneration, 322, 327–336 passim; as successional species, 322, 334

yellow sedge *(Carex flava)*, 124